MATRIX-GEOMETRIC SOLUTIONS IN STOCHASTIC MODELS

An Algorithmic Approach

MARCEL F. NEUTS

DOVER PUBLICATIONS, INC.
NEW YORK

Copyright

Published in Canada by General Publishing Company, Ltd., 30 Lesmill Road, Don Mills, Toronto, Ontario.
Published in the United Kingdom by Constable and Company, Ltd., 3 The Lanchesters, 162-164 Fulham Palace Road, London W6 9ER.

Bibliographical Note

This Dover edition, first published in 1994, is an unabridged and corrected republication of the work first published by The Johns Hopkins University Press, Baltimore, Maryland, in 1981 in *The Johns Hopkins Series in the Mathematical Sciences*.

Library of Congress Cataloging-in-Publication Data

Neuts, Marcel F.
Matrix-geometric solutions in stochastic models : an algorithmic approach / Marcel F. Neuts.
p. cm.
Originally published: Baltimore : Johns Hopkins University Press, 1981, in series: Johns Hopkins series in the mathematical sciences ; 2.
Includes bibliographical references and index.
ISBN 0-486-68342-7 (pbk.)
1. Markov processes. 2. Queuing theory. 3. Matrices. I. Title.
QA274.7.N48 1994
003'.76'0113351—dc20 94-36940
CIP

Manufactured in the United States of America
Dover Publications, Inc., 31 East 2nd Street, Mineola, N.Y. 11501

It is unworthy of excellent men to lose hours like slaves in the labor of calculation which could safely be relegated to anyone else if machines were used.

Gottfried Wilhelm Leibniz

Contents

Preface

This book is an outgrowth of a series of lectures I gave at The Johns Hopkins University, Baltimore, Maryland, from July 20 to July 24, 1979. It contains a variety of results on queues and other stochastic models with the unifying feature of ready algorithmic implementation. The material discussed in this book comes under the broader heading of *computational probability,* a subject area sufficiently young to require definition.

As I perceive it, computational probability is not primarily concerned with the algorithmic questions raised by the direct numerical computation of existing analytic solutions. Such questions are best considered within the framework of classical numerical analysis. It is the concern of the probabilist, however, to ensure that the solutions he obtains are in the best, most natural, form for numerical computation. The expression of this concern is of recent date. Before the era of the modern computer, much of the best effort in applied mathematics was aimed at obtaining insight into the behavior of formal models, while avoiding the drudgery of computation by primitive machinery. On the other hand, the early difficulty of computation has also allowed the development of a large number of formal solutions from which few, if any, qualitative conclusions may be drawn, and whose appropriateness for algorithmic implementation has not been seriously considered. There is, in fact, an attitude that still pervades most of the teaching and the research literature on applied probability today and that does not view algorithmic implementation as an integral, challenging part of the solution process. We view this attitude as a legacy of history, but not as a constructive one.

We therefore define computational probability as the study of stochastic models with a genuine added concern for algorithmic feasibility over a wide, realistic range of parameter values. We have imposed upon our own work and that of our students the requirement that careful and exhaustive computer studies be performed before research results are proposed as the solution to a given problem. The limitations on a specific algorithm are as real as are restrictive conditions on the validity of a theorem. They offer, as we shall see, the same challenge and stimulus to further research.

This self-imposed constraint has not been felt as a burden. It has, on the contrary, led us to seek alternatives to certain classical analytic methods that are often difficult and risky in their numerical implementation. Unifying structural properties of Markov chains have been identified, which, through purely probabilistic arguments, lead to highly stable numerical procedures. Rather than proceeding through purely formal manipulations, the analysis of many problems now runs parallel to the steps of the algorithms. The latter then acquire a significance, which is often of independent interest in the interpretation of numerical results. A number of specific algorithms and the interpretation of their intermediate steps are discussed in this book.

The reader who is acquainted with the literature on stochastic models may briefly miss in our treatment the prevalent transform solutions and the familiar arguments based on methods of complex analysis. The reason for this will soon be clear. Though useful tools in analytic investigations, these techniques may often be replaced by others that are better suited for numerical computation. For the models discussed in this book, these will be iterative matrix methods and classical systems of differential equations.

The oft-lamented *Laplacian curtain,* which covers the solution and hides the structural properties of many interesting stochastic models, is now effectively lifted, at least for broad classes of problems in the theory of queues. We feel that at this point the real difficulties of a detailed analysis of such models emerge. These are nearly all related to the vast increases in dimensionality which accompany even small increases in the complexity of the physical description of most models. The *curse of dimensionality,* so aptly named by R. Bellman in his book *Dynamic Programming* [13], also weighs heavily on applied probability.

About this matter, at least two things may be said. It has been stated that numerical procedures are always closely linked to the available computer technology and that matters of implementation will eventually be settled by those interested in specific answers. The implication that such issues need not preoccupy the theoretical investigator is usually unstated, yet very clear. The truism of the first statement hides the utter fallacy of the second. It is clear that with faster computers that are endowed with larger central memories we can handle problems of higher dimension and compute their solution in more detail and with greater ease. The dimensionality of the problems we would like to solve, however, far outruns even the most optimistic forecasts of the capabilities of future computers. There is also little evidence that persons working on specific applications can afford the time or the mathematical effort required to construct efficient and appropriate algorithms for their solutions. The algorithmic aspects of models usually lead to a complete rethinking of their mathematical structure and present challenges and technical difficulties on a par with those of the classical analytic procedures. They also lead to a more realistic view of what constitutes a meaningful and satisfactory solution. A case in point is the analysis of the $GI/PH/c$ queue, which is briefly discussed in Chapter 4. Our general theorems give a complete characterization of the stationary probability vector of this queue, but in terms of the solution of a nonlinear matrix equation involving matrices of order ν^c. This allows the computation of that vector for very small values of ν and c, but does not constitute in our view a meaningful and satisfactory solution to this problem. In this case, the limitations of the algorithm are brought about by the inherent complexity of the multi-server queue. The theoretical examination of these limitations will, we hope, shed light on qualitative results or approximate procedures that are capable of extending our study of systems with many servers and more versatile service time distributions.

The high dimensionality of most interesting problems offers the opportunity for investigations of a different nature. As in mathematical programming, many problems of practical significance have special structural properties that may be used to construct workable algorithms, even when the problem is inherently of high dimension. Where these models have been treated earlier, such special, simplifying structures have usually been obscured by the formalism of transform methods. In this book, there are many instances where I discuss how particular structural features can be utilized to simplify and expedite numerical computation. These examples are far from the final word on this subject. They should serve to illustrate the type of simplifying structure to look for in specific problems. Such examples further illustrate the need for the probabilist to be very closely involved with the algorithmic analysis of a problem. One may hardly expect simplifications, which require a thorough understanding of the problem, to be discovered by a programmer. Such an involvement is also necessary for another reason. Statements regarding the feasibility or limitations of algorithms have the same standing as theorems and should be subject to the same scrutiny and scientific critique. The profusion of trivial numerical examples and of unwarranted algorithmic claims in the "applied" literature on stochastic models suggests, however, that such standards are not yet widely held.

This book must not be regarded as a systematic treatment of all of computational probability, or even of its applications in the theory of queues. Insofar as it represents my work and that of my associates, it deals only with results obtained since 1975. A significant portion of its contents are of very recent date, and some useful algorithms are presented for the first time here. In order to keep its length within reasonable bounds, we have limited the book to the discussion of one class of structured Markov chains only, and to a broad selection of its applications. We have given detailed references to procedures for other types of Markov chains, whose structure we have found to be very useful for algorithmic purposes. The presentation of the book is also intentionally open-ended. Such important approaches as, for example, the use of interactive computation in the design of stochastic systems, are represented only by simple, but illustrative, examples. We are confident, however, that as the algorithmic approach gains wider currency, much more will be heard about this methodology.

The discussion in this book is also intentionally limited to the algorithmic aspects of the *stable versions in steady-state* of the stochastic models under consideration. This was done for several reasons. The mathematical structure of the models is particularly useful in the discussion of steady-state features. The computation of transient solutions is, even for simple models, a belabored task that may only rarely be simplified. It requires in the best instances the solution of very large systems of differential equations. Both the computational effort and the interpretation of numerical results of transient solutions depend crucially on the choice of the initial conditions. It is next to impossible to give a unified discussion of all types of transient behavior which may arise. In steady-state analysis, the consideration of initial conditions may, of course, be avoided. Provided it is carried out in sufficient detail, the steady-state solution sheds ample light on many behavioral features of the model. In such cases where transient solutions to, for example, a stable queue are desirable for a variety of initial conditions, it is still advisable to obtain detailed information on the steady-state distributions. In performing the belabored computations for the transient phase, the latter will then serve to indicate when the effect of the initial conditions has been attenuated to the point that further numerical solution of the time-dependent equations becomes uninformative.

The organization of this book is as follows. Chapter 1 contains a systematic discussion of the general properties of a class of Markov chains and processes that in the positive recurrent case have a matrix-geometric invariant probability vector. Chapter 2 is a self-contained treatment of the properties of phase type distributions. These distributions arise from a generalization of Erlang's method of stages in a form that is particularly well-suited for numerical computation. Chapter 3 deals with the case of block-tridiagonal transition probability matrices. For such matrices, which frequently arise in applications, much more detailed results may be obtained. Chapter 4 treats the *GI/PH/*1 queue and several of its variants and generalizations. A lengthy discussion of semi-Markovian arrival processes and their applications is given in section 4.2. Chapters 5 and 6 deal with an eclectic variety of models suggested by diverse applications. In these chapters, we concentrate primarily on the formulation and formalization of models and on the interpretation of numerical results.

In concluding this preface, I fulfill the pleasant task of acknowledging the support of several institutions and the cooperation of many persons, both in the research effort and in the writing effort that have led to this book.

Purdue University and, since 1976, the University of Delaware have provided me with an environment and an atmosphere conducive to learning and scholarly pursuit. The Air Force Office of Scientific Research, through grant numbers AFOSR-72-2350 and AFOSR-77-3236, and the National Science Foundation, through grant numbers APR 74-15256 and ENG-7908351, have generously supported my research and that of research associates and graduate students.

Lecture series presented during short-term visits to the Indian Institute of Management, Calcutta, and to Clemson University, Clemson, South Carolina, prior to the delivery of the Mathematical Sciences Lectures at The Johns Hopkins University in 1979 have helped me in finding greater unity and conciseness in much diverse material.

To Professor Moshe Barash of Purdue University go special thanks for broadening my views of applications in technology, notably in manufacturing processes. Dr. Guy Latouche of the Free University of Brussels has generously shared his insights into the problems of computer modeling. To Dr. V. Ramaswami of Drexel University, and to Professor Ohoe Kim of Towson State University, I am grateful for many constructive suggestions and comments on the mathematical results in this book. Dr. Alan F. Karr of The Johns Hopkins University has ably organized the lecture series given there in July, 1979.

D. M. Lucantoni, S. Chakravarthy, and S. Kumar, currently graduate students at the University of Delaware, have lightened our task through their enthusiasm for the subject and through a careful reading of the manuscript. The typescript was prepared by Karen L. Tanner, whose conscientious work is deeply appreciated.

Chapter 1
Matrix-Geometric Invariant Vectors

1.1. INTRODUCTION

The detailed study of many stochastic models is made possible by the presence of embedded Markov chains, which have particular structures. This is well recognized in the theory of queues—see, for example, D. G. Kendall [166-68]—and also in inventory models and branching processes. Each of these areas of applied probability theory has provided several interesting classes of Markov chains, whose analysis has been the basis for many extensions and generalizations. These particular Markov chains are well known to the student of probability models. They have become classical textbook examples. Two instances, which are relevant to the material in this book, are the embedded Markov chains of the elementary $M/G/1$ and $GI/M/1$ queues. The transition probability matrices P_1 and P_2 of these chains are respectively given by

$$P_1 = \begin{vmatrix} a_0 & a_1 & a_2 & a_3 & a_4 & \cdots \\ a_0 & a_1 & a_2 & a_3 & a_4 & \cdots \\ 0 & a_0 & a_1 & a_2 & a_3 & \cdots \\ 0 & 0 & a_0 & a_1 & a_2 & \cdots \\ 0 & 0 & 0 & a_0 & a_1 & \cdots \\ \vdots & \vdots & \vdots & \vdots & \vdots & \end{vmatrix} \tag{1.1.1}$$

and

$$P_2 = \begin{vmatrix} b_0 & a_0 & 0 & 0 & 0 & \cdots \\ b_1 & a_1 & a_0 & 0 & 0 & \cdots \\ b_2 & a_2 & a_1 & a_0 & 0 & \cdots \\ b_3 & a_3 & a_2 & a_1 & a_0 & \cdots \\ b_4 & a_4 & a_3 & a_2 & a_1 & \cdots \\ \vdots & \vdots & \vdots & \vdots & \vdots & \end{vmatrix}. \tag{1.1.2}$$

The scalars a_ν, $\nu \geq 0$, are the terms of a probability density on the nonnegative integers and $b_\nu = 1 - \Sigma_{k=0}^{\nu} a_k$, for $\nu \geq 0$.

The infinite stochastic matrix P_3, given by

$$P_3 = \begin{vmatrix} b_0 & a_0 & 0 & 0 & 0 & \cdots \\ a_2 & a_1 & a_0 & 0 & 0 & \cdots \\ 0 & a_2 & a_1 & a_0 & 0 & \cdots \\ 0 & 0 & a_2 & a_1 & a_0 & \cdots \\ 0 & 0 & 0 & a_2 & a_1 & \cdots \\ \vdots & \vdots & \vdots & \vdots & \vdots & \end{vmatrix}, \tag{1.1.3}$$

is of the types P_1 and P_2. As a discrete parameter Markov chain, it arises in the homogeneous random walk on $\{0, 1, 2, \ldots\}$ with a "barrier" at 0. Its continuous-parameter analogue describes, among others, the simple $M/M/1$ queue.

In specific applications, the quantities a_ν are themselves expressed in terms of the parameters of the system under consideration. For example, for the $M/G/1$ queue, a_ν is given by

$$a_\nu = \int_0^\infty e^{-\lambda u} \frac{(\lambda u)^\nu}{\nu!} \, dF(u), \quad \text{for} \quad \nu \geq 0,$$

in terms of the service time distribution $F(\cdot)$ and the Poisson arrival rate λ of the queue.

In the construction of algorithmic solutions, it is overwhelmingly clear that the *structure* of the embedded Markov chain is of paramount importance. The specific analytic form of the elements of the transition probability matrix is of far less consequence, so that, for example, many simple variations on the $M/G/1$ and $GI/M/1$ queues which are available in the literature have the structure of the matrices P_1 or P_2, but with different expressions for the elements a_ν.

The structural simplicity of the embedded Markov chains of many models for queues, inventories, and dams is frequently obscured by the introduction at an early stage of probability generating functions or Laplace-Stieltjes transforms. The specific definition of the elements of the transition probability matrices often leads to spurious analytic simplifications whose merit is quite clearly lost as soon as one tries to recover numerical results from any one of the abundantly available transform solutions.

The introduction of transform methods also leads the analysis of such models away from probabilistic arguments and, in a natural manner, to methodology suggested by the theory of analytic functions. This has indeed been the case in the development of the extensive literature on stochastic models. As we shall discuss at greater length herein, this methodological preference has had serious consequences in limiting the feasibility of numerical implementation and, in consequence, the practical applicability of many of the proposed results.

In 1970, and after having proposed some formidable transform solutions of our own, we developed a concern for solution methods that are implementable in a general and numerically stable manner and offer detailed information on at least some of the more complex models encountered in the study of practical queueing systems. These are usually rather far removed from the elementary models such as $M/G/1$ and $GI/M/1$.

As if to strengthen our trust in the importance of structural properties, it soon became apparent that many widely differing stochastic models have embedded Markov chains, which are indeed generalizations of the paradigms P_1 and P_2 for the $M/G/1$ and $GI/M/1$ queues respectively.

We first defined the *Markov chains of the M/G/1 type,* which have block-partitioned stochastic matrices P of the form

$$P = \left| \begin{array}{cccccc} B_0 & B_1 & B_2 & B_3 & B_4 & \cdots \\ C_0 & A_1 & A_2 & A_3 & A_4 & \cdots \\ 0 & A_0 & A_1 & A_2 & A_3 & \cdots \\ 0 & 0 & A_0 & A_1 & A_2 & \cdots \\ 0 & 0 & 0 & A_0 & A_1 & \cdots \\ \vdots & \vdots & \vdots & \vdots & \vdots & \end{array} \right|, \tag{1.1.4}$$

where the elements A_ν, $\nu \geq 0$, B_ν, $\nu \geq 1$, B_0, and C_0 are finite, nonnegative matrices of dimensions $m \times m$, $n \times m$, $n \times n$, and $m \times n$ respectively. It was found that for such Markov chains and the related stochastic models, it is possible to give a purely probabilistic analysis and to construct algorithmic solutions, which involve real arithmetic only. A discussion of the mathematical results and of the applications of Markov chains of the $M/G/1$ type would far exceed the bounds we have set for ourselves in writing the present book. A general survey of the results available up to 1977, along with a detailed list of references, is given in the Technical Report [194] by D. M. Lucantoni and M. F. Neuts.

In this book, we shall limit our attention to the *Markov chains of the GI/M/1 type,* first considered from the present viewpoint in M. F. Neuts [227]. In chapter 3, in dealing with block-Jacobi matrices that are of both types, we shall present some of the essential ideas underlying the analysis of Markov chains of the $M/G/1$ type, albeit in a somewhat restricted setting.

The transition probability matrix $\tilde{P}$ of a Markov chain of the $GI/M/1$ type is of the (canonical) form

$$\tilde{P} = \left|\begin{matrix} B_0 & A_0 & 0 & 0 & 0 & \cdots \\ B_1 & A_1 & A_0 & 0 & 0 & \cdots \\ B_2 & A_2 & A_1 & A_0 & 0 & \cdots \\ B_3 & A_3 & A_2 & A_1 & A_0 & \cdots \\ B_4 & A_4 & A_3 & A_2 & A_1 & \cdots \\ \vdots & \vdots & \vdots & \vdots & \vdots & \end{matrix}\right|, \tag{1.1.5}$$

in which all elements are $m \times m$ nonnegative matrices. The form (1.1.5) is called the *canonical* form because the essential results follow from its structure. In section 1.5 we shall modify these results in an elementary manner to yield solutions for several variants that arise in applications. Such variants will also be said to be of the $GI/M/1$ type.

Since the matrix $\tilde{P}$ is stochastic, we clearly have

$$B_k e + \sum_{\nu=0}^{k} A_\nu e = e, \quad \text{for} \quad k \geq 0, \tag{1.1.6}$$

where e is the column vector with all its components equal to one. (The vector e is so defined throughout this book.) The substochastic matrix $\sum_{\nu=0}^{\infty} A_\nu$ will be denoted by A. In most applications, A is *stochastic* and *irreducible*. The discussion in M. F. Neuts [227] was restricted to the case where these two properties hold. In the interest of generality, no such assumptions will be imposed until later, when their utility will be clear.

We do assume that the Markov chain $\tilde{P}$ is *irreducible*. In well-formulated, specific models, this is nearly always obvious, although no easily verifiable criterion for irreducibility is available. In the course of our discussion, some useful necessary conditions for irreducibility will be obtained.

The states of the Markov chain $\tilde{P}$ are denoted by (i, j), $i \geq 0$, and $1 \leq j \leq m$, and are ordered in the lexicographic order, that is, $(0, 1)$, $\ldots$, $(0, m)$, $(1, 1)$, $\ldots$, $(1, m)$, $(2, 1)$, $(2, 2)$, $\ldots$. The set of states $\{(i, 1), \ldots, (i, m)\}$, $i \geq 0$, will be called the *level* i. A pictorial representation of the state space is given in figure 1.1. The structure of the matrix $\tilde{P}$ implies that in a single transition the chain can move upwards only to the next higher level. In moving from the state (i, j) to a state $(i + k, \nu)$, with $k \geq 1$, the chain must visit all intermediate levels at least once. For such Markov chains, J. Keilson [162] has coined the descriptive term *skipfree to the right*. The chains under consideration are skipfree to the right *for levels*; within a level the possible transitions are unrestricted.

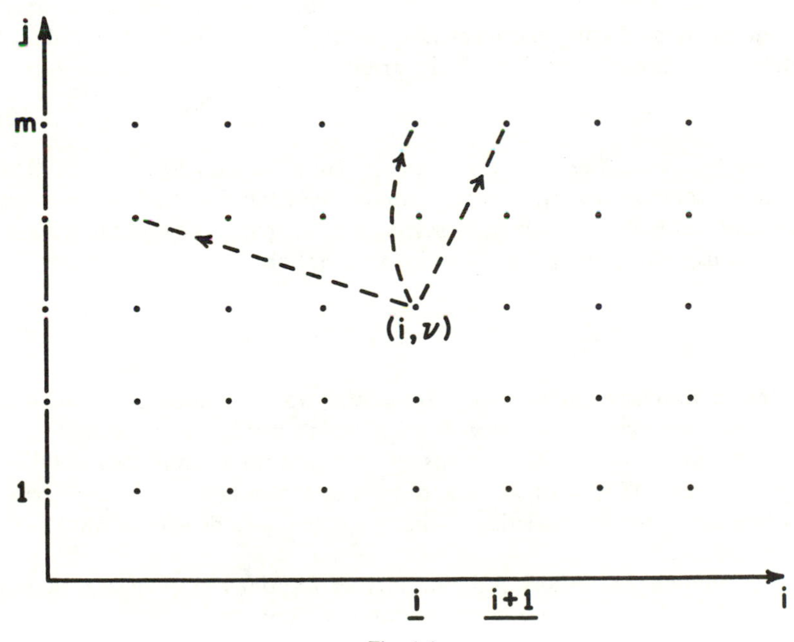

Fig. 1.1

We conclude the introduction by reviewing the earlier results, which were harbingers of the main theorem, which is discussed in the next section.

The earliest result, due to A. Ya. Khinchine [169], states that the Markov Chain P_2 of formula (1.1.2) is irreducible, provided $a_0 > 0$, $a_0 + a_1 < 1$, and *positive recurrent* if and only if

$$\sum_{\nu=1}^{\infty} \nu a_\nu > 1. \tag{1.1.7}$$

The equation

$$z = \sum_{\nu=0}^{\infty} a_\nu z^\nu, \tag{1.1.8}$$

then, has a unique solution ξ in (0, 1), and the invariant probability vector $\boldsymbol{x} = [x_0, x_1, x_2, \ldots]$ of P_2 is given by

$$x_k = (1 - \xi)\xi^k, \quad \text{for} \quad k \geq 0. \tag{1.1.9}$$

In the context of the $GI/M/1$ queue, the stationary queue length prior to arrivals in the stable queue has a *geometric distribution* with parameter $p = 1 - \xi$.

C. B. Winsten [304] observed that for Markov chains obtained from P_2 by modifying the transition probabilities near the "boundary" state 0 in a suitable manner, the invariant probability vector $\boldsymbol{x}$ satisfies

$$x_k = x_{c-1}\xi^{k-c+1}, \quad \text{for} \quad k \geq c - 1, \tag{1.1.10}$$

where ξ is again the unique root of the equation (1.1.8) in (0, 1). The initial c components $x_0, \ldots, x_{c-1}$ may, then, for the positive recurrent case, be uniquely determined by solving a homogeneous system of c linear equations and by using the normalizing equation

$$\sum_{\nu=0}^{c-2} x_\nu + x_{c-1}(1 - \xi)^{-1} = 1. \tag{1.1.11}$$

Winsten also gave a heuristic explanation for the appearance of this *modified geometric stationary density,* which yields the solution to the $GI/M/c$ queue with single arrivals. Recast into the terminology of taboo probabilities, Winsten's argument provided the key to the approach presented in the next section. This approach was developed in M. F. Neuts [236].

A number of authors obtained further solutions to queueing models of the $GI/M/1$ type by expressing the components x_k of the invariant vector $\boldsymbol{x}$ as linear combinations of formal geometric distributions with complex parameters ξ_j and complex mixing coefficients. How these results may be interpreted will be discussed in section 1.6.

More directly in line with the present approach are the paper [89] by R. V. Evans and the Ph.D. thesis [302] of V. Wallace. They showed that for block-Jacobi generators of continuous-parameter Markov processes of the $GI/M/1$ type, called *quasi birth and death processes,* the stationary probability vector x may be partitioned into m-vectors x_k, $k \geq 0$, which are given by

$$x_k = x_0 R^k, \quad \text{for} \quad k \geq 0, \tag{1.1.12}$$

where the square matrix R is the nonnegative solution to a matrix-quadratic equation. This result and its applications are discussed in detail herein. A probability vector x, which satisfies (1.1.12), will be called a *matrix-geometric probability vector.*

1.2. THE MAIN THEOREM

In this section we study necessary and sufficient conditions for the positive recurrence of Markov chains of the type $\tilde{P}$. In doing so, we shall characterize the invariant probability vector x of $\tilde{P}$, which in the positive recurrent case is the unique solution to the infinite system of linear equations

$$x\tilde{P} = x, \qquad xe = 1. \tag{1.2.1}$$

It will be convenient to partition x as $[x_0, x_1, x_2, \ldots]$, where the row vectors x_k, $k \geq 0$, are of dimension m. The steady-state equations (1.2.1) may then be rewritten as

$$\begin{aligned} x_0 &= \sum_{\nu=0}^{\infty} x_\nu B_\nu, \\ x_k &= \sum_{\nu=0}^{\infty} x_{k+\nu-1} A_\nu, \quad \text{for} \quad k \geq 1, \\ \sum_{k=0}^{\infty} x_k e &= 1. \end{aligned} \tag{1.2.2}$$

We now define certain *taboo probabilities,* which play a fundamental role in the sequel. We recall that, given two states i and j and a set H of states of a Markov chain, the *taboo probability* ${}_H P_{ij}^{(n)}$ is the conditional probability that, given that the chain starts in i at time 0, it reaches j at time n without having visited the set H at any of the times $1, \ldots, n-1$. A comprehensive treatment of taboo probabilities is given in K. L. Chung [51, pp. 45–54].

For the Markov chain $\tilde{P}$, we consider the taboo probability ${}_iP^{(n)}_{i,j;i+k,\nu}$ that, starting in the state (i, j), the chain reaches $(i + k, \nu)$ at time n without returning to the level i in between. This probability is defined for $n \geq 0$, $i \geq 0$, $k \geq 1$, $1 \leq \nu \leq m$, and $1 \leq j \leq m$, and is clearly equal to zero for $n < k$. More importantly, for all $n \geq 0$, $k \geq 1$, $1 \leq j \leq m$, and $1 \leq \nu \leq m$, *its value does not depend* on i. This follows immediately from the particular structure of the matrix $\tilde{P}$. A sample path going from (i, j) to $(i + k, \nu)$ which does not reenter the level i cannot visit any state (r, j') with $r \leq i$, $1 \leq j' \leq m$. The taboo probability ${}_iP^{(n)}_{i,j;i+k,\nu}$ therefore depends solely on the submatrix of $\tilde{P}$ obtained by deleting all rows and columns with indices (r, j'), $r \leq i$, $1 \leq j' \leq m$. These submatrices are identical for all $i \geq 0$.

The quantities $R_{j\nu}{}^{(k)}$, defined by

$$R_{j\nu}{}^{(k)} = \sum_{n=0}^{\infty} {}_iP^{(n)}_{i,j;i+k,\nu}, \quad \text{for} \quad k \geq 1, i \geq 0, 1 \leq j \leq m, 1 \leq \nu \leq m, \tag{1.2.3}$$

are of basic interest. $R_{j\nu}{}^{(k)}$ is the expected number of visits to the state $(i + k, \nu)$ before the first return to the level i, given that the chain $\tilde{P}$ starts in the state (i, j). The square matrix with elements $R_{j\nu}{}^{(k)}$, $1 \leq j \leq m$, $1 \leq \nu \leq m$, will be denoted by $R^{(k)}$ for $k \geq 1$. We shall agree to set $R^{(0)} = I$, the identity matrix. Also for $k = 1$, the superscript (1) will be suppressed, and the matrix $R = R^{(1)}$ will be called the *rate matrix* of the Markov chain $\tilde{P}$.

Lemma 1.2.1. If the Markov chain $\tilde{P}$ is positive recurrent, the matrices $R^{(k)}$, $k \geq 1$, are finite.

Proof. By virtue of theorem 3, p. 47, in K. L. Chung [51].

Lemma 1.2.2. The matrix $R^{(k)}$ is the k-th power of the matrix R.

Proof. For $n \geq k + 1$, we have

$$\begin{aligned} {}_iP^{(n)}_{i,j;i+k+1,\nu} &= \sum_{h=1}^{m} \sum_{r=0}^{n} {}_iP^{(r)}_{i,j;i+k,h} \cdot {}_{i+k}P^{(n-r)}_{i+k,h;i+k+1,\nu} \\ &= \sum_{h=1}^{m} \sum_{r=0}^{n} {}_iP^{(r)}_{i,j;i+k,h} \cdot {}_iP^{(n-r)}_{i,h;i+1,\nu}. \end{aligned} \tag{1.2.4}$$

The first equality is obtained from the law of total probability, by conditioning on the time (r) of the *last* visit to the level $i + k$ and on the state $(i + k, h)$ of that visit, before the chain reaches the state $(i + k + 1, \nu)$ at time n. Clearly every path from (i, j) to $(i + k + 1, \nu)$ visits the level $i + k$ at least once. The second equality follows from the fact that ${}_iP_{i,h;i+1,\nu}$ does not depend on i. For $n \le k$, both sides of (1.2.4) are zero, so that the equality holds for all $n \ge 0$.

Summation on n now yields

$$R_{j\nu}{}^{(k+1)} = \sum_{h=1}^{m} \sum_{n=0}^{\infty} \sum_{r=0}^{n} {}_iP^{(r)}_{i,j;i+k,h} \cdot {}_iP^{(n-r)}_{i,h;i+1,\nu}$$

$$= \sum_{h=1}^{m} \sum_{r=0}^{\infty} {}_iP^{(r)}_{i,j;i+k,h} \sum_{n'=0}^{\infty} {}_iP^{(n')}_{i,h;i+1,\nu} = \sum_{h=1}^{m} R_{jh}{}^{(k)} R_{h\nu}, \quad (1.2.5)$$

so that $R^{(k+1)} = R^{(k)}R$, and hence $R^{(k)} = R^k$, for $k \ge 1$.

Lemma 1.2.3. If the Markov chain $\tilde{P}$ is positive recurrent, the matrix R satisfies the equation

$$R = \sum_{k=0}^{\infty} R^k A_k \quad (1.2.6)$$

and is the minimal nonnegative solution to the matrix equation

$$X = \sum_{k=0}^{\infty} X^k A_k.$$

Proof. Clearly ${}_iP^{(1)}_{i,j;i+1,\nu} = (A_0)_{j\nu}$. For $n \ge 2$, we have

$${}_iP^{(n)}_{i,j;i+1,\nu} = \sum_{h=1}^{m} \sum_{k=1}^{\infty} {}_iP^{(n-1)}_{i,j;i+k,h} \cdot (A_k)_{h\nu}, \quad (1.2.7)$$

by conditioning on the state $(i + k, h)$ from which the state $(i + 1, \nu)$ is entered at time n. Note that all terms on the right-hand side for which $k \ge n$ are zero, but it is convenient to write the summation as in (1.2.7). Summation on n clearly yields the stated equation (1.2.6).

Now consider the sequence $\{X(N), N \ge 0\}$ of matrices, obtained by performing successive substitutions in $X = \sum_{k=0}^{\infty} X^k A_k$, starting with $X(0) = 0$. It is readily verified by induction that the sequence $\{X(N)\}$ is entry-wise nondecreasing and that $X(N) \le R$, for $N \ge 0$. The sequence $\{X(N)\}$ converges monotonically to a nonnegative matrix X^*, which, by the dominated convergence theorem, satisfies the equation (1.2.6). X^* is called the *minimal nonnegative solution to that equation.* It is readily verified that any other nonnegative solution X^o satisfies $X^* \le X^o$. In particular, $X^* \le R$.

It now remains to show that $R \le X^*$. To this effect, we define the matrices $R(N, k)$ with elements

$$R_{j\nu}(N, k) = \sum_{n=1}^{N} {}_iP^{(n)}_{i,j;i+k,\nu}, \quad \text{for} \quad k \ge 1, 1 \le j \le m, 1 \le \nu \le m. \quad (1.2.8)$$

Adding the equations (1.2.7) for n ranging from 1 to N yields

$$R(N, 1) = A_0 + \sum_{k=1}^{\infty} R(N-1, k)A_k. \tag{1.2.9}$$

We also add the equations (1.2.4) for n ranging from 1 to $N-1$ and we obtain

$$\begin{aligned}\sum_{n=1}^{N-1} {}_iP^{(n)}_{i,j;i+k+1,\nu} &= [R(N-1, k+1)]_{j\nu} \\ &= \sum_{h=1}^{m} \sum_{n=0}^{N-1} \sum_{r=0}^{n} {}_iP^{(r)}_{i,j;i+k,h} \cdot {}_iP^{(n-r)}_{i,h;i+1,\nu} \\ &= \sum_{h=1}^{m} \sum_{r=0}^{N-1} \sum_{n=0}^{N-1-r} {}_iP^{(r)}_{i,j;i+k,h} \cdot {}_iP^{(n)}_{i,h;i+1,\nu} \qquad (1.2.10) \\ &\le \sum_{h=1}^{m} \sum_{r=0}^{N-1} {}_iP^{(r)}_{i,j;i+k,h} \sum_{n=0}^{N-1} {}_iP^{(n)}_{i,h;i+1,\nu} \\ &= \sum_{h=1}^{m} [R(N-1, k)]_{jh} [R(N-1, 1)]_{h\nu},\end{aligned}$$

or, in matrix form,

$$R(N-1, k+1) \le R(N-1, k)R(N-1, 1). \tag{1.2.11}$$

By induction, we obtain $R(N-1, k) \le [R(N-1, 1)]^k$, for $k \ge 1$, and upon substitution in (1.2.9),

$$R(N, 1) \le \sum_{k=0}^{\infty} [R(N-1, 1)]^k A_k, \quad \text{for} \quad N \ge 2. \tag{1.2.12}$$

We now have that $R(1, 1) = A_0 = X(1)$, so that $R(2, 1) \le X(2)$. By induction, formula (1.2.12) yields that $R(N, 1) \le X(N)$, for $N \ge 1$. The sequence of matrices $R(N, 1)$ is obviously nondecreasing and tends to R, so that the preceding inequality implies that $R \le X^*$, and therefore $R = X^*$.

Theorem 1.2.1. If the Markov chain $\tilde{P}$ is positive recurrent, then

a. for $i \ge 0$, we have

$$x_{i+1} = x_i R, \tag{1.2.13}$$

b. the eigenvalues of R lie *inside* the unit disk,

c. the matrix

$$B[R] = \sum_{k=0}^{\infty} R^k B_k \tag{1.2.14}$$

is stochastic, and

d. the vector x_0 is a positive, left invariant eigenvector of $B[R]$, normalized by

$$x_0(I - R)^{-1}e = 1. \tag{1.2.15}$$

Proof. By conditioning on the time and the state of the last visit to the set $\boldsymbol{i}$, if there is such a visit, we obtain the relation

$$P^{(n)}_{i+1,j;i+1,j} = {}_iP^{(n)}_{i+1,j;i+1,j} + \sum_{\nu=1}^{m} \sum_{r=0}^{n} P^{(r)}_{i+1,j;i,\nu} \cdot {}_iP^{(n-r)}_{i,\nu;i+1,j}, \quad \text{for} \quad n \geq 1. \tag{1.2.16}$$

We add these equations for n ranging from 1 to N and divide the resulting sums by N. As $N \to \infty$, the left-hand side tends to $x_{i+1,j}$, by virtue of the classical ergodic theorem for Markov chains. Since the sum $\Sigma_{n=1}^{\infty} {}_iP^{(n)}_{i+1,j;i+1,j}$ is finite, the first term on the right-hand side tends to zero. The second term,

$$\sum_{\nu=1}^{m} \frac{1}{N} \sum_{n=1}^{N} \sum_{r=0}^{n} P^{(r)}_{i+1,j;i,\nu} \cdot {}_iP^{(n-r)}_{i,\nu;i+1,j} = \sum_{\nu=1}^{m} \frac{1}{N} \sum_{r=0}^{N} P^{(r)}_{i+1,j;i,\nu} \sum_{n=0}^{N-r} {}_iP^{(n)}_{i,\nu;i+1,j},$$

tends to $\Sigma_{\nu=1}^{m} x_{i\nu}R_{\nu j}$, by an elementary summability argument, since $N^{-1} \Sigma_{r=0}^{N} P^{(r)}_{i+1,j;i,\nu}$ tends to $x_{i\nu}$ and $\Sigma_{n=0}^{N} {}_iP^{(n)}_{i,\nu;i+1,j}$ has the limit $R_{\nu j}$, for $1 \leq \nu \leq m$. This completes the proof of the first statement.

Next, we consider the expected number of transitions before the first return to the level $\mathbf{0}$, given that the chain starts in the state $(0, j)$. This quantity is finite if and only if the chain is positive recurrent. It is given by the j-th component of the vector $\Sigma_{k=1}^{\infty} R^k e$, which is finite if and only if the matrix $\Sigma_{k=0}^{\infty} R^k = (I - R)^{-1}$ is finite, or equivalently if all the eigenvalues of R lie *inside* the unit disk.

The statement (*c*) follows by direct calculation. We have by using formulas (1.1.6) and (1.2.6) that

$$\begin{aligned} B[R]e &= \sum_{k=0}^{\infty} R^k B_k e = (I - R)^{-1}e - \sum_{k=0}^{\infty} R^k \sum_{\nu=0}^{k} A_\nu e \\ &= (I - R)^{-1}e - \sum_{\nu=0}^{\infty} \sum_{k=\nu}^{\infty} R^k A_\nu e \\ &= (I - R)^{-1}[I - \sum_{\nu=0}^{\infty} R^\nu A_\nu]e = e. \end{aligned} \tag{1.2.17}$$

Substitution of (1.2.13) into the first equation (1.2.2) now shows that

$$x_0 = x_0 B[R]. \tag{1.2.18}$$

The normalizing equation $\Sigma_{k=0}^{\infty} x_k e = 1$, implies formula (1.2.15). The vector x_0 is strictly positive whenever $\tilde{P}$ is positive recurrent. It is therefore necessary that $B[R]$ have a positive left invariant vector.

Remark. The statements (*b*) and (*d*) provide essentially the sufficient conditions for the positive recurrence of the Markov chain $\tilde{P}$. In order to show this, we need the following structural results on the matrix R.

Lemma 1.2.4. If the matrix $\tilde{P}$ is irreducible, the matrix R cannot have columns that are identically zero, and R must have at least one positive eigenvalue.

Proof. If the ν-th column of R is identically zero, so is the ν-th column of R^k, for $k \geq 1$. This implies that ${}_0P^{(n)}_{0,j;k,\nu} = 0$, for $n \geq 1$, $k \geq 1$, $1 \leq j \leq m$. None of the states (k, ν), $k \geq 1$, can therefore be reached from any of the m states $(0, j)$, so that the Markov chain $\tilde{P}$ is reducible.

The nonnegative matrix R has at least one positive eigenvalue, or else all its eigenvalues are zero. In the latter case, the Cayley-Hamilton theorem yields that $R^m = 0$. This in turn implies that the Markov chain $\tilde{P}$ is reducible, since ${}_0P^{(n)}_{0,j;m,\nu} = 0$, for $n \geq 1$, $1 \leq j \leq m$, $1 \leq \nu \leq m$. No path starting in the level **0** can reach the level ***m***.

Theorem 1.2.2. If the Markov chain $\tilde{P}$ is irreducible, if the minimal nonnegative solution R of the matrix equation (1.2.6) has all its eigenvalues inside the unit disk, and if the stochastic matrix $B[R]$ has a positive left invariant vector, then the Markov chain $\tilde{P}$ is positive recurrent.

Proof. Let R have the stated properties, then $I - R$ is nonsingular and, by the same argument as in the proof of theorem 1.2.1, the matrix $B[R]$ is stochastic. Let x_0 be a positive left invariant vector of $B[R]$, normalized by $x_0(I - R)^{-1}e = 1$. Since R does not have zero columns, the vector $x = [x_0, x_0R, x_0R^2, \ldots]$ is a positive probability vector. By direct verification, it satisfies $x\tilde{P} = x$. The Markov chain $\tilde{P}$ therefore is positive recurrent, and the vector x is its unique stationary probability vector.

Remarks

a. When the matrix A is irreducible, an *explicit* necessary and sufficient condition for the matrix R to have all its eigenvalues inside the unit disk may be given. This condition is discussed in section 1.3.

b. For specific models, it is frequently the case that the matrix R has zero rows. As will be discussed herein, this is useful in simplifying the algorithmic solution of such models.

c. If the Markov chain $\tilde{P}$ is irreducible, the inverse $(I - A_1)^{-1}$ exists. The (j, j')-element of the matrix $\Sigma_{\nu=0}^{\infty} A_1{}^{\nu}$ is the conditional expected number of visits to the state (i, j') before leaving the level i, given that the chain $\tilde{P}$ starts in the state (i, j). All elements of this matrix are finite, unless the level i contains an irreducible subclass. If $\tilde{P}$ is irreducible, this is not possible.

The matrix equation (1.2.6) may therefore be equivalently written as

$$R = \left(\sum_{k=0, k \neq 1}^{\infty} R^k A_k \right) (I - A_1)^{-1}, \tag{1.2.19}$$

and may be solved by successive substitutions, starting with $R = 0$. These successive iterates exceed those in ordinary successive substitutions, unless $A_1 = 0$. By using the form (1.2.19), one usually saves a few iterations in the computation of R to within a given tolerance. One needs, however, to compute and store the inverse $(I - A_1)^{-1}$, and for large values of m the gain in computer time may not outweigh the cost of the added storage.

d. One of the advantages of a purely probabilistic solution to stochastic models is the common availability of one or more *internal accuracy checks*. These are (loosely) defined as easily verifiable equalities, which must be satisfied by some simple functions of the results of much more complicated computations. The fact that in the positive recurrent case the matrix $B[R]$ is *stochastic* provides us with a first internal accuracy check on R. Additional checks will be discussed herein.

Internal accuracy checks usually have a probabilistic interpretation. For example, the matrix $B[R]$ is stochastic in the (positive) recurrent case, because with probability one there is *some* state (k, j'), $k \geq 0$, $1 \leq j' \leq m$, which a chain $\tilde{P}$ will visit immediately prior to its first return to the level $\mathbf{0}$, given any initial state $(0, j)$, $1 \leq j \leq m$. Formally,

$$\sum_{j'=1}^{m} \left\{ (B_0)_{jj'} + \sum_{h=1}^{m} \sum_{n=1}^{\infty} \sum_{k=1}^{\infty} {}_{\mathbf{0}}P^{(n)}_{0,j;k,h} (B_k)_{hj'} \right\} = 1, \tag{1.2.20}$$

for $1 \leq j \leq m$. Upon recalling the definition of $R^{(k)}$ and by using lemma 1.2.2, formula (1.2.20) is seen to be equivalent to $B[R]\mathbf{e} = \mathbf{e}$.

e. The eigenvalue of R with largest modulus is real and *positive,* provided $\tilde{P}$ is irreducible, as will henceforth be tacitly assumed. It will be called the *spectral radius* $sp(R)$ of R. The condition that all the eigenvalues of R lie inside the unit disk will be written as $\text{sp}(R) < 1$.

f. The next theorem provides us with another internal accuracy check on the computation of R.

Theorem 1.2.3. If the matrix A is stochastic and $\mathrm{sp}(R) < 1$, then

$$A_0 e = \sum_{k=1}^{\infty} R^k \sum_{\nu=k+1}^{\infty} A_\nu e. \tag{1.2.21}$$

Proof. Replace A_1 by $A - A_0 - \sum_{\nu=2}^{\infty} A_\nu$ in equation (1.2.6) and postmultiply by e. Since $Ae = e$, the terms Re and RAe cancel, and we obtain

$$(I - R)A_0 e = \sum_{\nu=2}^{\infty} (R - R^\nu)A_\nu e.$$

The matrix series on the right converges absolutely, so that the arithmetic operations involved are permitted. Since $I - R$ is nonsingular, we obtain

$$A_0 e = \sum_{\nu=2}^{\infty} \sum_{r=1}^{\nu-1} R^r A_\nu e = \sum_{k=1}^{\infty} R^k \sum_{\nu=k+1}^{\infty} A_\nu e.$$

Remark. Formula (1.2.21) also has an interesting interpretation. We consider the stationary version of the Markov chain $\tilde{P}$ and focus our attention on the state (i, j), $i \geq 1$, $1 \leq j \leq m$. Clearly $(A_0 e)_j$ is the probability that, if the chain is in (i, j), a transition to the right of level i occurs. The j-th component of the vector in the right-hand side of (1.2.21) is the probability that a path, which last visited the level i in the state (i, j) and moved from there to the right of i, crosses back to the left of the level i. In the positive recurrent chain $\tilde{P}$, the set of sample paths moving at any time from (i, j) to the right of i is matched by a set of paths of equal probability, which, having moved to the right of i from (i, j), cross back to the left of i at that same time. The relation (1.2.21) therefore has the significance of a *conservation law*.

1.3. THE CASE WHERE THE MATRIX A IS IRREDUCIBLE

In most practical cases it follows naturally from the description of the model that the matrix A is *irreducible*. The discussion in M. F. Neuts [227], written before the probabilistic significance of the matrix R, obtained in [236], had been clarified, was indeed restricted to the case where A is irreducible. The material in this section deals primarily with an easily verifiable, necessary, and sufficient condition for $\mathrm{sp}(R) < 1$, and is developed by matrix analytic rather than probabilistic arguments. In somewhat special practical models, it does arise that the matrix A is reducible. The spectrum of the matrix R must then be studied by ad hoc arguments. A detailed example is due to D. M. Lucantoni [195] and involves a situation where A is upper triangular. This example is also briefly discussed in sections 1.4 and 6.5.

Let now A be *irreducible.* We denote $\mathrm{sp}(A)$ by ξ, $0 < \xi \leq 1$, and the corresponding left eigenvector by $\boldsymbol{\pi}$. The vector $\boldsymbol{\pi}$ is chosen to be strictly positive and is normalized by $\boldsymbol{\pi e} = 1$. When A is stochastic, then clearly $\xi = 1$.

We define the matrix $A^*(z)$ by

$$A^*(z) = \sum_{k=0}^{\infty} A_k z^k, \quad \text{for} \quad 0 \leq z \leq 1. \tag{1.3.1}$$

For $0 < z \leq 1$, the matrix $A^*(z)$ is then clearly an *irreducible,* nonnegative matrix. We denote its maximal eigenvalue by $\chi(z)$. The function $\chi(z)$ is analytic on $(0, 1)$ (by the implicit function theorem), is continuous at $z = 1$, and may be defined by continuity at $z = 0$. $\chi(0)$ is then the maximal eigenvalue of A_0.

The vector $\boldsymbol{\beta}^*$ is defined by

$$\boldsymbol{\beta}^* = \sum_{k=1}^{\infty} kA_k \boldsymbol{e}. \tag{1.3.2}$$

Some components of $\boldsymbol{\beta}^*$ may be infinite.

The following theorem, essential to the arguments that follow, is due to J. F. C. Kingman [170].

Theorem 1.3.1. For $s \geq 0$, the function $\log \chi(e^{-s})$ is convex.

Lemma 1.3.1. The minimal, nonnegative solution R of equation (1.2.6) satisfies $\mathrm{sp}(R) \leq \xi \leq 1$.

Proof. Consider the sequence of matrices $\{R(n), n \geq 0\}$, obtained by successive substitutions, starting with $R(0) = 0$, in equation (1.2.6). We then trivially have $\boldsymbol{\pi} R(0) < \xi \boldsymbol{\pi}$. If $\boldsymbol{\pi} R(n) \leq \xi \boldsymbol{\pi}$, then

$$\boldsymbol{\pi} R(n+1) = \boldsymbol{\pi} \sum_{k=0}^{\infty} R^k(n) A_k \leq \boldsymbol{\pi} \sum_{k=0}^{\infty} A_k \xi^k = \boldsymbol{\pi} A^*(\xi) \leq \boldsymbol{\pi} A = \xi \boldsymbol{\pi}.$$

It follows by induction that $\boldsymbol{\pi} R(n) \leq \xi \boldsymbol{\pi}$, for $n \geq 0$, and hence also $\boldsymbol{\pi} R \leq \xi \boldsymbol{\pi}$. This implies, since $\boldsymbol{\pi}$ is a positive vector, that $\mathrm{sp}(R) \leq \xi$.

Corollary 1.3.1. If A is irreducible, but not stochastic, then $\mathrm{sp}(R) < 1$. Provided $B[R]$ has a positive left invariant vector, the irreducible Markov chain $\tilde{P}$ is then always positive recurrent.

Remarks

a. Lemma 1.3.1 holds whenever A has a *positive* left eigenvector corresponding to ξ. A sufficient, but not necessary, condition for this is that A is irreducible. The detailed structure of reducible, nonnegative matrices, which have a maximal eigenvalue with a corresponding *positive* eigenvector is given by theorem 6, pp. 77–78, in F. R. Gantmacher [104].

If A has a positive left eigenvector $\boldsymbol{\pi}$ corresponding to ξ, all the matrices $R(n)$ belong to the compact set of all nonnegative matrices, satisfying $\boldsymbol{\pi}X \leq \xi\boldsymbol{\pi}$. The minimal solution R, then, clearly exists, and $\text{sp}(R) \leq \xi$.

b. In view of corollary 1.3.1, we may now restrict our attention to the case where A is *stochastic*. For irreducible Markov chains $\tilde{P}$, we have already shown that $\text{sp}(R) > 0$.

Lemma 1.3.2. If the maximal eigenvalue η of R is positive, then

$$\eta = \chi(\eta), \tag{1.3.3}$$

and the left eigenvector $\boldsymbol{u}^*$ of R, corresponding to η, may be chosen to be *positive*. The eigenvalue η is of geometric multiplicity one.

Proof. The vector $\boldsymbol{u}^*$ may clearly be chosen to be nonnegative and is not equal to $\mathbf{0}$. Premultiplying by $\boldsymbol{u}^*$ in equation (1.2.6) yields

$$\eta\boldsymbol{u}^* = \boldsymbol{u}^*A^*(\eta). \tag{1.3.4}$$

Since $A^*(\eta)$ is irreducible and $\eta > 0$, it follows from the definition of $\chi(z)$, that for some positive vector $\boldsymbol{u}(\eta)$, we have

$$\chi(\eta)\boldsymbol{u}(\eta) = \boldsymbol{u}(\eta)A^*(\eta). \tag{1.3.5}$$

Since $\boldsymbol{u}(\eta)$ is positive and $\boldsymbol{u}^*$ nonnegative, it follows from remark 3, pp. 63–64, in F. R. Gantmacher [104] that η must be the maximal eigenvalue of $A^*(\eta)$, and hence $\eta = \chi(\eta)$. The maximal eigenvalue of $A^*(\eta)$ is simple, and every corresponding, nonnegative eigenvector is strictly positive. The vector $\boldsymbol{u}^*$ is therefore positive.

Every left eigenvector of R, corresponding to η, is also an eigenvector of $A^*(\eta)$. Since η, as the maximal eigenvalue of $A^*(\eta)$, is of geometric multiplicity one, the matrix R cannot have two linearly independent eigenvectors corresponding to η.

Lemma 1.3.3. The left-hand derivative $\chi'(1-)$ of $\chi(z)$ at $z = 1$ is given by

$$\chi'(1-) = \boldsymbol{\pi}\boldsymbol{\beta}^*. \tag{1.3.6}$$

Proof. Let $\boldsymbol{u}(z)$ and $\boldsymbol{\nu}(z)$ be respectively left and right eigenvectors of $A^*(z)$, corresponding to $\chi(z)$, for $0 < z \leq 1$. The vectors $\boldsymbol{u}(z)$ and $\boldsymbol{\nu}(z)$ may be chosen so that their components are continuously differentiable functions of z and such that the normalizations

$$\boldsymbol{u}(z)\boldsymbol{\nu}(z) = \boldsymbol{u}(z)\boldsymbol{e} = 1, \qquad \boldsymbol{u}(1-) = \boldsymbol{\pi}, \qquad \boldsymbol{\nu}(1-) = \boldsymbol{e} \tag{1.3.7}$$

hold for $0 < z \leq 1$.

Differentiation of the expressions in the equation

$$A^*(z)\boldsymbol{\nu}(z) = \chi(z)\boldsymbol{\nu}(z) \tag{1.3.8}$$

leads to

$$[\chi(z)I - A^*(z)]\boldsymbol{\nu}'(z) = [A^{*\prime}(z) - \chi'(z)I]\boldsymbol{\nu}(z). \tag{1.3.9}$$

Premultiplying by $\boldsymbol{u}(z)$ in (1.3.9), we obtain

$$\chi'(z) = \boldsymbol{u}(z)A^{*\prime}(z)\boldsymbol{\nu}(z), \quad \text{for} \quad 0 < z \leq 1. \tag{1.3.10}$$

Letting z tend to one and noting that $\boldsymbol{\beta}^* = A^{*\prime}(1-)\boldsymbol{e}$, we obtain (1.3.6).

Remark. The derivatives of $\chi(z)$, $\boldsymbol{u}(z)$, and $\boldsymbol{\nu}(z)$ at $z = 1-$ play an important role in the analysis of Markov chains of the $M/G/1$ type. Lemma 1.3.3 is only a simple, particular case of results used in M. F. Neuts [224] and in D. M. Lucantoni and M. F. Neuts [194].

Lemma 1.3.4. The equation

$$z = \chi(z), \quad \text{for} \quad 0 < z \leq 1, \tag{1.3.11}$$

has at most two roots, one of which is $z = 1$. If $\boldsymbol{\pi\beta}^* \leq 1$, then $z = 1$ is the only root. If $\boldsymbol{\pi\beta}^* > 1$, there is a second root $z = \eta$, $0 < \eta < 1$, provided that for some z_0, $0 < z_0 < 1$, we have

$$z\chi'(z) < \chi(z), \quad \text{for} \quad 0 < z < z_0. \tag{1.3.12}$$

If $\chi(0) > 0$, condition (1.3.12) is always satisfied.

Proof. By setting $z = e^{-s}$, for $s \geq 0$, the equation (1.3.11) becomes

$$s = -\log \chi(e^{-s}), \quad \text{for} \quad s \geq 0. \tag{1.3.13}$$

Since the matrix A is stochastic, that equation is satisfied by $s = 0$. By Kingman's theorem 1.3.1, the right-hand side is concave for $s \geq 0$. It is also positive and increasing for $s > 0$. The derivative of the right-hand side at $s = 0+$ is equal to $\chi'(1-) = \boldsymbol{\pi\beta}^*$.

If $\boldsymbol{\pi\beta}^* \leq 1$, the graph of the right-hand side lies everywhere below the bisectrix line. There cannot be a second point of intersection with that line, and therefore $s = 0$ is the unique solution of (1.3.13).

If $\pi\beta^* > 1$, and $\chi(0) > 0$, the graph of the right-hand side swings above the bisectrix line and then approaches a horizontal asymptote as $s \to +\infty$. There will be a unique second intersection at some point $s_0 > 0$, and η is then given by $\eta = e^{-s_0} < 1$.

If $\pi\beta^* > 1$, and $\chi(0) = 0$, there will be a unique second intersection at some point $s_0 > 0$, provided the derivative of the right-hand side is eventually less than one. This is equivalent to the condition (1.3.12).

Lemma 1.3.5. If the maximal eigenvalue η of R is positive, it is the *smallest positive* solution of the equation (1.3.11).

Proof. Let η^o be the smallest positive solution of equation (1.3.11), and denote by $\mathbf{u}^o > \mathbf{0}$ a left eigenvector of the irreducible matrix $A^*(\eta^o)$, corresponding to $\chi(\eta^o) = \eta^o$. Let the matrices $R(n)$ be defined as in the proof of lemma 1.3.1. Clearly $\mathbf{u}^o R(0) < \eta^o \mathbf{u}^o$. Moreover, if $\mathbf{u}^o R(n) \leq \eta^o \mathbf{u}^o$, then

$$\mathbf{u}^o R(n+1) = \mathbf{u}^o \sum_{k=0}^{\infty} R^k(n) A_k \leq \mathbf{u}^o A^*(\eta^o) = \eta^o \mathbf{u}^o.$$

All the matrices $R(n)$, $n \geq 0$, and hence also R, therefore, belong to the compact set of matrices $X \geq 0$, satisfying $\mathbf{u}^o X \leq \eta^o \mathbf{u}^o$. This implies that $0 < \eta \leq \eta^o$. Since η is also a solution to (1.3.11), it follows that $\eta = \eta_0$.

Corollary 1.3.2. If $\pi\beta^* \leq 1$, the Markov chain $\tilde{P}$ cannot be positive recurrent.

Proof. If the Markov chain $\tilde{P}$ is positive recurrent, the maximal eigenvalue, $0 < \eta < 1$, of R must be a solution to equation (1.3.11). By lemma 1.3.4, this is impossible.

Corollary 1.3.3. If $\chi(0) = 0$, $\pi\beta^* > 1$, and if condition (1.3.12) fails to hold, the Markov chain $\tilde{P}$ cannot be irreducible and positive recurrent.

Proof. By the same argument as in corollary 1.3.2.

Remark. The conditions of corollary 1.3.3 can only be satisfied when the Markov chain $\tilde{P}$ is highly reducible. The successive iterates $R(n)$ then all have spectral radius zero, and so does the matrix R. The condition $\pi\beta^* > 1$ may still be satisfied, but becomes irrelevant, since by lemma 1.2.4, $\tilde{P}$ is reducible.

The preceding discussion may now be summarized as follows:

Theorem 1.3.2. If the matrix A is irreducible, the irreducible Markov chain $\tilde{P}$ is positive recurrent if and only if $\pi\beta^* > 1$, and the stochastic matrix $B[R]$ has a strictly positive left invariant vector. The partitioned invariant probability vector x of $\tilde{P}$ is then given by

$$x_i = x_0 R^i, \quad \text{for} \quad i \geq 0,$$
$$x_0 = x_0 B[R], \tag{1.3.14}$$
$$x_0(I - R)^{-1}e = 1,$$

where R is the minimal nonnegative solution to the matrix equation

$$R = \sum_{k=0}^{\infty} R^k A_k. \tag{1.3.15}$$

Remarks

a. In practice, $\pi\beta^* > 1$ is the only condition requiring detailed verification. It is usually obvious that $B[R]$ is irreducible, so its invariant vector is positive. No specific models where $B[R]$ is reducible are known to date.

b. The following is an example where $\chi(0) = 0$. Let C_0, C_1, C_2, and C_3 be positive square matrices of order m and let $C = C_0 + C_1 + C_2 + C_3$ be stochastic. We define the matrices A_k, $k \geq 0$, of order $2m$ by

$$A_0 = \begin{vmatrix} 0 & 0 \\ C_0 & 0 \end{vmatrix}, \qquad A_1 = \begin{vmatrix} C_1 & C_0 \\ C_2 & C_1 \end{vmatrix}, \qquad A_2 = \begin{vmatrix} C_3 & C_2 \\ 0 & C_3 \end{vmatrix},$$

and $A_k = 0$, for $k \geq 3$. All the eigenvalues of A_0 are zero, so that $\chi(0) = 0$. The matrix A, given by

$$A = \begin{vmatrix} C_1 + C_3 & C_0 + C_2 \\ C_0 + C_2 & C_1 + C_3 \end{vmatrix},$$

is positive. The vector π is given by $\frac{1}{2}[\gamma, \gamma]$, where γ is the invariant probability vector of C. The vector β^* is the *direct sum*† of the vectors $(C_0 + C_1)e + 2(C_2 + C_3)e$ and $(C_1 + C_2)e + 2C_3e$. The condition $\pi\beta^* > 1$ is, after elementary calculations, seen to be equivalent to

$$\sum_{\nu=1}^{3} \nu\gamma C_\nu e > 1.$$

†The *direct sum* of $\nu \geq 2$ column vectors $u_1, \ldots, u_\nu$ of dimensions $m_1, \ldots, m_\nu$ is the column vector u of dimension $m_1 + \cdots + m_\nu$ obtained by stringing out the components of $u_1, \ldots, u_\nu$ into a single vector. The analogous definition holds for row vectors. This construction is frequently used throughout this book.

If this condition is satisfied, the minimal nonnegative solution S of the equation

$$S = \sum_{\nu=0}^{3} S^\nu C_\nu$$

is positive, and $\mathrm{sp}(S) < 1$. A straightforward calculation shows that the matrix R, the minimal nonnegative solution of $R = R^2 A_2 + RA_1 + A_0$, is given by

$$R = \begin{vmatrix} 0 & 0 \\ S & S^2 \end{vmatrix},$$

so that $\mathrm{sp}(R) = [\mathrm{sp}(S)]^2 < 1$.

The iterate $R(2)$, given by

$$R(2) = \begin{vmatrix} 0 & 0 \\ C_0 C_1 + C_0 & C_0{}^2 \end{vmatrix},$$

already has a positive spectral radius. Note, however, that the successive iterates are, in general, not of the form

$$\begin{vmatrix} 0 & 0 \\ U & U^2 \end{vmatrix}.$$

c. The following is a simple example where the iterates $R(n)$ remain stuck in the set of matrices of spectral radius zero. Let the matrices A_k, $k \geq 0$, of order $m = 2$, be given by

$$A_0 = \begin{vmatrix} 0 & 0 \\ b & 0 \end{vmatrix}, \quad A_1 = \begin{vmatrix} 0 & 0 \\ 0 & 0 \end{vmatrix}, \quad A_2 = \begin{vmatrix} 1-a & a \\ 0 & 1-b \end{vmatrix},$$

and $A_k = 0$, for $k \geq 3$. With $0 < a < 1$, $0 < b < 1$, the matrix A is positive. Since $A_0{}^2 = 0$, $R = R(n)$, for all $n \geq 1$. The inequality $\pi\beta^* > 1$ is equivalent to $a + b > 2ab$, and holds in particular for $a = b = 1/2$. In the latter case, $\chi(z) = 1/2(z^2 + z)$, and $\chi'(z) = 1/2(2z + 1)$. Since $z\chi'(z) - \chi(z) = 1/2 z^2 \geq 0$, condition (1.3.12) does not hold. Notice that regardless of the definition of B_0, the Markov chain $\tilde{P}$ with the A_k-matrices as above is always reducible.

Theorem 1.3.3 (Uniqueness Theorem). The matrix R is the *unique* nonnegative matrix with spectral radius η which satisfies the equation (1.2.6).

Proof. By lemma 1.3.2, the matrix R has a positive left eigenvector $\boldsymbol{u}^*$ corresponding to η. Let S be a nonnegative solution to (1.2.6) with $\text{sp}(S) = \eta$ and nonnegative left eigenvector $\boldsymbol{u}^o$. The same argument as in lemma 1.3.2 shows that $\boldsymbol{u}^o$ is a left eigenvector of $A^*(\eta)$, corresponding to η. Since $A^*(\eta)$ is irreducible, $\boldsymbol{u}^o = k\boldsymbol{u}^*$, for some $k > 0$. We may choose $\boldsymbol{u}^o = \boldsymbol{u}^* > \mathbf{0}$.

Since R is the minimal solution, it follows that $S \geq R$, and also $\boldsymbol{u}^*(S - R) = \mathbf{0}$. If any element of S exceeds the corresponding element of R, the preceding equality cannot hold. It is therefore clear that $S = R$.

Theorem 1.3.4. If A is irreducible, the matrix R is irreducible if and only if every row of the matrix A_0 has at least one positive element.

If, in addition, the matrix $\sum_{k=1}^{\infty} R^{k-1}A_k$ is irreducible, the matrix R is positive.

Proof. If A_0 has a zero row, the corresponding row in all the matrices $R(n)$ and hence in R is zero. This shows that the condition is necessary.

If A_0 has no vanishing rows, neither does R, since $R \geq A_0$. If R is reducible, we may find an irreducible square submatrix of order r, $1 \leq r < m$, such that, upon an appropriate common permutation of the row and column indices, the matrix R may be written as

$$R = \begin{vmatrix} R_1 & 0 \\ R_3 & R_4 \end{vmatrix}, \tag{1.3.16}$$

where R_1 is irreducible and of order r. The same common permutation of the row and column indices is performed on the coefficient matrices A_k, $k \geq 0$. The matrices A_k, $k \geq 0$, are then partitioned in the same manner as R and written as

$$A_k = \begin{vmatrix} A_k(1) & A_k(2) \\ A_k(3) & A_k(4) \end{vmatrix}, \quad \text{for} \quad k \geq 0.$$

The equation (1.2.6) leads to

$$R_1 = \sum_{k=0}^{\infty} R_1{}^k A_k(1), \qquad \sum_{k=0}^{\infty} R_1{}^k A_k(2) = 0. \tag{1.3.17}$$

Let R_1 have the maximal eigenvalue η_1 with corresponding *positive* left eigenvector $\boldsymbol{u}$; then the second equation in (1.3.17) leads to

$$\sum_{k=0}^{\infty} \eta_1{}^k \boldsymbol{u} A_k(2) = \mathbf{0}.$$

This implies that all the matrices $A_k(2)$, $k \geq 0$, are zero. The matrix A is therefore reducible. It follows that if A is irreducible, so is the matrix R.

To prove the second statement, we write the equation (1.2.6) as

$$R = A_0\left(I - \sum_{k=1}^{\infty} R^{k-1}A_k\right)^{-1}. \tag{1.3.18}$$

If $\sum_{k=1}^{\infty} R^{k-1}A_k$ is irreducible and the inverse in (1.3.18) exists, the inverse is positive, since a nonnegative matrix C of order m is irreducible if and only if $I + C + \cdots + C^{m-1}$ is positive. The inverse $(I - C)^{-1}$ is strictly larger than that matrix. Since A_0 has no zero rows, the matrix R is then clearly positive.

To show that the inverse exists, consider a positive left eigenvector $\boldsymbol{u}^*$ of R, corresponding to η. We then have

$$\boldsymbol{u}^* \sum_{k=1}^{\infty} R^{k-1}A_k = \eta^{-1}\boldsymbol{u}^*[A^*(\eta) - A_0] = \boldsymbol{u}^* - \eta^{-1}\boldsymbol{u}^*A_0 \leq \boldsymbol{u}^*,$$

with strict inequality for some components. This, together with the irreducibility of $\sum_{k=1}^{\infty} R^{k-1}A_k$, implies that this matrix is of spectral radius less than one, by corollary 2.3, p. 551, in S. Karlin and H. M. Taylor [156].

Remarks

a. The irreducibility of $\sum_{k=1}^{\infty} R^{k-1}A_k$ is not necessary for the existence of the inverse in formula (1.3.18).

b. In many practical models, it is clear at a glance that R has to be positive. Often the matrix A_0 has no zero rows and the matrix A_1 is irreducible. By considering formula (1.2.19), it is clear that R is then positive.

c. From a computational viewpoint, it is not desirable that R is positive, since the dimension m is the main limiting factor on the algorithmic utility of the solutions based on our results. In some models, to be discussed herein, the matrix A_0, and hence also R, has many zero rows. This may be exploited to make computation feasible for large values of m.

The following theorem is given for the sake of completeness and will not be used in the sequel.

Theorem 1.3.5. If the Markov chain $\tilde{P}$ is null-recurrent, the matrix $B[R]$ is stochastic. If $B[R]$ has a positive, invariant vector $\boldsymbol{x}_0$, then the vector $\boldsymbol{x} = [\boldsymbol{x}_0, \boldsymbol{x}_1, \boldsymbol{x}_2, \ldots]$ with $\boldsymbol{x}_k = \boldsymbol{x}_0R^k$, $k \geq 1$, is a positive *stationary measure* for $\tilde{P}$. If R is irreducible, the vector $\boldsymbol{x}$ is bounded.

Proof. If the Markov chain $\tilde{P}$ is null-recurrent, then any path starting in $(0, j)$, $1 \leq j \leq m$, eventually returns to the level $\mathbf{0}$ with probability one. The same argument as that in the remark (d) following theorem 1.2.2 shows that $B[R]$ is stochastic. The matrix R now has spectral radius one. The vector x satisfies the equation $x = x\tilde{P}$, but its components do not have a finite sum.

If R is irreducible, $x_0 R^k$ converges to a finite vector (when R is aperiodic) and remains bounded in any case.

Remark. By an argument related to Markov chains of $M/G/1$ type, it is possible to show that the chain can be null-recurrent only if $\pi\beta^* = 1$. As this topic is peripheral to the subject matter of this book, we do not present the detailed argument here.

The transience (including the null-recurrent case) of the irreducible Markov chain $\tilde{P}$, whenever $\pi\beta^* \leq 1$, also follows from criteria established by F. G. Foster [99]; see also J. W. Cohen [61, pp. 21–31].

1.4. A CASE WHERE THE MATRIX A IS REDUCIBLE

In rare applications where the matrix A is reducible, an ad hoc argument is needed to identify conditions under which the matrix R has all its eigenvalues inside the unit disk. As an example, we consider the case where all A_k, $k \geq 0$, and hence also A, are *upper triangular.* This situation arises, for example, in the study of a $GI/M/c$ queue in which customers who do not have to wait are served at a different rate from those who do have to wait before entering service. A detailed discussion of this model is given in D. M. Lucantoni [195], and a brief presentation is given in section 6.5 herein. We only consider the case where A is *stochastic.*

Each of the matrices $R(n)$, $n \geq 0$, and hence also R, is upper triangular. The eigenvalues of R are its diagonal elements R_{jj}, $1 \leq j \leq m$. Introducing the generating functions

$$\phi_j(z) = \sum_{k=0}^{\infty} (A_k)_{jj} z^k, \quad \text{for} \quad 1 \leq j \leq m, \tag{1.4.1}$$

the equation (1.2.6) implies that

$$R_{jj} = \phi_j(R_{jj}), \quad \text{for} \quad 1 \leq j \leq m. \tag{1.4.2}$$

Each R_{jj} is the smallest nonnegative solution to the corresponding equation.

There are three possibilities. If $\phi_j(0) = 0$, then $R_{jj} = 0$. If $\phi_j(0) > 0$ and $\phi_j(1) < 1$, the convex nondecreasing graph of $\phi_j(z)$ has a unique intersection at some $z = R_{jj}$, $0 < R_{jj} < 1$, with the bisectrix line $y = z$. If $\phi_j(0) > 0$ and $\phi_j(1) = 1$, the equation $z = \phi_j(z)$ has a unique solution $z = R_{jj}$, $0 < R_{jj} < 1$, if and only if $\phi_j'(1-) > 1$. If $\phi_j'(1-) \leq 1$, the smallest solution R_{jj} is equal to one.

The following theorem is now obvious.

Theorem 1.4.1. The eigenvalues of the matrix R all lie in the interval (0, 1) if and only if $\phi_j'(1-) > 1$, for every index j, $1 \leq j \leq m$, for which $A_{jj} = 1$, and $(A_0)_{jj} > 0$.

Remarks

a. Notice that $A_{jj} = 1$ is necessary for $\phi_j(1) = 1$. This always holds for $j = m$, but may also hold for some other indices j.

b. Once the quantities R_{jj}, $1 \leq j \leq m$, are known, the special structure of the matrix R makes it possible to evaluate successively the quantities $R_{j,j+1}$, $1 \leq j \leq m - 1$, $R_{j,j+2}$, $1 \leq j \leq m - 2$, and finally $R_{1,m}$ by solving systems of *linear* equations. This is simple in principle, but quite belabored in practice and not always numerically stable. It may still be preferable in this case to compute R iteratively, starting with the matrix $R(0) = \text{diag}[R_{11}, \ldots, R_{mm}]$.

c. This example illustrates the importance of the boundary states. Many states in such a Markov chain $\tilde{P}$ can be reached from a given state (i, j) only via a path passing through the boundary states. It now becomes a matter of prime concern to verify that the matrix $B[R]$ has a positive left invariant vector.

1.5. COMPLEX BOUNDARY BEHAVIOR

As in the scalar case $m = 1$, discussed by C. B. Winsten [304], we may modify the matrix $\tilde{P}$ to have a much more complicated structure near the lower boundary. In this section we discuss a general instance of such boundary behavior, which leads to the consideration of *modified matrix-geometric invariant vectors.*

Assume that the stochastic matrix $\tilde{P}$ may be partitioned as

$$\tilde{P} = \begin{vmatrix} B_{00} & B_{01} & 0 & 0 & 0 & \cdots \\ B_{10} & B_{11} & A_0 & 0 & 0 & \cdots \\ B_{20} & B_{21} & A_1 & A_0 & 0 & \cdots \\ B_{30} & B_{31} & A_2 & A_1 & A_0 & \cdots \\ B_{40} & B_{41} & A_3 & A_2 & A_1 & \cdots \\ \vdots & \vdots & \vdots & \vdots & \vdots & \end{vmatrix}, \tag{1.5.1}$$

where the matrices B_{00} and B_{01} are of dimensions $(m_1 - m) \times (m_1 - m)$ and $(m_1 - m) \times m$ respectively. The matrices B_{k0}, $k \geq 1$, are of dimensions $m \times (m_1 - m)$. All other blocks B_{k1}, $k \geq 1$, and A_k, $k \geq 0$, are square matrices of order m. The first m_1 states are called the *boundary states,* and $m_1 \geq m$. Since $m_1 = m$ corresponds to the canonical form, we need only consider the case $m_1 > m$ here.

We shall assume that the matrix $\tilde{P}$ is irreducible and limit our discussion to the case where $A = \sum_{k=0}^{\infty} A_k$ is stochastic. The vector x is partitioned into an $(m_1 - m)$-vector x_0 and m-vectors x_k, $k \geq 1$. The matrix R is defined as before.

The matrix $B[R]$ is defined by

$$B[R] = \begin{vmatrix} B_{00} & B_{01} \\ \sum_{k=1}^{\infty} R^{k-1}B_{k0} & \sum_{k=1}^{\infty} R^{k-1}B_{k1} \end{vmatrix}. \tag{1.5.2}$$

Lemma 1.5.1. If $\text{sp}(R) < 1$, the matrix $B[R]$ is stochastic.

Proof. It suffices to verify that

$$\sum_{k=1}^{\infty} R^{k-1}(B_{k0} + B_{k1})e = \sum_{k=1}^{\infty} R^{k-1}\left(e - \sum_{\nu=0}^{k} A_\nu e\right) = e.$$

This is done by direct calculation as in the proof of theorem 1.2.1.

Theorem 1.5.1. The irreducible Markov chain $\tilde{P}$ is positive recurrent if and only if $\text{sp}(R) < 1$ and the stochastic matrix $B[R]$ has a positive left invariant m_1-vector $[x_0, x_1]$. Normalizing the vector $[x_0, x_1]$ by

$$x_0 e + x_1(I - R)^{-1}e = 1, \tag{1.5.3}$$

the invariant probability vector x of $\tilde{P}$ is given by

$$x_k = x_1 R^{k-1}, \quad \text{for} \quad k \geq 1. \tag{1.5.4}$$

Proof. It suffices to verify that the stated vector x satisfies the equations $x = x\tilde{P}$, $xe = 1$, and this is straightforward.

Remarks

a. The matrix $\tilde{P}$ in (1.5.1) differs from the canonical form (1.1.5) only in the definition of the transition probabilities from the boundary states. The probabilistic significance of the matrix R therefore remains the same, but applies only to nonboundary states.

b. The vector $x = [x_0, x_1, x_1R, x_1R^2, \ldots]$ is called a *modified matrix-geometric invariant vector.* As we shall see in many instances that follow, the vector x_0 may frequently be further partitioned for ease of computation.

c. We note that in the matrix $\tilde{P}$ in (1.5.1), the submatrix $[B_{00}, B_{01}]$, by which the first m_1 columns protrude above the regular pattern formed by the other columns, must have fewer rows than columns. If this were not the case, the Markov chain $\tilde{P}$ would be reducible.

d. There are some forms of partitioned matrices that are superficially similar to the form (1.5.1), but that need to be appropriately repartitioned before our methods of analysis can be applied to them.

A matrix of the form

$$\tilde{P} = \begin{vmatrix} B_0 & A_0 & 0 & 0 & \cdots \\ B_1 & A_1 & A_0 & 0 & \cdots \\ B_2 & A_2 & A_1 & A_0 & \cdots \\ B_3 & A_3 & A_2 & A_1 & \cdots \\ \vdots & \vdots & \vdots & \vdots & \end{vmatrix}, \tag{1.5.5}$$

where the matrices B_ν, $\nu \geq 0$, are $m \times m_1$ with $m_1 < m$ is actually of the canonical form, but is inappropriately partitioned. By repartitioning the matrix into $m \times m$ blocks, a matrix of the canonical form is obtained. Although a partition such as (1.5.5) may be suggested by the specifics of the model, it is not the correct one to identify the matrix-geometric solution.

In the study of queues with bulk arrivals in groups of bounded size, we encounter stochastic matrices, partitioned, for example, as follows:

$$\tilde{P} = \begin{vmatrix} B_{00} & B_{01} & B_{02} & A_0 & 0 & 0 & 0 & 0 & 0 & \cdots \\ B_{10} & B_{11} & B_{12} & A_1 & A_0 & 0 & 0 & 0 & 0 & \cdots \\ B_{20} & B_{21} & B_{22} & A_2 & A_1 & A_0 & 0 & 0 & 0 & \cdots \\ B_{30} & B_{31} & B_{32} & A_3 & A_2 & A_1 & A_0 & 0 & 0 & \cdots \\ B_{40} & B_{41} & B_{42} & A_4 & A_3 & A_2 & A_1 & A_0 & 0 & \cdots \\ B_{50} & B_{51} & B_{52} & A_5 & A_4 & A_3 & A_2 & A_1 & A_0 & \cdots \\ \vdots & \vdots & \vdots & \vdots & \vdots & \vdots & \vdots & \vdots & \vdots & \end{vmatrix}, \tag{1.5.6}$$

where all blocks are square matrices of order m. This matrix becomes of the canonical form when it is repartitioned into blocks of order $3m$.

Denoting these larger blocks by A_k^* and B_k^*, $k \geq 0$, we see that the A_k^* are then given by

$$A_0^* = \begin{vmatrix} A_0 & 0 & 0 \\ A_1 & A_0 & 0 \\ A_2 & A_1 & A_0 \end{vmatrix}, \quad A_k^* = \begin{vmatrix} A_{3k} & A_{3k-1} & A_{3k-2} \\ A_{3k+1} & A_{3k} & A_{3k-1} \\ A_{3k+2} & A_{3k+1} & A_{3k} \end{vmatrix}, \tag{1.5.7}$$

for $k \geq 1$. This particular structure of the matrices A_k^* has only minor computational advantages, which lie in the reduced storage—since only the matrices A_k and not the A_k^* need to reside in memory. No particular, useful structure of the R-matrix follows, in general, from the form of the A_k^*. The matrix $A^* = \sum_{k=0}^{\infty} A_k^*$, is a block-circulant matrix given by

$$A^* = \begin{vmatrix} C_0 & C_2 & C_1 \\ C_1 & C_0 & C_2 \\ C_2 & C_1 & C_0 \end{vmatrix}, \tag{1.5.8}$$

where $C_r = \sum_{\nu=0}^{\infty} A_{3\nu+r}$, for $r = 0, 1, 2$. As we shall see, this simplifies the calculation of the quantity $\pi\beta^*$ somewhat. The structure (1.5.6) is readily generalized to the case where the matrices A_k^* are of order Km, $K \geq 2$. The case $K = 3$ was chosen for ease of notation.

We see that relatively simple modifications of queueing models, such as group arrivals or services, may result in computational problems of considerably increased dimension, even though the basic structure of the Markov chain $\tilde{P}$ is preserved. For a discussion of this point, we refer to M. F. Neuts [233].

1.6. THE APPROACH USING COMPLEX ANALYSIS

Some of the specific stochastic models discussed in the following have been treated in the literature by the use of methods from complex analysis. This approach involves a number of steps without a probabilistic interpretation. It also presents a number of substantial technical difficulties, which have only rarely been discussed in full detail. Related to these mathematical difficulties are certain intermediate steps, which can create pitfalls in numerical implementations. As we shall point out in what follows, these hazards are brought about by the *method* of solution and are not inherent to the problem itself. In this section we present a brief outline of the complex variable method as it applies to matrices $\tilde{P}$ of the canonical form.

For such matrices, one sets out to construct an invariant probability vector $\boldsymbol{x} = [\boldsymbol{x}_0, \boldsymbol{x}_1, \boldsymbol{x}_2, \ldots]$ of the form

$$\boldsymbol{x}_i = \sum_{k=1}^{m} \boldsymbol{a}_k \eta_k{}^i, \quad \text{for} \quad i \geq 0, \tag{1.6.1}$$

where the (usually complex) numbers η_k, $1 \leq k \leq m$, are of modulus less than one. These quantities and the row vectors $\boldsymbol{a}_k$, $1 \leq k \leq m$, are to be determined.

Before proceeding, we note that when $\tilde{P}$ is irreducible and the rate matrix R is *diagonalizable,* the matrix-geometric solution is immediately seen to be equivalent to the form (1.6.1). The quantities η_k are then clearly the *eigenvalues of the matrix R,* and the vectors $\boldsymbol{a}_k$ are readily expressed in terms of $\boldsymbol{x}_0$ and the left and right eigenvectors of R. Even then, the form (1.6.1) is, because of potential loss of significance, far less desirable for numerical work than the highly stable matrix-geometric form.

Substitution of the expressions in (1.6.1) into the system of linear equations $\boldsymbol{x} = \boldsymbol{x}\tilde{P}$ leads after elementary calculations to the equations

$$\sum_{k=1}^{m} \boldsymbol{a}_k \eta_k{}^{i-1}[\eta_k I - A^*(\eta_k)] = \boldsymbol{0}, \quad \text{for} \quad i \geq 1, \tag{1.6.2}$$

for the nonboundary states, and to

$$\sum_{k=1}^{m} \boldsymbol{a}_k = \sum_{k=1}^{m} \boldsymbol{a}_k \sum_{\nu=0}^{\infty} B_\nu \eta_k{}^\nu \tag{1.6.3}$$

for the boundary states of the level $\boldsymbol{0}$. The matrix $A^*(z)$ is defined, as in (1.3.1), by $A^*(z) = \sum_{\nu=0}^{\infty} A_\nu z^\nu$, for $|z| \leq 1$.

The normalizing equation $\boldsymbol{xe} = 1$ leads to

$$\sum_{k=1}^{m} (1 - \eta_k)^{-1} \boldsymbol{a}_k \boldsymbol{e} = 1. \tag{1.6.4}$$

One now examines the zeros in the unit disk of the function

$$\phi(z) = \det[zI - A^*(z)]. \tag{1.6.5}$$

It is usually relatively easy to show that the function $\phi(z)$ has exactly m zeros η_k, $1 \leq k \leq m$, satisfying $|\eta_k| \leq 1$. Some cases, usually corresponding to the critical queue, must be handled with special care due to the appearance of additional multiple zeros at $z = 1$. The customary method of proof is based on *Rouché's theorem,* which is used with great care, for example, in L. Takács [293].

A much more detailed argument is now required to ascertain under which conditions all m zeros η_k actually lie *inside* the unit disk. For a particular model, this is ascertained in full detail by examining the specific form of the function $\phi(z)$, for example, in the paper by A. B. Clarke [58]. Other discussions have followed in essence the argument based on the convexity of the function $\log \chi(e^{-s})$, $s \geq 0$, which we have presented in section 1.3. In this manner, the condition $\pi\beta^* > 1$ is obtained, although in most papers it appears in a specific explicit form, appropriate only to the model at hand.

At this point, having identified conditions for $|\eta_k| < 1$, for $1 \leq k \leq m$, it is necessary to examine the multiplicities of the zeros η_k. In cases where multiple zeros are present, the trial solution (1.6.1) needs to be modified. Since $\phi(z)$ is usually a transcendental function, or at best a polynomial of high degree, it is possible to carry out this examination only in the rarest of cases. Proofs of the distinctness of the zeros η_k are given for such cases, for example, in N. T. J. Bailey [9] and in P. E. Boudreau, J. S. Griffin, and M. Kac [32]. In E. Çinlar [52], a set of matrix-algebraic conditions are given under which the analysis of a specific queueing model (the *M/SM/*1 queue) can be carried through in the presence of multiple zeros. These conditions would, however, require explicit knowledge of the zeros η_k and can hence only rarely be actually verified.

Further discussions are usually restricted to the case where the roots η_k are simple. The vectors $\mathbf{a}_k$ are now in principle determined each up to a multiplicative constant θ_k by requiring that

$$\mathbf{a}_k[\eta_k I - A^*(\eta_k)] = \mathbf{0}, \quad \text{for} \quad 1 \leq k \leq m. \tag{1.6.6}$$

The equations (1.6.2) are then satisfied. The remaining m constants θ_k may be found, again in principle, by solving the m homogeneous linear equations, derived from (1.6.3), and the additional inhomogeneous equation, obtained from (1.6.4). The coefficients of these equations are, in general, complex numbers.

There is some similarity between this method and the classical spectral decomposition method for the solution of systems of linear differential equations with constant coefficients. Both methods are algorithmically unattractive for the same reasons. These are:

a. Even if the parameters of the solution (1.6.1) can be accurately computed, the *form* of that solution may easily cause loss of significance when implemented.

b. For large values of m, it is difficult to compute all the η_k accurately, even using techniques which exploit the fact that complex η_k must occur in conjugate pairs.

c. The matrix R often has the eigenvalue zero with high multiplicity. For indices k, for which $\eta_k = 0$, the vector $\mathbf{a}_k$ may be set equal to zero, and the method needs to be appropriately modified. For particular choices of the parameters of the model, the zeros η_k may be multiple, and a solution of the form (1.6.1) then fails altogether. Although one might argue that this will not arise "in practice," the occurrence of distinct, but closely clustered, η_k is common and just as pernicious. The systems of equations for the $\boldsymbol{a}_k$ then have coefficient matrices that are "close" to matrices of rank lower than $m - 1$, and the system of linear equations for the θ_k is close to singular. The accurate computation of the $\boldsymbol{a}_k$ is then exceedingly delicate.

d. Since there are no good internal accuracy checks, the only warnings of grave numerical errors are given by the appearance of negative or otherwise clearly erroneous values for the components of $\boldsymbol{x}_i$ or by the failure of the computed vector $\boldsymbol{x}$ to satisfy the steady-state equations to within a satisfactory tolerance. Backsubstitution of $\boldsymbol{x}$ requires much computational effort when m is large, and if the vector $\boldsymbol{x}$ fails this test, there is no clear way to recovery and a more accurate solution.

All these difficulties, both mathematical and numerical, are greatly compounded in the case where $\boldsymbol{x}$ is modified matrix-geometric, particularly when m_1 is large, as is frequently the case in applications.

1.7. CONTINUOUS-PARAMETER MARKOV PROCESSES

There is an entirely analogous theory for continuous-parameter Markov processes, whose infinitesimal generator $\tilde{Q}$ may be partitioned similarly to the matrix $\tilde{P}$ in (1.5.1). The earliest examples of matrix-geometric solutions arose, in fact, for Markov processes, rather than for chains, in the work of R. V. Evans [89] and V. Wallace [302].

In this section we shall restate the principal results obtained up to this point for the case of a continuous-parameter Markov process with generator $\tilde{Q}$ of the form

$$\tilde{Q} = \begin{vmatrix} B_0 & A_0 & 0 & 0 & 0 & \cdots \\ B_1 & A_1 & A_0 & 0 & 0 & \cdots \\ B_2 & A_2 & A_1 & A_0 & 0 & \cdots \\ B_3 & A_3 & A_2 & A_1 & A_0 & \cdots \\ B_4 & A_4 & A_3 & A_2 & A_1 & \cdots \\ \vdots & \vdots & \vdots & \vdots & \vdots & \end{vmatrix}. \tag{1.7.1}$$

The off-diagonal elements of $\tilde{Q}$ are nonnegative. Its diagonal elements are all (strictly) negative and

$$\sum_{\nu=0}^{k} A_\nu e + B_k e = \mathbf{0}, \quad \text{for} \quad k \geq 0. \tag{1.7.2}$$

We shall limit our discussion to the canonical form $\tilde{Q}$, displayed in (1.7.1). The elementary modifications, needed to handle more involved boundary behavior, are entirely similar to those discussed in section 1.5. In the subsequent chapters, we shall have occasion to illustrate these by many specific examples.

We assume that $\tilde{Q}$ is *irreducible.* Necessary conditions for this are that the $m \times m$ matrices B_0 and A_1 are *nonsingular.* The inverses B_0^{-1} and A_1^{-1} are then *nonpositive.* The matrix $A = \sum_{k=0}^{\infty} A_k$ has negative diagonal and nonnegative off-diagonal elements. Its row sums are nonpositive. We shall limit our attention to the case where $Ae = \mathbf{0}$. The matrix A is then called a *conservative, semi-stable matrix.* Such matrices are the analogues of stochastic matrices, and most of their properties may indeed be derived from those of stochastic matrices by an elementary transformation, which we shall use in the proof of theorem 1.7.1. We shall for brevity refer to a finite or infinite conservative, semi-stable matrix as a *generator.* Most of the classical properties of such matrices are listed in R. Bellman [14], M. Marcus and H. Minc [200], or F. R. Gantmacher [104].

Positive recurrence of the process $\tilde{Q}$ is equivalent to the existence of a positive probability vector $\boldsymbol{x} = [\boldsymbol{x}_0, \boldsymbol{x}_1, \boldsymbol{x}_2, \ldots]$, satisfying the equations

$$\sum_{\nu=0}^{\infty} \boldsymbol{x}_\nu B_\nu = \mathbf{0},$$

$$\sum_{\nu=0}^{\infty} \boldsymbol{x}_{k+\nu-1} A_\nu = \mathbf{0}, \quad \text{for} \quad k \geq 1. \tag{1.7.3}$$

Let τ be any real number satisfying

$$\tau \geq \max_j \max[-(B_0)_{jj}, -(A_1)_{jj}] > 0, \tag{1.7.4}$$

then the equations (1.7.3) may be rewritten

$$\boldsymbol{x}_0 = \sum_{\nu=0}^{\infty} \boldsymbol{x}_\nu B_\nu',$$

$$\boldsymbol{x}_k = \sum_{\nu=0}^{\infty} \boldsymbol{x}_{k+\nu-1} A_\nu', \quad \text{for} \quad k \geq 1, \tag{1.7.5}$$

where $B_\nu' = \delta_{\nu 0}I + \tau^{-1}B_\nu$, and $A_\nu' = \delta_{\nu 1}I + \tau^{-1}A_\nu$, for $\nu \geq 0$. The equations (1.7.5) are entirely similar to the equations (1.2.2). The recurrence properties of the process $\tilde{Q}$ may therefore be deduced from those of the chain $\tilde{P}'$ with

$$\tilde{P}' = \begin{vmatrix} B_0' & A_0' & 0 & 0 & \cdots \\ B_1' & A_1' & A_0' & 0 & \cdots \\ B_2' & A_2' & A_1' & A_0' & \cdots \\ B_3' & A_3' & A_2' & A_1' & \cdots \\ \vdots & \vdots & \vdots & \vdots & \end{vmatrix}. \tag{1.7.6}$$

Theorem 1.7.1. The irreducible Markov process $\tilde{Q}$ is *positive recurrent* if and only if the minimal nonnegative solution R of the equation

$$\sum_{k=0}^{\infty} R^k A_k = 0 \tag{1.7.7}$$

has $\mathrm{sp}(R) < 1$, and if there exists a positive vector x_0 such that

$$x_0 B[R] = \mathbf{0}. \tag{1.7.8}$$

The matrix $B[R] = \sum_{k=0}^{\infty} R^k B_k$ is a generator.

The stationary probability vector x, satisfying $x\tilde{Q} = \mathbf{0}$, $xe = 1$, is then given by

$$x_k = x_0 R^k, \quad \text{for} \quad k \geq 0, \tag{1.7.9}$$

and x_0 is normalized by

$$x_0(I - R)^{-1}e = 1. \tag{1.7.10}$$

The matrix R has a positive maximal eigenvalue η. If the generator A is irreducible, the left eigenvector u^* of R, corresponding to η, is determined up to a multiplicative constant and may be chosen to be positive. The matrix R then satisfies $\mathrm{sp}(R) < 1$, if and only if

$$\pi A_0 e < \sum_{k=2}^{\infty} (k-1)\pi A_k e, \tag{1.7.11}$$

where π is given by $\pi A = \mathbf{0}$, $\pi e = 1$.

Whenever $\eta = \mathrm{sp}(R) < 1$, the equality

$$A_0 e = \sum_{k=1}^{\infty} R^k \sum_{\nu=k+1}^{\infty} A_\nu e \tag{1.7.12}$$

holds.

Proof. The statements of this theorem follow immediately from the results for the discrete case, applied to the matrix $\tilde{P}'$. It suffices to replace the matrices A_ν' and B_ν' by their definitions in terms of A_ν and B_ν, for $\nu \geq 0$. Only minor details then remain to be verified.

Since $B'[R] = \sum_{\nu=0}^{\infty} R^\nu B_\nu' = I + \tau^{-1}B[R]$ is stochastic, it follows that $B[R]e = \mathbf{0}$. Since the off-diagonal elements of $B[R]$ are nonnegative, its diagonal elements are clearly nonpositive. If any diagonal element of $B[R]$ were zero, the corresponding row of $B[R]$ and hence of B_0 would vanish. Since B_0 is nonsingular, this is impossible.

In order to prove (1.7.11), we observe that π is also the invariant probability vector of $A' = \sum_{k=0}^{\infty} A_k'$, and that $\beta^* = e + \tau \sum_{k=1}^{\infty} kA_k e$. The inequality $\pi\beta^* > 1$ is therefore equivalent to $\sum_{k=1}^{\infty} k\pi A_k e > 0$. Replacing $A_1 e$ by $-A_0 e - \sum_{k=2}^{\infty} A_k e$, we obtain formula (1.7.11). The remaining formula (1.7.12) is obtained from (1.2.21) by direct translation.

Remark. Formula (1.7.11), the key condition for positive recurrence, has an appealing interpretation. Defining the average transition rates to the left and to the right from the level i, $i > 0$, by the right and left sides of (1.7.11) respectively, the condition says that the rate to the left should exceed that to the right. The components of the vector π provide the correct weights for the appropriate definitions of the average transition rates.

In order to identify the probabilistic significance of the matrix R for the continuous parameter case, we define a second discrete parameter chain $\tilde{P}^o$, of the form

$$\tilde{P}^o = \begin{vmatrix} B_0^o & A_0^{o\prime} & 0 & 0 & \cdots \\ B_1^o & A_1^o & A_0^o & 0 & \cdots \\ B_2^o & A_2^o & A_1^o & A_0^o & \cdots \\ B_3^o & A_3^o & A_2^o & A_1^o & \cdots \\ \vdots & \vdots & \vdots & \vdots & \end{vmatrix}, \tag{1.7.13}$$

where

$$\begin{aligned} B_0^o &= \Delta_0^{-1}B_0 + I, \\ A_0^{o\prime} &= \Delta_0^{-1}A_0, \\ B_\nu^o &= \Delta^{-1}B_\nu, \qquad &&\text{for} \quad \nu \geq 1, \\ A_\nu^o &= \Delta^{-1}A_\nu + \delta_{1\nu}I, \qquad &&\text{for} \quad \nu \geq 0, \\ \Delta &= -\text{diag}(A_1), \\ \Delta_0 &= -\text{diag}(B_0). \end{aligned} \tag{1.7.14}$$

The matrix $\tilde{P}^o$, which is in general not of the canonical form, is the transition probability matrix of the Markov chain obtained by considering the process $\tilde{Q}$ immediately after its successive transitions. The matrix $\tilde{P}^o$ could also have served in the proof of theorem 1.7.1, but would have required more belabored calculations, as well as an argument from Markov renewal theory.

The matrix $A^o = \Sigma_{\nu=0}^{\infty} A_\nu{}^o$ is given by

$$A^o = I + \Delta^{-1}A, \tag{1.7.15}$$

and is stochastic if and only if $A\boldsymbol{e} = \boldsymbol{0}$. Similarly A^o is irreducible if and only if A is, and its invariant vector $\boldsymbol{\pi}^o$ is then given by

$$\boldsymbol{\pi}^o = (\boldsymbol{\pi}\Delta\boldsymbol{e})^{-1}\boldsymbol{\pi}\Delta. \tag{1.7.16}$$

The vector $\boldsymbol{\beta}^o = \Sigma_{\nu=1}^{\infty} \nu A_\nu \boldsymbol{e}$ is given by

$$\boldsymbol{\beta}^o = \boldsymbol{e} + \Delta^{-1} \sum_{\nu=1}^{\infty} \nu A^o{}_\nu \boldsymbol{e}, \tag{1.7.17}$$

and a ready calculation shows that $\boldsymbol{\pi}^o\boldsymbol{\beta}^o > 1$, if and only if the inequality (1.7.11) holds.

Let R^o now be the minimal nonnegative solution of the matrix equation $R^o = \Sigma_{\nu=0}^{\infty} R^{o\nu} A_\nu{}^o$, then the matrices R and R^o are related by the following theorem.

Theorem 1.7.2. We have that

$$R = \Delta R^o \Delta^{-1}, \tag{1.7.18}$$

and $\text{sp}(R) = \text{sp}(R^o)$. In the positive recurrent case, the invariant probability vector $\boldsymbol{x}^o = [\boldsymbol{x}_0{}^o, \boldsymbol{x}_1{}^o, \boldsymbol{x}_2{}^o, \ldots]$ of $\tilde{P}^o$ is related to the stationary probability vector $\boldsymbol{x}$ of $\tilde{Q}$ by

$$\boldsymbol{x}_0{}^o = h\boldsymbol{x}_0\Delta_0,$$

$$\boldsymbol{x}_k{}^o = h\boldsymbol{x}_0 R^k \Delta = \boldsymbol{x}_1{}^o R^{ok-1}, \quad \text{for} \quad k \geq 1, \tag{1.7.19}$$

where $h^{-1} = \boldsymbol{x}_0\Delta_0\boldsymbol{e} + \boldsymbol{x}_0 R(I - R)^{-1}\Delta\boldsymbol{e}$.

Proof. Equation (1.7.7) is clearly equivalent to

$$\Delta^{-1}R\Delta = \sum_{\nu=0}^{\infty} (\Delta^{-1}R\Delta)^\nu \Delta^{-1}(A_\nu + \delta_{1\nu}\Delta) = \sum_{\nu=0}^{\infty} (\Delta^{-1}R\Delta)^\nu A_\nu{}^o,$$

so that $R^o = \Delta^{-1}R\Delta$. The matrices R and R^o trivially have the same eigenvalues. The expressions (1.7.19) are verified by direct substitution into the equations $\boldsymbol{x}^o\tilde{P}^o = \boldsymbol{x}^o$, $\boldsymbol{x}^o\boldsymbol{e} = 1$.

Remarks

a. The elementary probability of a transition in the time interval $(t, t + dt)$ in the stationary version of the process $\tilde{Q}$ is clearly given by $h^{-1}dt$. The quantity h^{-1} may therefore be interpreted as the *stationary transition rate* of the process $\tilde{Q}$.

b. Writing (1.7.18) as

$$R_{j\nu} = \left(\frac{R_{j\nu}{}^o}{\Delta_{\nu\nu}}\right)\Delta_{jj}, \quad \text{for} \quad 1 \le j \le m,\ 1 \le \nu \le m, \tag{1.7.20}$$

we see that $R_{j\nu}$ is Δ_{jj} times the expected time spent in the state $(i + 1, \nu)$ before the first return to the level i, $i \ge 0$, given that the process $\tilde{Q}$ starts in the state (i, j). Equivalently $R_{j\nu}$ is the expected time spent in the state $(i + 1, \nu)$ before the first return to i, expressed in the time unit $\Delta_{jj}{}^{-1}$, given the initial state (i, j). $\Delta_{jj}{}^{-1}$ is of course the mean sojourn time in the state (i, j), for $i \ge 1$.

1.8. MARGINAL AND CONDITIONAL PROBABILITIES, MOMENTS

Whenever the vector x is matrix-geometric, easily computable formulas for several marginal and conditional probabilities may be obtained. The corresponding expressions for the modified matrix-geometric case are somewhat more involved, but are derived with similar ease.

The sum $x^* = \sum_{k=0}^{\infty} x_k$ is given by

$$x^* = x_0(I - R)^{-1}. \tag{1.8.1}$$

Its component $x_j{}^*$ is the stationary probability that the chain $\tilde{P}$ is in the set $\{(i, j), i \ge 0\}$, and for specific models $x_j{}^*$ usually has a useful practical interpretation. The vector x^* is easily computed by solving the equation $x^* - x^*R = x_0$, and provides us with a stopping criterion for the computation of the vectors x_k, $k \ge 0$.

The marginal density $\{q_k, k \ge 0\}$, given by

$$q_k = x_k e = x_0 R^k e, \qquad k \ge 0, \tag{1.8.2}$$

is, in most applications in queueing theory, the steady-state density of the queue length.

The conditional probability densities $\{q_k(j), k \ge 0\}$, for $1 \le j \le m$, given by

$$q_k(j) = (x_j{}^*)^{-1} x_{kj}, \qquad k \ge 0, \tag{1.8.3}$$

play a useful role in many applications. They show, for example, how the queue length varies with the indices j and so serve to shed light on the fluctuations of the queue length. We shall return to the interpretation of these densities for specific numerical examples in chapter 6.

Factorial moments of the marginal and conditional densities are readily expressed in convenient closed forms by using the formula

$$L(r) = \sum_{k=r}^{\infty} \frac{k!}{(k-r)!} x_k = r! x_0 R^r (I - R)^{-1-r}. \tag{1.8.4}$$

Whenever tabulation of the probability distributions, rather than the densities $\{q_k\}$ or $\{q_k(j)\}$, is preferred, it is more efficient and slightly more accurate to use the formula

$$\sum_{\nu=0}^{k} x_\nu = x^* - x^* R^{k+1}, \quad \text{for} \quad k \geq 0, \tag{1.8.5}$$

rather than to compute and then sum the vectors x_ν.

1.9. COMPUTATIONAL PROCEDURES

For specific models, some overhead computations are usually required to evaluate the matrices A_k and B_k, $k \geq 0$, in terms of the parameters of the model. These computations may involve a substantial amount of effort, but may often be efficiently performed by taking special features of the model into account. This will be illustrated by a number of examples that follow.

Limiting our attention to the common case where the stochastic matrix A is irreducible, we next evaluate the vectors π and β^*. For many models, these vectors are available in an explicit or easily computable form. If they are not, then β^* is best evaluated by truncating the sum

$$\beta^* = \sum_{k=0}^{\infty} \left[e - \sum_{\nu=0}^{k} A_\nu e \right] \tag{1.9.1}$$

at a sufficiently high index k.

The vector π requires the solution of the system of linear equations $\pi A = \pi$, $\pi e = 1$. This may best be done by the use of library routines, which contain safeguards against the occasional ill-conditioning problems of stochastic matrices. It is customary to assign one of the components, say π_1, a temporary value such as $\pi_1 = 1$, and to solve the system of $m - 1$ linear equations, obtained from $\pi A = \pi$, by setting $\pi_1 = 1$ throughout and deleting one of the equations. This system then uniquely determines π up to a multiplicative constant, which may be found by use of the normalization $\pi e = 1$.

There is an elegant trick, obtained in the M.Sc. thesis by P. G. Wachter [301] and discussed in C. C. Paige, G. P. H. Styan, and P. G. Wachter [247], by which the normalizing condition is used to eliminate one of the unknowns, say π_m. The equations for $\boldsymbol{\pi}$ may be rewritten as

$$\sum_{\nu=1}^{m} \pi_\nu(\delta_{\nu j} - A_{\nu j} + A_{mj}) = A_{mj}, \quad \text{for} \quad 1 \leq j \leq m$$

$$\pi_m = 1 - (\pi_1 + \cdots + \pi_{m-1}). \qquad (1.9.2)$$

The first $m - 1$ equations do not involve π_m and may be shown to form a nonsingular system. They are used to determine $\pi_1, \ldots, \pi_{m-1}$, and π_m is then obtained from the final equation.

If $\boldsymbol{\pi}\boldsymbol{\beta}^* > 1$, we proceed. In many problems the matrices A_k are zero for some K satisfying $k \geq K \geq 3$, but when this is not the case, the problem of truncating the summation in the equation $R = \sum_{k=0}^{\infty} R^k A_k$ needs to be considered. A complete error analysis of the computation of R has not yet been carried out at this time, but we have, with some theoretical justification and on the basis of a considerable amount of computational experience, arrived at the recommendation to truncate that sum and the sequence $\{A_k\}$ at the smallest index K for which

$$\max_j \left\{\boldsymbol{\beta}^* - \sum_{k=0}^{K} \left(\mathbf{e} - \sum_{\nu=0}^{k} A_\nu \mathbf{e}\right)\right\}_j < \epsilon, \qquad (1.9.3)$$

where $\epsilon = 10^{-8}$ was found to be entirely adequate.

In all computations, we have found it expedient to compute R by *successive substitutions,* after rewriting the equation (1.2.6) in the form (1.2.19), whenever the dimension of the problem made the evaluation and storage of the inverse $(I - A_1)^{-1}$ convenient.

When K is small, typically $K \leq 4$, and m is small, there is some merit in computing the approximating sequence $R'(n)$, defined by

$$R'(n+1) = A_0\left(I - \sum_{\nu=1}^{\infty} R'^{\nu-1}(n) A_\nu\right)^{-1}, \qquad R'(0) = 0, \qquad (1.9.4)$$

which also converges increasingly to R. This procedure was recommended for matrix-quadratic equations in V. Wallace [302]. The need to perform a matrix inversion at each iteration soon causes much larger processing times, in spite of a reduced number of iterations.

Methods based on the Newton-Kantorovich approximation, discussed, for example, in J. M. Ortega and W. C. Rheinboldt [245], suffer from the same problem. The number of iterations is typically greatly reduced, but the amount of computation required by each iteration is so vastly increased that the unsophisticated approach of (modified) successive substitutions generally outperforms those more advanced methods by a factor of two or three in the processing times for sample computations. For a comparison of various methods for a related matrix equation, we refer to M. F. Neuts [223].

It is clearly advantageous to evaluate matrix polynomials of the form $\Sigma_{k=0}^{K} R^k A_k$ by *Horner's algorithm,* that is, by successively computing the matrices

$$X_K = A_K,$$

$$X_{K-\nu} = A_{K-\nu} + RX_{K-\nu+1}, \quad \text{for} \quad 1 \leq \nu \leq K. \tag{1.9.5}$$

This algorithm minimizes both the number of matrix multiplications and the number of matrix additions needed to evaluate $\Sigma_{k=0}^{K} R^k A_k$.

Iterates $R(n)$ are now computed until

$$\max_{i,j}[R_{ij}(n+1) - R_{ij}(n)] < \epsilon. \tag{1.9.6}$$

The last iterate R^* is then subject to the accuracy checks dictated by $B[R]e = e$ and by formula (1.2.21). If these are not satisfactorily met, additional iterations may be performed with an appropriate reduction of ϵ. We have only rarely found this necessary in practice; both sides of the terms arising in the accuracy tests typically agree up to six or seven decimal places.

The computation of x_0, or of $[x_0, x_1]$, in the modified matrix-geometric case, is now routine. In the latter case, when m_1 is large, we may frequently exploit particular features of the boundary matrices to perform this computation with the least amount of effort. This clearly requires separate consideration for each model.

It is usually advisable to have the interactive computer program first report a number of easily computed quantities, such as moments, which are expressed in x_0 and R only. More belabored computations, such as the densities $\{q_k\}$ and $\{q_k(j)\}$, may be performed by subroutines, callable at the user's discretion. This makes it feasible to vary parameters of the model and to examine some features of the probability distributions of interest without expending unnecessary computational effort and without being flooded by an unwieldy amount of output.

The Spectral Radius η: Elsner's Algorithm. The spectral radius $\text{sp}(R) = \eta$ of the matrix R rarely needs to be explicitly computed, but it does play a role in the asymptotic form of certain waiting time distributions, as discussed in section 3.9. We shall consider two cases, depending on whether the matrix R is explicitly available or not. In the former case, we limit our attention to the cases where R is irreducible or where, because of vanishing rows, we can easily identify the irreducible principal submatrix of R with spectral radius η. The second case, for which it suffices that A is irreducible, is based on the numerical solution of the equation $\eta = \chi(\eta)$, obtained in section 1.3.

Both cases utilize iterative methods for the computation of the spectral radius of an irreducible, nonnegative matrix C. A number of such methods are known, but we favor the following method, due to L. Elsner [86, 87], for the ease with which it may be implemented and for its generally rapid convergence.

We consider a sequence of positive vectors $\boldsymbol{u}^{(n)}$, satisfying $\boldsymbol{u}^{(n)}\boldsymbol{e} = 1$, for $n \geq 0$. The vector $\boldsymbol{u}^{(o)}$ is arbitrary. The quantities $S_j^{(n)}$ are defined by

$$S_j^{(n)} = (\boldsymbol{u}^{(n)}C)_j / u_j^{(n)}, \quad \text{for} \quad 1 \leq j \leq m, \tag{1.9.7}$$

and the indices ν_n and μ_n are chosen so that

$$S_{\nu_n}^{(n)} \leq S_j^{(n)} \leq S_{\mu_n}^{(n)}, \quad \text{for} \quad 1 \leq j \leq m. \tag{1.9.8}$$

For a fixed α, $0 < \alpha < 1$, we define d_n by

$$d_n = (S_{\nu_n}^{(n)} - C_{\nu_n \nu_n})(S_{\nu_n}^{(n)} - C_{\nu_n \nu_n} + \alpha(S_{\mu_n}^{(n)} - S_{\nu_n}^{(n)}))^{-1}. \tag{1.9.9}$$

The vector $\boldsymbol{u}^{(n+1)}$ is then defined in terms of $\boldsymbol{u}^{(n)}$ by

$$\begin{aligned} u_j^{(n+1)} &= (1 - (1 - d_n)u_{\nu_n}^{(n)})^{-1} u_j^{(n)}, \qquad \text{for} \quad j \neq \nu_n, \\ &= (1 - (1 - d_n)u_{\nu_n}^{(n)})^{-1} d_n u_{\nu_n}^{(n)}, \quad \text{for} \quad j = \nu_n. \end{aligned} \tag{1.9.10}$$

L. Elsner has shown that for all $n \geq 0$, we have

$$S_{\nu_n}^{(n)} \leq \mathrm{sp}(C) \leq S_{\mu_n}^{(n)}, \tag{1.9.11}$$

and that

$$\lim_{n \to \infty} S_{\mu_n}^{(n)} = \lim_{n \to \infty} S_{\nu_n}^{(n)} = \mathrm{sp}(C), \qquad \boldsymbol{u}^{(n)} \to \boldsymbol{u}, \tag{1.9.12}$$

where $\boldsymbol{u}$ is a left eigenvector corresponding to the maximal eigenvalue $\mathrm{sp}(C)$.

This procedure converges for all α, $0 < \alpha < 1$, but the number of iterations until $S_{\mu_n}^{(n)} - S_{\nu_n}^{(n)} < \epsilon$ is, for any $\epsilon > 0$, quite sensitive to the choice of α. For lower order matrices, Elsner has, on experimental grounds, recommended choosing α in the interval [0.5, 0.7]. Much additional computer experimentation by students in my seminar on computational probability at the University of Delaware has confirmed the merits of this recommendation. For higher order matrices, α should be chosen somewhat larger with $\alpha = 0.75$, a good overall choice.

If the matrix R is known and irreducible, the maximal eigenvalue η and the corresponding left eigenvector $\boldsymbol{u}^*$ may readily be computed by Elsner's algorithm.

We may also compute η from the matrix $A^*(z)$ by solving the equation $z = \chi(z)$ on (0, 1] by applying an elementary procedure such as the *bisection* or the *secant* method. For any trial value z', it is not necessary to compute $\chi(z')$ to high accuracy, but only to determine whether $\chi(z') \geq z'$ or $\chi(z') < z'$. The inequalities (1.9.11) are handy in doing so, since $S_{\nu_n}^{(n)} > z'$ clearly implies that $\chi(z') > z'$, and similarly for the other inequality. It is also indicated to choose the starting vectors $\boldsymbol{u}^{(o)}(z')$ appropriately. If, for example, we have determined that $0 < z_1 < \eta < z_2 < 1$, and try the trial value $z' = (z_1 + z_2)/2$ (as in the bisection method), then we may choose $\boldsymbol{u}^{(o)}(z')$ to be the average of the last approximation vectors $\boldsymbol{u}_1$ and $\boldsymbol{u}_2$ used in testing the trial values z_1 and z_2. This typically results in appreciable savings in computer time.

Remark. All comments on numerical computation have been stated for the discrete parameter case and for nonnegative matrices. They all apply mutatis mutandis to the case of Markov processes $\tilde{Q}$ and to stable matrices.

For example, if C is an irreducible matrix with negative diagonal elements and nonnegative off-diagonal elements, such that $C\boldsymbol{e} \leq \boldsymbol{0}$, then C has a simple, nonpositive eigenvalue $-\lambda$, such that for all other eigenvalues λ_j of C, we have $\text{Re}(\lambda_j) < -\lambda$.

The eigenvalue $-\lambda$ may be computed by considering the nonnegative matrix $C_1 = \tau I + C$, where $\tau \geq \max_j(-C_{jj})$. It is then clear that $-\lambda = \text{sp}(C_1) - \tau$.

Chapter 2
Probability Distributions of Phase Type

2.1. INTRODUCTION

Many probability models that have matrix-geometric solutions involve in one way or another the probability distributions to be discussed in this chapter. The class of probability distributions *of phase type* is also of considerable independent interest. This chapter is therefore written as a self-contained, comprehensive treatment of the properties of phase type distributions known to date. Most of these properties will be used in the context of specific models, discussed in the subsequent chapters.

Before proceeding to the formal discussion, we review the reasons for introducing this highly versatile class of probability distributions. It is well known that the main barrier to the explicit solution of even very simple stochastic models is the increasing complexity of the conditional probability distributions that arise in their analysis. The pervasiveness in stochastic modeling of the *exponential distribution* and of the related *Poisson process* is rarely due to persuasive empirical evidence in support of their assumption, but, far more so, to the ease of conditioning which results from the *lack-of-memory* property.

To require that, for example, arrival processes to a queue be Poisson or that the service times be exponential is to impose severe restrictions on the qualitative nature of the interarrival or service times being modeled. For truly complex models, as arise in networks of queues, one is fortunate to obtain tractable steady-state results at all, and exponential assumptions are then considered a small price to pay to obtain some information to guide highly costly and time-consuming simulations.

For relatively simple models it remains desirable to obtain exact and detailed results whose numerical implementation is easy and inexpensive, even under less restrictive distributional assumptions. There are many good reasons for this. First among these is the importance of bringing the detailed information about the *behavior* of stochastic models, which is inherent in the theories of queues, inventories, and dams, much closer to the practitioner than is done by the formal variations upon the $M/M/1$ queue, which tend to be the principal ones to make their way into practically oriented texts. That such is necessary is quite clear from some of the strongly critical remarks on the merits of queueing theory, expressed in a debate carried in *Interfaces* [11, 28, 46, 173, 300]. My views on this were expressed earlier [222].

The second reason, particularly for the examination of models under nonexponential assumptions, lies in questions related to the ***robustness or insensitivity*** to distributional assumptions of certain features of stochastic models. A number of general results, primarily obtained in a unified way by R. Schassberger [277–80], show that some steady-state averages, such as mean queue lengths or waiting times, often depend only on the first few moments and not on the detailed form of, for example, the service time distribution. The following property of the $M/G/\infty$ queue is an example of an elementary insensitivity result with important consequences in network theory. If the Poisson arrival rate to that queue is λ and its service time distribution is $H(\cdot)$ of finite mean α, then the stationary queue length distribution is Poisson with parameter $\lambda\alpha$ and the stationary process of departures is a Poisson process of rate λ. These two features of the stationary $M/G/\infty$ queue are *insensitive* to the service time distribution $H(\cdot)$.

In view of the practical and mathematical appeal of insensitivity results, it is also important to be keenly aware of what they *do not* imply. Their practical uses should be strictly limited to those inferences which are warranted by their precise mathematical statements. Some features of a stochastic model may be insensitive, while others of equal or even greater practical significance are definitely not. Here again, it becomes important to extend the range of assumptions under which detailed results may be computed. The robustness of insensitivity results may then be numerically examined, and highly sensitive features of the model can be identified. For two discussions in this spirit, we refer the reader to V. Ramaswami and M. F. Neuts [262] and to V. Ramaswami and D. M. Lucantoni [259]. Both papers are based on the theory of phase type distributions. In the former, the sensitivity of the $M/G/\infty$ queue to deviations from Poisson arrivals is examined. In the latter, it is shown that a proposed approximation to the mean duration of the busy period in the $GI/G/1$ queue is not to be recommended.

A third reason, which may ultimately be the most compelling, is the growing importance of *qualitative modeling*. It is highly desirable to model queues with such features as fluctuating arrival rates, breakdowns of servers, rare exceptionally long service times, overflow, and customers of several types or priority rules, as well as inventories or dams with seasonal patterns to their inputs or demands. Analytic approaches to such models under general distributional assumptions either fail or become so complicated as to be essentially intractable. An alternative is to study probability distributions and point processes, which remain mathematically elementary, yet which are sufficiently versatile and computationally tractable that they may be used to reflect the essential qualitative features of the model and to provide, through the interpretation of numerical results, much useful information on its physical behavior. The appeal of phase type distributions for this purpose will be amply demonstrated in the sequel.

The fourth reason, a mathematician's delight in unifying a large number of special results and in the formal elegance of the results based on phase type distributions, is a personal one. If the reader shares it with us, we shall be truly gratified.

The earliest generalization of the exponential distribution, which preserved much of the analytic tractability of the latter, is due to A. K. Erlang [88] and led to the familiar Erlang distributions. Apart from being underdispersed—their coefficient of variation is always less than one (except in the exponential case, where it is one)—the ordinary Erlang distributions have a disturbingly special property. In queueing theory, they allow us to "count stages, rather than customers." This familiar trick has, in our opinion, delayed the recognition of the beautiful general formalism, which underlies their use. It has also led to a profusion of papers dealing with very special models.

For analyses by transform methods, it was recognized that the key property of Erlang distributions is the fact that their Laplace-Stieltjes transform is *rational*. This was systematically discussed by D. R. Cox [72, 73], and led to the widespread use of probability distributions with rational Laplace-Stieltjes transforms and the corresponding method of phases (or stages) in stochastic modeling. See, for example, J. W. Cohen [61] or R. Schassberger [276]. This generalization of Erlang's approach relies heavily on the use of complex variable methods and is closely related to the *spectral method* for the solution of systems of linear differential equations with constant coefficients. Its numerical implementation frequently presents the same difficulties as the latter.

The class of probability distributions of phase type is a proper subset of those discussed by D. R. Cox [72]. They are, however, related to finite-state Markov processes, have an appealing algebraic formalism, and can be implemented by using real arithmetic only. The resulting expressions may often be given useful interpretations. Their algorithmic value was noted in M. F. Neuts [220], where they were introduced as a natural probabilistic generalization of Erlang's approach. The relationship between Erlang's method of stages and absorption time distributions in finite-state Markov processes was already noted earlier, in the monograph by A. Jensen [153], but its useful algorithmic consequences were not emphasized.

In order to overcome the limitations of the classical Erlang distributions while preserving some of their analytic tractability, many authors have considered mixtures of such distributions. Without the unifying matrix formalism to be developed in this chapter, the use of *hyperexponential* or (more generally) *mixed Erlang distributions* leads to highly complicated analytic expressions. Notably in the theory of queues, it has led to a large number of papers dealing with classical models under slightly varying distributional assumptions. Examples of the use of mixtures of Erlang distributions may be found in, among others, K. L. Arora [4, 5], M. L. Chaudhry [49], D. P. Gaver [105], D. Gross and C. M. Harris [115], S. K. Gupta and J. K. Goyal [117, 123], S. K. Gupta [118–22], T. L. Healy [131], P. Hokstad [145], R. R. P. Jackson [148, 149], A. S. Kapadia [155], G. Luchak [196–98], P. M. Morse [210, 211], T. Nishida [242], R. K. Rana [264], and D. M. G. Wishart [305, 306].

2.2. DEFINITIONS AND CLOSURE PROPERTIES

There are two parallel discussions; one for distributions on $[0, \infty)$, and the other for discrete distributions on the nonnegative integers. We shall only discuss the former in detail, stating results for the latter without proofs.

We consider a Markov process on the states $\{1, \ldots, m + 1\}$ with infinitesimal generator

$$Q = \begin{vmatrix} T & T^0 \\ \mathbf{0} & 0 \end{vmatrix}, \tag{2.2.1}$$

where the $m \times m$ matrix T satisfies $T_{ii} < 0$, for $1 \le i \le m$, and $T_{ij} \ge 0$, for $i \ne j$. Also $T\boldsymbol{e} + T^0 = \mathbf{0}$, and the initial probability vector of Q is given by $(\boldsymbol{\alpha}, \alpha_{m+1})$, with $\boldsymbol{\alpha e} + \alpha_{m+1} = 1$. We assume that the states $1, \ldots, m$ are all transient, so that absorption into the state $m + 1$, from any initial state, is certain. A useful equivalent condition is given by the following lemma.

Lemma 2.2.1. The states $1, \ldots, m$ are transient if and only if the matrix T is nonsingular.

Proof. Let a_i, $1 \leq i \leq m$, be the probability of eventual absorption into the state $m + 1$, starting in the state i. The probabilities a_i satisfy the system of linear equations

$$a_i = (-T_i^o T_{ii}^{-1}) + \sum_{\nu \neq i} (-T_{i\nu} T_{ii}^{-1}) a_\nu, \quad \text{for} \quad 1 \leq i \leq m, \tag{2.2.2}$$

or, equivalently,

$$T\boldsymbol{a} + \boldsymbol{T}^o = \boldsymbol{0}. \tag{2.2.3}$$

If T is nonsingular, the solution $\boldsymbol{a} = \boldsymbol{e}$ is unique, so that absorption from any state i is certain.

We shall now show that if absorption from every state is certain, the matrix T is nonsingular.

If the matrix T is singular, there exists a nonnegative, nonzero vector $\boldsymbol{y}$, such that $\boldsymbol{y}T = \boldsymbol{0}$. This implies that the inner product $\boldsymbol{y} \exp(Tt)\boldsymbol{e} = \boldsymbol{ye}$ is positive for all $t \geq 0$. Furthermore, the vector $\boldsymbol{u} = \lim_{t\to\infty} \exp(Tt)\boldsymbol{e}$ does not vanish, since $\boldsymbol{yu} = \boldsymbol{ye} > 0$. This shows that there is at least one initial state $i \in \{1, \ldots, m\}$ from which the probability $1 - u_i$ of eventual absorption is not equal to one. It now follows by contraposition that certain eventual absorption from every initial state implies that T is nonsingular.

Lemma 2.2.2. The probability distribution $F(\cdot)$ of the time until absorption in the state $m + 1$, corresponding to the initial probability vector $(\boldsymbol{\alpha}, \alpha_{m+1})$, is given by

$$F(x) = 1 - \boldsymbol{\alpha} \exp(Tx)\boldsymbol{e}, \quad \text{for} \quad x \geq 0. \tag{2.2.4}$$

Proof. The unconditional probabilities $\nu_j(x)$, that the process is in the state j at time x, satisfy the system of differential equations

$$\boldsymbol{\nu}'(x) = \boldsymbol{\nu}(x)T, \quad \text{for} \quad x > 0, \tag{2.2.5}$$

with initial conditions $\boldsymbol{\nu}(0) = \boldsymbol{\alpha}$. Its solution is given by $\boldsymbol{\nu}(x) = \boldsymbol{\alpha} \exp(Tx)$, so that $F(x) = 1 - \boldsymbol{\nu}(x)\boldsymbol{e}$.

Definition. A probability distribution $F(\cdot)$ on $[0, \infty)$ is a *distribution of phase type* (*PH-distribution*) if and only if it is the distribution of the time until absorption in a finite Markov process of the type defined in (2.2.1). The pair $(\boldsymbol{\alpha}, T)$ is called a *representation* of $F(\cdot)$.

The following statements are readily verified.

a. The distribution $F(\cdot)$ has a jump of height α_{m+1} at $x = 0$, and its density portion $F'(x)$ on $(0, \infty)$ is given by $F'(x) = \boldsymbol{\alpha} \exp(Tx) T^o$.

b. The Laplace-Stieltjes transform $f(s)$ of $F(\cdot)$ is given by

$$f(s) = \alpha_{m+1} + \boldsymbol{\alpha}(sI - T)^{-1} T^o, \quad \text{for} \quad \operatorname{Re} s \geq 0. \tag{2.2.6}$$

c. The noncentral moments μ_i' of $F(\cdot)$ are all finite and given by

$$\mu_i' = (-1)^i i!(\boldsymbol{\alpha} T^{-i} \boldsymbol{e}), \quad \text{for} \quad i \geq 0. \tag{2.2.7}$$

Examples. The (generalized) Erlang distribution of order m with parameters $\lambda_1, \ldots, \lambda_m$ has the representation $\boldsymbol{\alpha} = (1, 0, \ldots, 0)$ and

$$T = \begin{vmatrix} -\lambda_1 & \lambda_1 & & & \\ & -\lambda_2 & \lambda_2 & & \\ & & \cdots & & \\ & & & -\lambda_{m-1} & \lambda_{m-1} \\ & & & & -\lambda_m \end{vmatrix}.$$

The hyperexponential distribution $F(x) = \sum_{\nu=1}^{m} \alpha_\nu (1 - e^{-\lambda_\nu x})$ has the representation $\boldsymbol{\alpha} = (\alpha_1, \ldots, \alpha_m)$ and $T = \operatorname{diag}(-\lambda_1, \ldots, -\lambda_m)$.

The Discrete Case

Discrete PH-distributions are defined by considering an $(m + 1)$-state Markov chain P of the form

$$P = \begin{vmatrix} T & T^o \\ \mathbf{0} & 1 \end{vmatrix}, \tag{2.2.8}$$

where T is a substochastic matrix, such that $I - T$ is nonsingular. The initial probability vector is $(\boldsymbol{\alpha}, \alpha_{m+1})$. The probability density $\{p_k\}$ of phase type is given by

$$p_0 = \alpha_{m+1},$$
$$p_k = \boldsymbol{\alpha} T^{k-1} T^o, \quad \text{for} \quad k \geq 1. \tag{2.2.9}$$

Its probability generating function $P(z) = \alpha_{m+1} + z\boldsymbol{\alpha}(I - zT)^{-1} T^o$ and its factorial moments are given by $P^{(k)}(1) = k!\boldsymbol{\alpha} T^{k-1}(I - T)^{-k} \boldsymbol{e}$.

The natural discrete analogue of the (generalized) Erlang distribution is the (generalized) *negative binomial distribution.* We shall use this example to illustrate a property of discrete PH-distributions that is not shared by the continuous ones. It is easy to see that any right shift of a discrete PH-distribution is again of phase type. In order to shift the density in (2.2.9), r places to the right, we consider the representation $(\boldsymbol{\beta}, S)$ of dimension $m + r$, given by $\boldsymbol{\beta} = (1, 0, \ldots, 0)$, and

$$
S = \begin{array}{r|cccccccccc|}
1 & 0 & 1 & 0 & \cdots & 0 & & & & \\
2 & 0 & 0 & 1 & \cdots & 0 & & & & \\
3 & 0 & 0 & 0 & \cdots & 0 & & & & \\
\vdots & \vdots & \vdots & \vdots & & \vdots & & & & \\
r-1 & 0 & 0 & 0 & \cdots & 1 & & & & \\
r & 0 & 0 & 0 & \cdots & 0 & \alpha_1 & \alpha_2 & \cdots & \alpha_m \\
r+1 & & & & & & T_{11} & T_{12} & \cdots & T_{1m} \\
\vdots & & & & & & T_{21} & T_{22} & \cdots & T_{2m} \\
 & & & & & & \vdots & \vdots & & \vdots \\
r+m & & & & & & T_{m1} & T_{m2} & \cdots & T_{mm}
\end{array}
$$

In an $(m + 1)$-state Markov chain, if a state can be reached from another state, it can be reached in m steps or less. Not all of the terms $p_0, p_1, \ldots, p_m$ in a discrete PH-density, therefore, can be zero. Suppose that p_r, with $1 \leq r \leq m$, is the first nonzero term of a PH-density, then that density may be shifted up to r steps to the left simply by changing the initial probability vector. The shifted densities are still of phase type. To illustrate this, we consider the density with $q_0 = \boldsymbol{\alpha} T^{r-1} \mathbf{T}^o = p_r$, and

$$q_k = p_{k+r} = \boldsymbol{\alpha} T^r \cdot T^{k-1} \mathbf{T}^o, \quad \text{for} \quad k \geq 1.$$

The density $\{q_k\}$ is of phase type with representation $(\boldsymbol{\beta}, T)$, with $\boldsymbol{\beta} = \boldsymbol{\alpha} T^r$. It is easily verified that $p_j = 0$, $0 \leq j \leq r - 1$, implies that $\boldsymbol{\alpha} T^{r-1} \mathbf{T}^o + \boldsymbol{\alpha} T^r \mathbf{e} = 1$. The left shifts of the negative binomial density on $\{m, m + 1, \ldots\}$ are therefore also of phase type.

The same argument shows that if $\{p_k\}$ is of phase type, then so is the density $q_0 = \sum_{\nu=0}^{r} p_\nu$, $q_k = p_{k+r}$, for $k \geq 1$. Elementary though these properties are, they are useful in modeling discrete versions of the classical queues. It is also readily verified that every density with finite support on the nonnegative integers is of phase type.

Irreducible Representations

The following construction will be very useful in the sequel. Suppose that upon absorption into the state $m + 1$, we instantaneously perform independent multinomial trials with probabilities $\alpha_1, \ldots, \alpha_m, \alpha_{m+1}$, until one of the alternatives $1, \ldots, m$ occurs. Restarting the process Q in the corresponding state, we consider the time of the next absorption and repeat the same procedure there. It is easy to see that by continuing this procedure indefinitely we construct a new Markov process in which the state $m + 1$ is an instantaneous state. When that process enters the instantaneous state, the probability that it stays there for $r \geq 1$ transitions is given by $(1 - \alpha_{m+1})\alpha_{m+1}^{r-1}$, and the numbers of visits to the instantaneous state are independent, with that common geometric distribution.

If we further consider the version of that Markov process in which the path functions are right-hand continuous, we obtain a Markov process on $\{1, \ldots, m\}$ with the infinitesimal generator

$$Q^* = T + T^o A^o, \tag{2.2.10}$$

where T^o is an $m \times m$ matrix with identical columns $\mathbf{T}^o$ and $A^o = (1 - \alpha_{m+1})^{-1} \operatorname{diag}(\alpha_1, \ldots, \alpha_m)$. The trivial case $\alpha_{m+1} = 1$ will henceforth be excluded. The matrix $T^o A^o$ may also be written as

$$T^o A^o = (1 - \alpha_{m+1})^{-1} \mathbf{T}^o \cdot \boldsymbol{\alpha}, \tag{2.2.11}$$

a property that will frequently be used.

It is readily verified that the successive visits to the instantaneous state form a renewal process with underlying distribution $F(\cdot)$, given by (2.2.4). The point process so obtained is a *PH-renewal process.*

Definition. The representation $(\boldsymbol{\alpha}, T)$ is called *irreducible* if and only if the matrix Q^* is irreducible.

We now show that we may always restrict our attention to irreducible representations.

Lemma 2.2.3. Each component of $\boldsymbol{\nu}(t) = \boldsymbol{\alpha} \exp(Tt)$, $t \geq 0$, is either strictly positive for all $t > 0$, or is identically zero for $t \geq 0$. In the latter case, the matrix Q^* is reducible.

Proof. Since $T_{ij} \geq 0$, for $i \neq j$, the matrix $\exp(Tt)$ is nonnegative for all $t \geq 0$. See, for example, R. Bellman [14, p. 172]. It follows that $\boldsymbol{\nu}(t) \geq \mathbf{0}$, but since $\boldsymbol{\alpha} \neq \mathbf{0}$, $\boldsymbol{\nu}(t)$ does not vanish identically.

Assume now that at $t_0 > 0$, some components of $\boldsymbol{\nu}(t_0)$ are zero.

We may relabel the indices so that $v_i(t_0) > 0$, for $1 \leq i \leq m_1 < m$, and $v_i(t_0) = 0$, for $m_1 + 1 \leq i \leq m$. We have

$$v_j'(t_0) = \sum_{i=1}^{m_1} v_i(t_0)T_{ij}, \quad \text{for} \quad m_1 + 1 \leq j \leq m.$$

If any of the elements T_{ij}, $1 \leq i \leq m_1$, $m_1 + 1 \leq j \leq m$, are positive, then for some j, $m_1 + 1 \leq j \leq m$, $v_j'(t_0) > 0$, and hence $v_j(t_0 - \epsilon) < 0$, $\epsilon > 0$, which is a contradiction. The matrix T is therefore of the form

$$T = \begin{vmatrix} T_1 & 0 \\ T_3 & T_4 \end{vmatrix}.$$

We now partition $\boldsymbol{\alpha}$ and $\boldsymbol{\nu}(t)$ accordingly into $[\boldsymbol{\alpha}_1, \boldsymbol{\alpha}_2]$ and $[\boldsymbol{\nu}_1(t), \boldsymbol{\nu}_2(t)]$ respectively. It then follows that

$$\boldsymbol{\nu}_2(t) = \boldsymbol{\alpha}_2 \exp(T_4 t), \quad \text{for} \quad t \geq 0.$$

The matrix $\exp(T_4 t)$ is nonsingular. See R. Bellman [14, p. 166]. Since $\boldsymbol{\nu}_2(t_0) = \mathbf{0}$, it follows that $\boldsymbol{\alpha}_2 = \mathbf{0}$, and hence that $\boldsymbol{\nu}_2(t) = \mathbf{0}$, for $t \geq 0$. This in turn implies that Q^* is reducible and that

$$F(t) = 1 - \boldsymbol{\alpha}_1 \exp(T_1 t)\boldsymbol{e}_1, \quad \text{for} \quad t \geq 0.$$

Theorem 2.2.1. If the matrix Q^* is reducible, we may delete rows and columns of Q corresponding to a subset of indices of $\{1, \ldots, m\}$ to obtain a smaller, irreducible representation $(\boldsymbol{\alpha}_1, T_1)$ of $F(\cdot)$.

Proof. If Q^* is reducible, it may be written (after relabeling if necessary) into the form

$$\begin{vmatrix} T_1 + T_1{}^oA_1{}^o & T_2 + T_1{}^oA_2{}^o \\ T_3 + T_2{}^oA_1{}^o & T_4 + T_2{}^oA_2{}^o \end{vmatrix},$$

where $T_1 + T_1{}^oA_1{}^o$ and $T_4 + T_2{}^oA_2{}^o$ are of orders m_1 and $m - m_1$ respectively, and $T_2 + T_1{}^oA_2{}^o = 0$. Since the off-diagonal elements of Q^* are nonnegative, it follows that $T_2 = 0$, and $T_1{}^oA_2{}^o = 0$.

If $\alpha_j > 0$, for some j with $m_1 + 1 \leq j \leq m$, it follows that $T_1{}^o = \mathbf{0}$. Since T is of the form

$$\begin{vmatrix} T_1 & 0 \\ T_3 & T_4 \end{vmatrix},$$

this would imply that the matrix T_1 and hence also the matrix T are singular. As this contradicts an earlier assumption, we have $\boldsymbol{\alpha}_2 = \mathbf{0}$.

Upon evaluation of the distribution $F(\cdot)$, we now obtain

$$F(x) = 1 - (\alpha_1, \mathbf{0}) \exp\left\{x \begin{vmatrix} T_1 & 0 \\ T_3 & T_4 \end{vmatrix}\right\}\begin{pmatrix} e_1 \\ e_2 \end{pmatrix}$$
$$= 1 - \alpha_1 \exp(T_1 x) e_1,$$

so that $F(\cdot)$ also has the smaller representation (α_1, T_1). It is clear that this reduction may be continued until an irreducible representation is obtained.

Corollary 2.2.1. For an irreducible representation (α, T), the vectors $\alpha \exp(Tt)$ and $\exp(Tt) T^o$ are strictly positive for $t > 0$.

Proof. Only the positivity of $\exp(Tt) T^o$ needs to be shown. The same argument as in lemma 2.2.3 shows that each component of that vector is either identically zero or positive for all $t > 0$. Were the i-th component zero, the probability of absorption in the process Q, starting in the state i, would be zero. This is a contradiction.

Corollary 2.2.2. For $s \geq 0$, the vectors $\alpha(sI - T)^{-1}$ and $(sI - T)^{-1} T^o$ are positive.

Henceforth we shall always assume that a given representation (α, T) of a PH-distribution is *irreducible.*

Remarks

a. Even irreducible representations are far from unique. One should therefore be cautious in giving a physical interpretation to the m phases of the process Q.

b. A difficult unsolved problem is that of characterizing *minimal* irreducible representations for a given PH-distribution. In those, the order of the matrix T is as small as possible. An extreme example is the following. Let $F(\cdot)$ be a PH-distribution with representation (α, T), and let T be irreducible. The eigenvalue of maximal real part of T is denoted by $-\xi < 0$; the corresponding left eigenvector is positive. If α happens to be a left eigenvector of T corresponding to $-\xi$, then clearly

$$F(x) = 1 - \alpha \exp(Tx) e = 1 - \exp(-\xi x), \quad \text{for} \quad x \geq 0.$$

For this choice of α and T, $F(\cdot)$ is exponential and has a minimal representation of order one. Any such pair (α, T) is nevertheless an irreducible representation of the exponential distribution $F(\cdot)$.

c. As discussed in R. Syski [290], the instantaneous resetting used to define the Markov process Q^* is an example of the *compensation method* due to J. Keilson [162] and may be used to study a variety of problems in Markov processes. The theory of PH-distributions is also related to the *Laguerre transform,* whose algorithmic uses in stochastic modeling were treated in J. Keilson and W. R. Nunn [163].

Closure Properties

A number of operations on PH-distributions lead again to distributions of phase type. In each case, a representation for the new distribution is obtained. This is algorithmically useful in that somewhat involved operations, which in general require numerical integrations, may here be replaced by matrix operations.

We first make some notational conventions. If T^o is an m-vector and $\boldsymbol{\beta}$ an n-vector, we denote by T^oB^o the $m \times n$ matrix $T^o \cdot \boldsymbol{\beta}$, with elements $T_i^o\beta_j$, $1 \leq i \leq m$, $1 \leq j \leq n$. By the symbol I, we always denote an identity matrix of the dimension appropriate to the formula in which it appears.

Theorem 2.2.2. If $F(\cdot)$ and $G(\cdot)$ are both continuous (or both discrete) PH-distributions with representations $(\boldsymbol{\alpha}, T)$ and $(\boldsymbol{\beta}, S)$ of orders m and n respectively, then their convolution $F*G(\cdot)$ is a PH-distribution with representation $(\boldsymbol{\gamma}, L)$, given by (in the continuous case):

$$\boldsymbol{\gamma} = [\boldsymbol{\alpha}, \alpha_{m+1}\boldsymbol{\beta}],$$

$$L = \begin{vmatrix} T & T^oB^o \\ 0 & S \end{vmatrix}. \tag{2.2.12}$$

Proof. It is readily seen that the Laplace-Stieltjes transform of the PH-distribution with representation $(\boldsymbol{\gamma}, L)$ is given by

$$\alpha_{m+1}\beta_{n+1} + \beta_{n+1}\boldsymbol{\alpha}(sI - T)^{-1}T^o + \alpha_{m+1}\boldsymbol{\beta}(sI - S)^{-1}S^o$$
$$+ \boldsymbol{\alpha}(sI - T)^{-1}T^oB^o(sI - S)^{-1}S^o.$$

The final term simplifies to $\boldsymbol{\alpha}(sI - T)^{-1}T^o \cdot \boldsymbol{\beta}(sI - S)^{-1}S^o$, and the entire expression is therefore equal to

$$[\alpha_{m+1} + \boldsymbol{\alpha}(sI - T)^{-1}T^o][\beta_{n+1} + \boldsymbol{\beta}(sI - S)^{-1}S^o].$$

Remark. It is interesting to note the relation of the Markov process with generator

$$\begin{vmatrix} T & T^oB^o \\ S^oA^o & S \end{vmatrix}, \qquad (2.2.13)$$

where we assume for convenience that $\alpha_{m+1} = \beta_{n+1} = 0$, to the *alternating renewal process* with the probability distributions $F(\cdot)$ and $G(\cdot)$. If at time t the Markov process is in the set of states $\{1, \ldots, m\}$, the point t is covered by an interval with distribution $F(\cdot)$. A similar interpretation clearly holds for sojourns in the set $\{m+1, \ldots, m+n\}$. Transitions from the set $\{1, \ldots, m\}$ to the set $\{m+1, \ldots, m+n\}$, and vice versa, correspond to renewals.

This construction and theorem 2.2.2 may clearly be extended to several PH-distributions.

Before we state the next theorem, we define the vector $\boldsymbol{\pi}$ as the stationary probability vector of the *irreducible* generator $Q^* = T + T^oA^o$, defined in (2.2.10). Clearly the positive vector $\boldsymbol{\pi}$ is the unique solution to the equations

$$\boldsymbol{\pi}(T + T^oA^o) = \mathbf{0}, \qquad \boldsymbol{\pi}\mathbf{e} = 1. \qquad (2.2.14)$$

The following simple formulas will be repeatedly used in the sequel.

$$\boldsymbol{\pi} = -\boldsymbol{\pi}T^oA^oT^{-1} = -(1-\alpha_{m+1})^{-1}(\boldsymbol{\pi}T^o)\boldsymbol{\alpha}T^{-1}, \qquad (2.2.15)$$

and

$$\boldsymbol{\pi}T^o = (1-\alpha_{m+1})\mu_1'^{-1}, \qquad (2.2.16)$$

where μ_1' is the mean of $F(\cdot)$.

Theorem 2.2.3. If $F(\cdot)$ is a PH-distribution with (irreducible) representation $(\boldsymbol{\alpha}, T)$, then

$$F^*(x) = \frac{1}{\mu_1'}\int_0^x [1 - F(u)]\,du, \quad \text{for} \quad x \geq 0, \qquad (2.2.17)$$

is a PH-distribution with representation $(\boldsymbol{\pi}, T)$.

Proof

$$F^*(x) = \frac{1}{\mu_1'}\int_0^x \boldsymbol{\alpha}\exp(Tu)\,du\,\mathbf{e} = \frac{1}{\mu_1'}\boldsymbol{\alpha}T^{-1}[\exp(Tx) - I]\mathbf{e}$$
$$= 1 - \boldsymbol{\pi}\exp(Tx)\mathbf{e},$$

by (2.2.7) and (2.2.15).

The moments μ_i^* of $F^*(\cdot)$ are given by

$$\mu_i^* = (-1)^i i! \pi T^{-i} e = (-1)^{i+1} \frac{i!}{\mu_1'} \alpha T^{-i-1} e$$

$$= \frac{\mu_{i+1}'}{(i+1)\mu_1'}, \quad \text{for} \quad i \geq 1. \tag{2.2.18}$$

Theorem 2.2.4. A finite mixture of PH-distributions is a PH-distribution. If $(p_1, \ldots, p_k)$ is the mixing density and $F_j(\cdot)$ has the representation $[\alpha(j), T(j)]$, $1 \leq j \leq k$, then the mixture has the representation $\alpha = [p_1\alpha(1), p_2\alpha(2), \ldots, p_k\alpha(k)]$, and

$$T = \begin{vmatrix} T(1) & 0 & \cdots & 0 \\ 0 & T(2) & \cdots & 0 \\ \cdot & \cdot & & \cdot \\ \cdot & \cdot & & \cdot \\ \cdot & \cdot & & \cdot \\ 0 & 0 & \cdots & T(k) \end{vmatrix}.$$

Proof. Obvious.

Infinite mixtures of PH-distributions are generally not of phase type. An important and useful exception is given by the next theorem. In its proof, we require some properties of the *Kronecker product* of matrices, which we now review. See also R. Bellman [14] or M. Marcus and H. Minc [200].

Definition. If L and M are rectangular matrices of dimensions $k_1 \times k_2$ and $k_1' \times k_2'$, their Kronecker product $L \otimes M$ is the matrix of dimensions $k_1k_1' \times k_2k_2'$, written in block-partitioned form as

$$\begin{vmatrix} L_{11}M & L_{12}M & \cdots & L_{1k_2}M \\ \cdot & \cdot & & \cdot \\ \cdot & \cdot & & \cdot \\ \cdot & \cdot & & \cdot \\ L_{k_1 1}M & L_{k_1 2}M & \cdots & L_{k_1 k_2}M \end{vmatrix}.$$

A useful property of the Kronecker product that is repeatedly used in the sequel is the following:

Product Property. If L, M, U, and V are rectangular matrices such that the ordinary matrix products LU and MV are defined, then

$$(L \otimes M)(U \otimes V) = LU \otimes MV. \tag{2.2.19}$$

Theorem 2.2.5. Let $\{s_\nu\}$ be a discrete PH-density with representation (β, S) of order n and $F(\cdot)$ a continuous PH-distribution with representation (α, T) of order m, then the mixture $\Sigma_{\nu=0}^{\infty} s_\nu F^{(\nu)}(\cdot)$, of the successive convolutions of $F(\cdot)$, is of phase type with representations (γ, L) of order mn, given by

$$\gamma = \alpha \otimes \beta(I - \alpha_{m+1}S)^{-1},$$
$$L = T \otimes I + (1 - \alpha_{m+1})T^o A^o \otimes (I - \alpha_{m+1}S)^{-1}S. \qquad (2.2.20)$$

The height of the jump γ_{mn+1} at zero and the vector L^o are given by

$$\gamma_{mn+1} = \beta_{n+1} + \alpha_{m+1}\beta(I - \alpha_{m+1}S)^{-1}S^o,$$
$$L^o = T^o \otimes (I - \alpha_{m+1}S)^{-1}S^o. \qquad (2.2.21)$$

Proof. The probability generating function $P(z)$ of $\{s_\nu\}$ and the Laplace-Stieltjes transform $f(s)$ of $F(\cdot)$ are given by $P(z) = \beta_{n+1} + z\beta(I - zS)^{-1}S^o$ and $f(s) = \alpha_{m+1} + \alpha(sI - T)^{-1}T^o$ respectively. The Laplace-Stieltjes transform of the mixture is given by

$$P[f(s)] = \beta_{n+1} + f(s)\cdot\beta[I - f(s)S]^{-1}S^o. \qquad (2.2.22)$$

Letting $s \to \infty$, we obtain for $\gamma_{mn+1} = \lim_{s\to\infty} P[f(s)]$, the expression stated in (2.2.21).

It suffices to verify that

$$\gamma(sI - L)^{-1}L^o = P[f(s)] - \gamma_{mn+1}.$$

We first check that the diagonal elements of L are negative, that its off-diagonal elements are nonnegative, and that its row sums are non-positive.

Clearly $L^o \geq \mathbf{0}$, and it readily follows from (2.2.20) that the off-diagonal elements of L are nonnegative. Now

$$\begin{aligned} Le + L^o &= [T \otimes I + (1 - \alpha_{m+1})T^o A^o \otimes (I - \alpha_{m+1}S)^{-1}S](e_m \otimes e_n) \\ &\quad + T^o \otimes (I - \alpha_{m+1}S)^{-1}S^o \\ &= -T^o \otimes e_n + (1 - \alpha_{m+1})T^o \otimes (I - \alpha_{m+1}S)^{-1}(e_n - S^o) \\ &\quad + T^o \otimes (I - \alpha_{m+1}S)^{-1}S^o \\ &= T^o \otimes (I - \alpha_{m+1}S)^{-1}[S^o + (1 - \alpha_{m+1})(e_n - S^o) \\ &\qquad\qquad - e_n + \alpha_{m+1}Se_n] = \mathbf{0}, \end{aligned}$$

since the term in brackets vanishes. We shall see in the following that L is nonsingular, so that no row of L can be identically zero. The preceding calculation then also shows that the diagonal elements are negative.

Let U be the inverse of $sI - L$, then

$$sU - (T \otimes I)U - [(1 - \alpha_{m+1})T^o A^o \otimes (I - \alpha_{m+1}S)^{-1}S]U = I.$$

The matrix $sI - T \otimes I = (sI - T) \otimes I$ is clearly nonsingular, so that the preceding equation may be written as

$$\{I - [(sI - T)^{-1} \otimes I][(1 - \alpha_{m+1})T^o A^o \otimes (I - \alpha_{m+1}S)^{-1}S]\}U$$
$$= (sI - T)^{-1} \otimes I. \quad (2.2.23)$$

We shall now show that the coefficient matrix of U in (2.2.23) is nonsingular. Since

$$(sI - T)^{-1}T^o A^o \otimes (I - \alpha_{m+1}S)^{-1}S$$
$$= \int_0^\infty e^{-su}\{\exp(Tu)T^o A^o \otimes (I - \alpha_{m+1}S)^{-1}S\}du, \quad \text{for} \quad s \geq 0,$$

it suffices to verify the nonsingularity of the coefficient matrix for $s = 0$. Now

$$-T^{-1}(1 - \alpha_{m+1})T^o A^o \otimes (I - \alpha_{m+1}S)^{-1}S = A^{oo} \otimes (I - \alpha_{m+1}S)^{-1}S,$$

where $A_{ij}{}^{oo} = \alpha_j$, for $1 \leq i, j \leq m$. If the matrix $I - A^{oo} \otimes (I - \alpha_{m+1}S)^{-1}S$ were singular, there would exist m n-vectors $\xi_1, \ldots, \xi_m$, not all zero, so that

$$\xi_i = \sum_{\nu=1}^{m} \alpha_\nu (I - \alpha_{m+1}S)^{-1}S\xi_\nu, \quad \text{for} \quad 1 \leq i \leq m.$$

This implies that $\xi_i = \xi$, for $1 \leq i \leq m$, and hence that

$$\xi = (1 - \alpha_{m+1})(I - \alpha_{m+1}S)^{-1}S\xi.$$

This in turn implies that $(I - S)\xi = \mathbf{0}$, which is impossible unless $\xi = \mathbf{0}$, since $I - S$ is nonsingular.

It follows that the coefficient matrix of U, and hence U itself, are nonsingular. We can now write

$$(sI - L)^{-1} = \sum_{\nu=0}^{\infty} [(sI - T)^{-1}T^o\boldsymbol{\alpha} \otimes (I - \alpha_{m+1}S)^{-1}S]^\nu[(sI - T)^{-1} \otimes I]$$

$$= \sum_{\nu=0}^{\infty} [(sI - T)^{-1}T^o\boldsymbol{\alpha}]^\nu (sI - T)^{-1} \otimes [(I - \alpha_{m+1}S)^{-1}S]^\nu,$$

so that

$$\boldsymbol{\gamma}(sI - L)^{-1}L^o = \sum_{\nu=0}^{\infty} \boldsymbol{\alpha}[(sI - T)^{-1}T^o\boldsymbol{\alpha}]^\nu (sI - T)^{-1}T^o$$
$$\otimes \boldsymbol{\beta}[(I - \alpha_{m+1}S)^{-1}S]^\nu (I - \alpha_{m+1}S)^{-2}S^o.$$

The Kronecker product in this expression is clearly the product of two scalars. Moreover, a simple matrix calculation shows that

$$\alpha[(sI - T)^{-1}T^o\alpha]^\nu(sI - T)^{-1}T^o = [\alpha(sI - T)^{-1}T^o]^{\nu+1}$$
$$= [f(s) - \alpha_{m+1}]^{\nu+1}.$$

It follows that

$$\gamma(sI - L)^{-1}L^o$$
$$= [f(s) - \alpha_{m+1}]\beta\{I - [f(s) - \alpha_{m+1}](I - \alpha_{m+1}S)^{-1}S\}^{-1}$$
$$\cdot (I - \alpha_{m+1}S)^{-2}S^o$$
$$= [f(s) - \alpha_{m+1}]\beta[I - f(s)S]^{-1}(I - \alpha_{m+1}S)^{-1}S^o.$$

An elementary matrix calculation now shows that the last expression is equal to $P[f(s)] - \gamma_{mn+1}$. This completes the proof of the theorem.

A corresponding result holds when $F(\cdot)$ is a *discrete* PH-distribution. We state the discrete analogue of theorem 2.2.5 without proof.

Theorem 2.2.6. Let $\{s_\nu\}$ and $\{p_k\}$ be discrete PH-densities with representations (β, S) and (α, T) of orders n and m respectively. The mixture $\Sigma_{\nu=0}^{\infty} s_\nu\{p_k\}^{(\nu)}$ is of phase type with representation (γ, L) of order mn, given by

$$\gamma = \alpha \otimes \beta(I - \alpha_{m+1}S)^{-1},$$
$$L = T \otimes I + (1 - \alpha_{m+1})T^oA^o \otimes (I - \alpha_{m+1}S)^{-1}S. \qquad (2.2.24)$$

Remark. Theorem 2.2.6 readily implies that if the number of offspring of each individual in the classical *Galton-Watson process* has a PH-density, the same is true for the number of k-th generation descendants of an individual. The representation obtained by repeated application of that theorem would, however, involve a matrix of order m^k, so that the theorem is not useful for computational purposes. For $m = 1$, the number of descendants of an individual is geometric, possibly with a modified first term. The preservation of this distribution under functional iteration of the probability generating function is well known. See, for example, S. Karlin and H. M. Taylor [156, pp. 402–4]. Some computational problems related to the Galton-Watson process are discussed in M. F. Neuts [235].

Theorem 2.2.5 has a very useful application in the theory of queues. Consider an $M/G/1$ queueing model with arrival rate λ and service time distribution $F(\cdot)$ of mean μ_1'. If $\rho = \lambda\mu_1' < 1$, the queue is *stable,* and the waiting time of the n-th customer as well as the virtual waiting time at time t converge in distribution to the same probability distribution $W(\cdot)$, whose Laplace-Stieltjes transform $w(s)$ is given by the classical *Pollachek-Khinchine formula*

$$w(s) = \frac{(1-\rho)s}{s - \lambda + \lambda f(s)} = \sum_{k=0}^{\infty} (1-\rho)\rho^k \left(\frac{1-f(s)}{\mu_1{}'s}\right)^k. \tag{2.2.25}$$

See, for example, L. Takács [293, p. 68].

In the $M/G/1$ queue, where $F(\cdot)$ is of phase type with the (irreducible) representation $(\boldsymbol{\beta}, S)$, we obtain the following computationally useful theorem.

Theorem 2.2.7. In the stable $M/G/1$ queue with service time distribution $F(\cdot)$, the stationary waiting time distribution $W(\cdot)$ is of phase type, with representation

$$\boldsymbol{\gamma} = \rho\boldsymbol{\theta}, \qquad L = S + \rho S^o \Theta^o, \tag{2.2.26}$$

where $\Theta^o = \operatorname{diag}(\theta_1, \ldots, \theta_n)$, and $\boldsymbol{\theta}$ is the stationary probability vector of $S + S^o B^o$.

Proof. Formula (2.2.25) shows that $W(\cdot)$ is a geometric mixture of the successive convolutions of the PH-distribution $F^*(\cdot)$ with representation $(\boldsymbol{\theta}, S)$. The geometric distribution is obviously of phase type. By applying theorem 2.2.5, the stated representation for $W(\cdot)$ is obtained.

A simple direct verification (for classroom use) goes as follows:

$$\begin{aligned} w(s) &= 1 - \rho + \rho\boldsymbol{\theta}(sI - S - \rho S^o\Theta^o)^{-1}(1-\rho)S^o \\ &= 1 - \rho + (1-\rho) \sum_{\nu=0}^{\infty} \rho\boldsymbol{\theta}[\rho(sI - S)^{-1}S^o\Theta^o]^\nu (sI - S)^{-1}S^o \\ &= 1 - \rho + (1-\rho) \sum_{\nu=0}^{\infty} \rho^{\nu+1}[\boldsymbol{\theta}(sI - S)^{-1}S^o]^{\nu+1}. \end{aligned}$$

This is clearly the same calculation as that used in the proof of theorem 2.2.5, but for the simple case of a geometric mixture.

Remarks

a. In discussing queues with interarrival or service time distributions of phase type, we shall consistently use the representations $(\boldsymbol{\alpha}, T)$ for interarrival times and $(\boldsymbol{\beta}, S)$ for service times.

b. Theorem 2.2.7 immediately yields a simple algorithm for the computation of $W(\cdot)$ and its density portion $W'(\cdot)$. Given $\boldsymbol{\beta}$ and S, we compute ρ and $\boldsymbol{\theta}$ and solve the system of differential equations

$$\nu'(t) = \nu(t)S + \rho[\nu(t)S^o]\boldsymbol{\theta}, \quad \text{for} \quad t \geq 0, \tag{2.2.27}$$

with $\nu(0) = \rho\boldsymbol{\theta}$. The distribution $W(\cdot)$ is then given by

$$W(t) = 1 - \nu(t)\boldsymbol{e}, \quad \text{for} \quad t \geq 0, \tag{2.2.28}$$

and

$$W'(t) = (1 - \rho)\nu(t)S^o, \quad \text{for} \quad t > 0. \tag{2.2.29}$$

We see that $W'(t)$ is obtained "free" in the calculation of $W(\cdot)$. It is also not necessary to compute $F(\cdot)$ in order to find $W(\cdot)$.

c. In the usual Kendall notation for queues, we shall use PH to denote an interarrival or service time distribution of phase type. Theorem 2.2.7 thus gives the stationary waiting time distribution for the $M/PH/1$ queue.

d. For the general $M/G/1$ queue, the distribution $W(\cdot)$ is most easily found by numerical solution of the Volterra equation

$$W(x) = 1 - \rho + \lambda \int_0^x [1 - F(x - y)]\, W(y)dy, \quad \text{for} \quad x \geq 0. \tag{2.2.30}$$

e. The simplifications, due to theorems 2.2.5 and 2.2.7, are particularly striking when we consider the stationary distribution $W(\cdot)$ of the *virtual* waiting time distribution for the case where customers arrive in groups. The group size density $\{p_\nu\}$ is assumed to be *of phase type* with representation $(\boldsymbol{\kappa}, K)$ of order m, with $\kappa_{m+1} = 0$. The mean group size is denoted by η_1'. The queue is *stable* if and only if $\rho = \lambda\eta_1'\mu_1' < 1$.

The stationary virtual waiting time distribution is clearly the same as in the $M/PH/1$ queue with single arrivals of rate λ and service time distribution $F_1(\cdot) = \sum_{\nu=1}^{\infty} p_\nu F^{(\nu)}(\cdot)$. By theorem 2.2.5, the representation $(\boldsymbol{\gamma}, L)$ of $F_1(\cdot)$ is given by

$$\boldsymbol{\gamma} = \boldsymbol{\beta} \otimes \boldsymbol{\kappa} \qquad L = S \otimes I + S^oB^o \otimes K, \tag{2.2.31}$$

and $L^o = S^o \otimes K^o$.

Denoting by Γ^o the diagonal matrix with the components of $\boldsymbol{\gamma}$ on its diagonal, we need the matrix $L + L^o\Gamma^o$ and its invariant probability vector $\boldsymbol{\delta}$. Let $\boldsymbol{\omega}$ be the invariant probability vector of the stochastic matrix $K + K^o\boldsymbol{\kappa}$, then we shall verify that $\boldsymbol{\delta} = \boldsymbol{\theta} \otimes \boldsymbol{\omega}$.

By direct calculation, we have $L^o\Gamma^o = S^o\boldsymbol{\beta} \otimes K^o\boldsymbol{\kappa}$, and therefore

$$\begin{aligned}(\boldsymbol{\theta} \otimes \boldsymbol{\omega})[S \otimes I + S^o\boldsymbol{\beta} \otimes K + S^o\boldsymbol{\beta} \otimes K^o\boldsymbol{\kappa}]& \\ &= \boldsymbol{\theta}S \otimes \boldsymbol{\omega} + (\boldsymbol{\theta}S^o)\boldsymbol{\beta} \otimes \boldsymbol{\omega}(K + K^o\boldsymbol{\kappa}) \\ &= \boldsymbol{\theta}(S + S^oB^o) \otimes \boldsymbol{\omega} = \mathbf{0} \otimes \boldsymbol{\omega} = \mathbf{0}.\end{aligned}$$

By theorem 2.2.7, we see that the representation of $W(\cdot)$ is given by $\rho\boldsymbol{\theta} \otimes \boldsymbol{\omega}$ and $S \otimes I + S^o B^o \otimes K + \rho S^o \boldsymbol{\beta} \otimes K^o \boldsymbol{\kappa}$. Although the computation of $W(\cdot)$ involves the solution of a system of mn differential equations, the coefficient matrix of that system has a highly special structure, which may be exploited through efficient programming.

f. For the $M/PH/1$ and related queues, it is possible to compute efficiently the stationary distribution of the waiting times under service in random order and under the last-come, first-served queue disciplines. In each case, the algorithm consists in solving a carefully truncated version of a highly structured infinite system of differential equations. For a detailed discussion, we refer to M. F. Neuts [225]. A similar approach may be used to evaluate the distribution of the busy period of the $M/PH/1$ queue. Details of that algorithm are given in section 3.4.

In many applications, we need to consider the number N of arrivals in a Poisson process of rate λ during an interval $[0, X]$, where X is a random variable independent of the Poisson process. If $F(\cdot)$ is the distribution of X, then

$$a_k = P\{N = k\} = \int_0^\infty e^{-\lambda u} \frac{(\lambda u)^k}{k!} dF(u), \quad \text{for} \quad k \geq 0, \tag{2.2.32}$$

and the probability generating function $A(z)$ of $\{a_k\}$ is given by $A(z) = f(\lambda - \lambda z)$, where $f(\cdot)$ is the Laplace-Stieltjes transform of $F(\cdot)$. The following theorem shows that if $F(\cdot)$ is a PH-distribution, then the density $\{a_k\}$ can be computed *without numerical integrations.*

Theorem 2.2.8. If $F(\cdot)$ is a PH-distribution with representation $(\boldsymbol{\alpha}, T)$, then $\{a_k\}$ is a discrete PH-density with representation $(\boldsymbol{\beta}, S)$, given by

$$\boldsymbol{\beta} = \lambda\boldsymbol{\alpha}(\lambda I - T)^{-1}, \qquad S = \lambda(\lambda I - T)^{-1}, \tag{2.2.33}$$

and

$$\beta_{m+1} = \alpha_{m+1} + \boldsymbol{\alpha}(\lambda I - T)^{-1}T^o, \qquad S^o = (\lambda I - T)^{-1}T^o.$$

Proof. Elementary matrix manipulations show that

$$\begin{aligned} A(z) &= \alpha_{m+1} + \boldsymbol{\alpha}[(\lambda - \lambda z)I - T]^{-1}T^o \\ &= \alpha_{m+1} + \boldsymbol{\alpha}(\lambda I - T)^{-1}T^o \\ &\quad + z\boldsymbol{\alpha}\lambda(\lambda I - T)^{-1}[I - \lambda z(\lambda I - T)^{-1}]^{-1}(\lambda I - T)^{-1}T^o. \end{aligned}$$

Clearly

$$\alpha_{m+1} + \boldsymbol{\alpha}(\lambda I - T)^{-1}T^o + \lambda\boldsymbol{\alpha}(\lambda I - T)^{-1}\mathbf{e} = 1,$$

since $T^o = -Te$. Also

$$\lambda(\lambda I - T)^{-1}e + (\lambda I - T)^{-1}T^o = e,$$

for the same reason. Finally, since T is nonsingular, the matrix

$$I - \lambda(\lambda I - T)^{-1} = -(\lambda I - T)^{-1}T$$

is nonsingular.

An efficient way of computing the density $\{a_k\}$ is to evaluate the vectors

$$\nu(0) = \frac{1}{\lambda}\alpha, \qquad \nu(k+1) = \nu(k)S, \quad \text{for} \quad k \geq 0,$$

and to note that $a_0 = \alpha_{m+1} + \nu(1)T^o$, and $a_k = \nu(k+1)T^o$, for $k \geq 1$. In some applications, we also need the probability density, defined by

$$\tilde{a}_k = \frac{1}{\lambda\mu_1{}'} \sum_{\nu=k+1}^{\infty} a_\nu, \quad \text{for} \quad k \geq 0. \tag{2.2.34}$$

This density is obtained with little extra effort, since

$$\sum_{\nu=k+1}^{\infty} a_\nu = \frac{1}{\lambda}\alpha \sum_{\nu=k+1}^{\infty} S^{\nu+1}T^o = \lambda\nu(k+1)e, \quad \text{for} \quad k \geq 0.$$

This is occasionally more efficient and more accurate than computing $\tilde{a}_k$ by direct summation.

If there are independent group arrivals at each Poisson event and if the group sizes have a PH-density, then theorem 2.2.6 immediately yields the corresponding result for that case.

We conclude our discussion of the closure properties of the PH-distributions by observing that if X and Y are independent random variables with PH-distributions $F(\cdot)$ and $G(\cdot)$, then the distributions $F_1(\cdot) = F(\cdot)G(\cdot)$ and $F_2(\cdot) = 1 - [1 - F(\cdot)][1 - G(\cdot)]$, corresponding to $\max(X, Y)$ and $\min(X, Y)$, are also of phase type. Although this result has not yet proved to be of computational utility, the construction of the representations of $F_1(\cdot)$ and $F_2(\cdot)$ is of interest.

Theorem 2.2.9. Let $F(\cdot)$ and $G(\cdot)$ have representations (α, T) and (β, S) of orders m and n respectively, then $F_1(\cdot)$ has the representation (γ, L) of order $mn + m + n$, given by

$$\gamma = [\boldsymbol{\alpha} \otimes \boldsymbol{\beta}, \beta_{n+1}\boldsymbol{\alpha}, \alpha_{m+1}\boldsymbol{\beta}],$$

$$L = \begin{vmatrix} T \otimes I + I \otimes S & I \otimes S^o & T^o \otimes I \\ 0 & T & 0 \\ 0 & 0 & S \end{vmatrix}, \tag{2.2.35}$$

and $F_2(\cdot)$ has the representation $[\boldsymbol{\alpha} \otimes \boldsymbol{\beta}, T \otimes I + I \otimes S]$.

Proof. The validity of the representations is readily ascertained by considering the behavior of the absorbing Markov chains. There is one point, however, which is of independent interest. That is to show the nonsingularity of the matrix $T \otimes I + I \otimes S$.

Suppose to the contrary that $T \otimes I + I \otimes S$ is singular. This is equivalent to the existence of m column vectors $\boldsymbol{u}_i$, $1 \leq i \leq m$, not all zero, such that

$$\sum_{\nu=1}^{m} T_{i\nu}\boldsymbol{u}_\nu + S\boldsymbol{u}_i = \mathbf{0}, \quad \text{for} \quad 1 \leq i \leq m.$$

Without loss of generality, we may assume that the vectors $\boldsymbol{u}_i$ are nonnegative, since the matrix $T \otimes I + I \otimes S$ is semi-stable. The eigenvalue zero would then be the eigenvalue with largest real part.

Let $-\phi$, with $\phi > 0$, be the rightmost eigenvalue of T, with $\boldsymbol{v}$ a corresponding right eigenvector, chosen so that $\boldsymbol{v} \geq \mathbf{0}$. The preceding equation then implies that

$$[-\phi I + S]\left(\sum_{\nu=1}^{m} v_\nu \boldsymbol{u}_\nu\right) = \mathbf{0}.$$

Since S is nonsingular and stable, it follows that $\sum_{\nu=1}^{m} v_\nu \boldsymbol{u}_\nu = 0$. If $\boldsymbol{v}$ is positive, it follows that $\boldsymbol{u}_\nu = \mathbf{0}$, for $1 \leq \nu \leq m$, and the nonsingularity is proved. If $\boldsymbol{v}$ is not strictly positive, then $v_\nu > 0$ implies that $\boldsymbol{u}_\nu = \mathbf{0}$.

Now let $\boldsymbol{u}_1, \ldots, \boldsymbol{u}_k$ have at least one positive component and let $\boldsymbol{u}_{k+1} = \cdots = \boldsymbol{u}_m = \mathbf{0}$, for some k with $1 \leq k < m$. We then have

$$\sum_{\nu=1}^{k} T_{i\nu}\boldsymbol{u}_\nu + S\boldsymbol{u}_i = \mathbf{0}, \quad \text{for} \quad 1 \leq i \leq k,$$

$$\sum_{\nu=1}^{k} T_{i\nu}\boldsymbol{u}_\nu = \mathbf{0}, \qquad \text{for} \quad k+1 \leq i \leq m. \tag{2.2.36}$$

The coefficients $T_{i\nu}$ in the second equation are nonnegative. Were any $T_{i\nu}$, $k + 1 \leq i \leq m$, $1 \leq \nu \leq k$, positive, the corresponding $\boldsymbol{u}_\nu$ would vanish. Since this would be a contradiction, it follows that T is of the form

$$T = \begin{vmatrix} T_1 & T_2 \\ 0 & T_3 \end{vmatrix},$$

where T_1 is a nonsingular $k \times k$ matrix. We now apply exactly the same argument as before to the first equation in (2.2.36). We denote by $\nu_1 \geq \mathbf{0}$ and by $-\phi_1 < 0$ a right eigenvector of T_1, corresponding to the eigenvalue $-\phi$ of largest real part of T_1. As before, we obtain that $\sum_{\nu=1}^{k} \nu_{1\nu} \boldsymbol{u}_\nu = \mathbf{0}$. Since at least one component of ν_1 is positive, at least one of the vectors $\boldsymbol{u}_\nu$, $1 \leq \nu \leq k$, vanishes. This is a contradiction, and the matrix $T \otimes I + I \otimes S$ is therefore nonsingular.

Remarks

a. In discussing the *PH/PH/*1 queue in M. F. Neuts [227], we expressed concern over the possible singularity of the matrix $T \otimes I + I \otimes S$. The preceding proof shows that this concern was unwarranted.

b. The nonsingularity of $T \otimes I + I \otimes S$ is also evident from the discussion on pp. 230–31 in R. Bellman [14].

2.3. ASYMPTOTIC BEHAVIOR OF PH-DISTRIBUTIONS

In this section, we prove a number of simple results on the asymptotically exponential nature of PH-distributions. These results are useful in the computation of the waiting time distribution for a number of queues.

Definition. A probability distribution $F(\cdot)$ is *asymptotically exponential* if and only if for some $K > 0$ and $\eta > 0$,

$$1 - F(x) = Ke^{-\eta x} + o(e^{-\eta x}), \quad \text{as} \quad x \to \infty. \tag{2.3.1}$$

Theorem 2.3.1. If the matrix T is irreducible, any PH-distribution $F(\cdot)$ with representation (α, T) is asymptotically exponential, and $-\eta$ is the eigenvalue with largest real part of T. The constant $K = \alpha\nu$, where ν is the positive right eigenvector of T, corresponding to $-\eta$, uniquely determined by the requirements that $\boldsymbol{u}\nu = \boldsymbol{u}\boldsymbol{e} = 1$, where $\boldsymbol{u}$ is a left eigenvector of T, associated with $-\eta$.

Proof. The matrix $\exp(Tt)$ is irreducible for $t > 0$, with maximal eigenvalue $\exp(-\eta t)$ and left and right eigenvectors $\boldsymbol{u}$ and ν, uniquely determined by the normalizing conditions $\boldsymbol{u}\nu = \boldsymbol{u}\boldsymbol{e} = 1$. The vectors $\boldsymbol{u}$ and ν are strictly positive.

A classical corollary of the Perron-Frobenius theorem states that

$$\exp(Tt) = e^{-\eta t}M + o(e^{-\eta t}), \quad \text{as} \quad t \to \infty.$$

The matrix M is given by $M_{ij} = v_i u_j$. The result follows upon substitution.

Corollary 2.3.1. A geometric mixture of the convolution powers of a PH-distribution is asymptotically exponential.

Proof. By theorem 2.2.5, the mixture $\Sigma_{\nu=1}^{\infty} (1 - r)r^{\nu-1}F^{(\nu)}(\cdot)$ is a PH-distribution. The matrix in its representation is given by $T_1 = T + rT^oA^o$, which is irreducible since $T + T^oA^o$ is. The decay parameter η_1 satisfies $0 < \eta_1 < \eta$.

Corollary 2.3.2. The stationary waiting time distribution $W(\cdot)$ in the $M/PH/1$ queue is asymptotically exponential. Its decay parameter η is the positive part of the eigenvalue with largest real part of the matrix $S + \rho S^o\Theta^o$.

Remarks. Theorem 2.3.1 holds under the more general condition that $-\eta$ is a simple pole of the Laplace-Stieltjes transform $f(s)$. As shown by the Erlang distribution, it does not hold when $-\eta$ is a multiple pole, although $e^{\eta_1 x}[1 - F(x)] \to 0$, as $x \to \infty$, for $\eta_1 < \eta$.

Corollary 2.3.2 is only a particular case of a general result on the $GI/PH/1$, proved in theorem 4.1.5.

2.4. PH-RENEWAL PROCESSES

To the m-state, irreducible Markov process with generator $Q^* = T + T^oA^o$, defined in (2.2.10), we may adjoin an instantaneous state $m + 1$, such that the elementary probability of at least one visit to $m + 1$ in the interval $(t, t + dt)$, given that the chain is in i at time t, is $T_i^o dt$. By choosing the return state to the set $\{1, \ldots, m\}$ according to multinomial trials, as described earlier, we see that the successive visits to the instantaneous state form a renewal process with underlying distribution $F(\cdot)$, a PH-distribution of representation $(\boldsymbol{\alpha}, T)$. In this section, we derive a number of useful formulas for PH-renewal processes. We first prove the following lemma.

Lemma 2.4.1. Let Q^* be the irreducible generator of a finite Markov process with stationary probability vector $\boldsymbol{\pi}$, then for any constant $\tau^* \geq \max_i(-Q_{ii}^*)$, the matrix $\tau^*\Pi - Q^*$, where $\Pi_{ij} = \pi_j$, is nonsingular and

$$\int_0^t \exp(Q^*u)du = \Pi t + [I - \exp(Q^*t)](\tau^*\Pi - Q^*)^{-1},$$

$$\text{for} \quad t \geq 0. \quad (2.4.1)$$

Proof. The first statement is equivalent to a well-known property of irreducible stochastic matrices. It is clear that the matrix $P_1 = I + (\tau^*)^{-1}Q^*$ is irreducible and stochastic. Its invariant probability vector is $\boldsymbol{\pi}$. As shown, for example, in J. Kemeny and J. L. Snell [164, p. 100], the matrix $I - P_1 + \Pi = (\tau^*)^{-1}(\tau^*\Pi - Q^*)$ is nonsingular.

Furthermore,

$$\int_0^t \exp(Q^*u)du(\tau^*\Pi - Q^*) = \sum_{\nu=0}^{\infty} \frac{t^{\nu+1}}{(\nu + 1)!} Q^{*\nu}(\tau^*\Pi - Q^*)$$

$$= \tau^*\Pi t + I - \exp(Q^*t).$$

Since $\tau^*\Pi = \Pi(\tau^*\Pi - Q^*)$, formula (2.4.1) follows.

Remarks

a. The matrix $(\tau^*\Pi - Q^*)^{-1}$ is a generalized inverse of Q^*.

b. The (i, j)-element of the matrix $\int_0^t \exp(Q^*u)du$ is the expected total sojourn time in the state j during $(0, t)$, given that the process Q^* starts in the state i.

Let now $N(t)$ be the number of renewals in $(0, t]$, then $\{N(t), t \geq 0\}$ is the *counting process* of the PH-renewal process. Denoting the Markov process with generator Q^* by $\{J(t), t \geq 0\}$, we introduce the matrices $P(n, t) = \{P_{ij}(n, t)\}$, for $n \geq 0$, by

$$P_{ij}(n, t) = P\{N(t) = n, J(t) = j \mid N(0) = 0, J(0) = i\}, \qquad (2.4.2)$$

for $t \geq 0, n \geq 0, 1 \leq i \leq m, 1 \leq j \leq m$.

A standard Chapman-Kolmogorov argument shows that the matrices $P(n, t), n \geq 0, t \geq 0$, satisfy the differential equations

$$P'(0, t) = TP(0, t) = P(0, t)T, \qquad (2.4.3)$$

$$P'(n, t) = TP(n, t) + (1 - \alpha_{m+1}) \sum_{\nu=1}^{n} \alpha_{m+1}^{\nu-1} T^o A^o P(n - \nu, t)$$

$$= P(n, t)T + (1 - \alpha_{m+1}) \sum_{\nu=1}^{n} \alpha_{m+1}^{\nu-1} P(n - \nu, t) T^o A^o,$$

with the initial conditions $P(n, 0) = \delta_{0n}I$, for $n \geq 0$.

Remarks. In most applications, $\alpha_{m+1} = 0$. The system (2.4.3) is then particularly simple. It is the natural analogue of the classical differential equations

$$P'(0, t) = -\lambda P(0, t), \qquad P'(n, t) = -\lambda P(n, t) + \lambda P(n-1, t), \qquad n \geq 1,$$

for the Poisson counting process. It is only for this simple case and for the Erlang process that it has an explicit analytic solution. The straightforward numerical solution of the system (2.4.3) is discussed in what follows.

The matrix $P^*(z, t) = \Sigma_{n=0}^{\infty} P(n, t)z^n$ is readily seen to satisfy the differential equation

$$\frac{\partial}{\partial t} P^*(z, t) = [T + (1 - \alpha_{m+1}z)^{-1}(1 - \alpha_{m+1})zT^oA^o]P^*(z, t),$$

with $P^*(z, 0) = I$. Its solution is given in matrix form by

$$P^*(z, t) = \exp\{[T + (1 - \alpha_{m+1}z)^{-1}(1 - \alpha_{m+1})zT^oA^o]t\}, \text{ for } \quad t \geq 0. \tag{2.4.4}$$

We recall that $A^o = (1 - \alpha_{m+1})^{-1}\operatorname{diag}(\alpha_1, \ldots, \alpha_m)$.

Lemma 2.4.2. The vector $\nu = (\tau^*\Pi - Q^*)^{-1}T^o$, satisfies the equation

$$\nu = (1 - \alpha_{m+1})\mu_1'^{-1}T^{-1}e + [1 + (1 - \alpha_{m+1})^{-1}(\alpha\nu)]e. \tag{2.4.5}$$

Proof. The vector of interest satisfies the equation

$$(\tau^*\Pi - T - T^oA^o)\nu = T^o,$$

which may be equivalently written as

$$\tau^*(\pi\nu)e - T\nu - (\alpha\nu)(1 - \alpha_{m+1})^{-1}T^o = T^o.$$

Premultiplying by π leads to

$$\tau^*(\pi\nu) = \pi T^o = (1 - \alpha_{m+1})\mu_1'^{-1},$$

by formula (2.2.16). Upon substitution and premultiplication by T^{-1}, we obtain (2.4.5). Note that the right-hand side of (2.4.5) still depends on the arbitrary constant $\alpha\nu$. This reflects the arbitrary choice of τ^*, but as we shall see, this is of no consequence to the formulas of interest.

Theorem 2.4.1. The expected number of renewals $H(t) = EN(t)$, in $(0, t]$, is given by

$$H(t) = \mu_1'^{-1}t - (1 - \alpha_{m+1})^{-1} + \frac{\sigma^2 + \mu_1'^2}{2\mu_1'^2} + \mu_1'^{-1}(1 - \alpha_{m+1})^{-1}\alpha[\Pi - \exp(Q^*t)]T^{-1}e, \tag{2.4.6}$$

where σ^2 is the variance of $F(\cdot)$.

Proof. Upon differentiating with respect to z in formula (2.4.4) and setting $z = 1$, we obtain

$$\left(\frac{\partial}{\partial z} P^*(z, t)\right)_{z=1} = (1 - \alpha_{m+1})^{-1} \sum_{n=1}^{\infty} \frac{t^n}{n!} \sum_{\nu=0}^{n-1} Q^{*\nu} T^o A^o Q^{*n-\nu-1},$$

and hence

$$\begin{aligned}\left(\frac{\partial}{\partial z} P^*(z, t)e\right)_{z=1} &= (1 - \alpha_{m+1})^{-1} \int_0^t \exp(Q^*u)du \cdot T^o \\ &= (1 - \alpha_{m+1})^{-1}(\pi T^o)te \\ &\quad + (1 - \alpha_{m+1})^{-1}[I - \exp(Q^*t)](\tau^*\Pi - Q^*)^{-1} T^o \\ &= \mu_1'^{-1} te + \mu_1'^{-1}[I - \exp(Q^*t)]T^{-1}e, \qquad (2.4.7)\end{aligned}$$

by formulas (2.2.16) and (2.4.5).

In the ordinary renewal process, the initial state of the process Q^* is chosen according to the probability vector $(1 - \alpha_{m+1})^{-1}\alpha$. Premultiplying in (2.4.7) by that vector, we obtain

$$H(t) = \mu_1'^{-1}t + \mu_1'^{-1}(1 - \alpha_{m+1})^{-1}\alpha[I - \exp(Q^*t)]T^{-1}e.$$

However,

$$\mu_1'^{-1}(1 - \alpha_{m+1})\alpha(I - \Pi)T^{-1}e = -(1 - \alpha_{m+1})^{-1} - \mu_1'^{-1}\pi T^{-1}e.$$

Recalling from formula (2.2.18) that $-\pi T^{-1}e = \frac{1}{2}\mu_1'^{-1}\mu_2'$, we obtain formula (2.4.6).

Remarks

a. Formula (2.4.6) is a somewhat more detailed version of the expression for $H(t)$, obtained in M. F. Neuts [226].

b. By choosing the initial probability vector of the process Q^* according to the vector π, we obtain the stationary version of the PH-renewal process. By premultiplying by π in (2.4.7), the familiar result $\tilde{H}(t) = \mu_1'^{-1}t$, for $t \geq 0$, is obtained.

c. Since $\lim_{t\to\infty} \exp(Q^*t) = \Pi$, we have the familiar asymptotic formula for the expected number of renewals in $[0, t]$:

$$H(t) + (1 - \alpha_{m+1})^{-1} = \mu_1'^{-1}t + \frac{\sigma^2 + \mu_1'^2}{2\mu_1'^2} + o(1), \quad \text{as} \quad t \to \infty.$$

d. Further differentiations with respect to z in (2.4.4) lead to expressions for the higher factorial moments of $N(t)$. The resulting expressions are complicated and will not be needed in this book. For the second moment of $N(t)$ and an expression for the covariance function of $N(t)$, we refer to M. F. Neuts [232], where a more general Markovian point process related to the PH-renewal process is discussed.

e. It is clear from (2.4.6) that the renewal function $H(t)$ and the renewal density $H'(t)$ for PH-renewal processes may be easily computed. It suffices to evaluate the vector $(1-\alpha_{m+1})^{-1}\boldsymbol{\alpha}[\Pi - \exp(Q^*t)] = \boldsymbol{\nu}(t)$, which is the solution of the system of differential equations

$$\boldsymbol{\nu}'(t) = \boldsymbol{\nu}(t)Q^*, \qquad \boldsymbol{\nu}(0) = \boldsymbol{\pi} - (1-\alpha_{m+1})^{-1}\boldsymbol{\alpha}. \tag{2.4.8}$$

This may serve to illustrate the behavior of $H(t)$ and $H'(t)$ for small values of t for a wide variety of probability distributions $F(\cdot)$.

The following result on the matrices $P(n, t)$, $n \geq 1$, is useful in the sequel.

Theorem 2.4.2. For $n \geq 1$, the matrices $P(n, t)$ are positive for $t > 0$.

Proof. For simplicity of notation only, we assume that $\alpha_{m+1} = 0$. The matrix $P(1, t)$, which may be written as

$$\begin{aligned} P(1, t) &= \int_0^t \exp(Tu)T^0A^0\exp[T(t-u)]du \\ &= \int_0^t \exp(Tu)\mathbf{T}^0\cdot\boldsymbol{\alpha}\exp[T(t-u)]du, \end{aligned}$$

is positive for $t > 0$, by virtue of corollary 2.2.1. Since for $n \geq 1$ we have

$$P(n, t) = \int_0^t \exp(Tu)\mathbf{T}^0\cdot\boldsymbol{\alpha}P(n-1, t-u)du, \quad \text{for} \quad t \geq 0,$$

the result follows by induction.

Remark. The matrix $P(0, t) = \exp(Tt)$ is positive for $t > 0$ only when T is irreducible. For example, in the case of the Erlang PH-renewal process, it is upper triangular.

Computation of the Matrices $P(n, t)$. The computation of the matrices $P(n, t)$, $n \geq 0$, $t \geq 0$, is laborious but elementary. It can be efficiently performed. Since in most situations of interest, $\alpha_{m+1} = 0$, we shall, for ease of discussion, limit our remarks to this case.

The system (2.4.3) may then be used in the simple form

$$\begin{aligned} &P'(0, t) = P(0, t)T, \\ &P'(n, t) = P(n, t)T + P(n-1, t)\mathbf{T}^0\cdot\boldsymbol{\alpha}, \quad \text{for} \quad n \geq 1. \end{aligned} \tag{2.4.9}$$

Note that in the second equation, only the vector $P(n-1, t)T^o$ appears. In large-scale computation, this is useful in the efficient use of central memory. The preceding "value" of $P(n-1, t)$ can be written on a remote storage medium, while only the vector $P(n-1, t)T^o$ is retained in central memory.

For any fixed n, $P(n, t) \to 0$, as $t \to \infty$. Similarly, for any fixed t, $P(n, t) \to 0$, as $n \to \infty$. Moreover for every $t \geq 0$, $\Sigma_{n=0}^{\infty} P(n, t) = \exp(Q^*t)$ is a stochastic matrix.

For every $t \geq 0$, we may compute the matrices $P(n, t)$ (approximately, of course) by adaptively truncating the system (2.4.9) at upper and lower indices $N_2'(t)$ and $N_1'(t)$ respectively.

In practice, this works as follows. We start by setting $N_1'(0) = 0$, $N_2'(0) = 1$, so that only the first differential equation is numerically integrated. Since $P(0, 0) = I$, a stochastic matrix, $P(0, t)$ will remain close to a stochastic matrix on some interval $(0, t_1)$. We can monitor this by, for example, keeping track of the quantity $\max_i |1 - [P(0, t)e]_i|$. When this quantity exceeds a given $\epsilon > 0$, say at $t = t_1$, we increase N_2' by setting $N_2'(t_1) = 1$. In order to alleviate to some extent the effect of the systematic error due to the truncation, we may define $P(1, t_1)$ by, for example,

$$[P(1, t_1)]_{ij} = [1 - (P(0, t_1)e)_i]\alpha_j,$$

so that at t_1, the sum of the computed matrices $P(0, t_1)$ and $P_1(1, t_1)$ is again stochastic.

This procedure of adaptively increasing $N_1'(t)$ and $N_2'(t)$ may be continued in an obvious manner for higher values of t. $N_2'(t)$, the upper index, is increased by one whenever the sum of the computed matrices differs by more than a tolerable amount from a stochastic matrix; the lower index $N_1'(t)$ is increased whenever the row sums of the computed matrix $P(N_1', t)$ become negligibly small.

It is clear that a rigorous study of the combined global errors due to truncation and to the numerical integration of the differential equations would be forbiddingly complicated. In practice, we monitor the deterioration in accuracy by a number of *internal accuracy checks*. For many problems of computational probability these are fortunately easy to construct. For obvious practical reasons, the effort in performing internal accuracy checks should be only a small fraction of the total computational effort.

For the present procedure, the renewal function $H(t)$ provides an accuracy check, as does the matrix $\exp(Q^*t)$. It is clear that

$$\alpha \sum_{n=1}^{\infty} nP(n, t)e = H(t), \quad \text{for} \quad t \geq 0.$$

For a more efficient and accurate computation, we write the left-hand side as

$$\sum_{n=1}^{\infty} \left(1 - \alpha \sum_{\nu=0}^{n-1} P(\nu, t)e\right).$$

The terms in this summation are readily accumulated in the course of evaluating the matrices $P(n, t)$. The accuracy check

$$\sum_{n=0}^{\infty} P(n, t) = \exp(Q^*t)$$

is performed in an obvious manner.

In many applications in queueing theory, we do not face the substantial storage problems arising in the computation of a large number of the matrices $P(n, t)$. As we shall see, many important queueing models require only the evaluation of matrices A_n and $\tilde{A}_n$, defined by

$$A_n = \int_0^{\infty} P(n, u)dK(u),$$

$$\tilde{A}_n = \int_0^{\infty} P(n, u) \frac{1 - K(u)}{\kappa_1'} du, \quad \text{for} \quad n \geq 0, \tag{2.4.10}$$

where $K(\cdot)$ is a probability distribution on $[0, \infty)$ with a finite, nonzero mean κ_1'.

In computing the matrices A_n and A_n, we can generally use a progressive integration method to accumulate the computed approximations to the integral. In most practical cases Simpson's rule or some variation based on splines perform very satisfactorily.

Again, there are a number of accuracy checks available. These involve quantities, which are anyway needed in the larger queueing theoretic algorithm so that their implementation does not involve any appreciable additional computer cost. They are:

a. The matrices $A = \sum_{n=0}^{\infty} A_n = \int_0^{\infty} \exp(Q^*u)dK(u)$ and $\tilde{A} = \sum_{n=0}^{\infty} \tilde{A}_n = \int_0^{\infty} \exp(Q^*u)[1 - K(u)]\kappa_1'^{-1}du$ are positive and stochastic.

b. With π defined as in (2.2.14), we have

$$\pi A = \pi \tilde{A} = \pi. \tag{2.4.11}$$

c. The equalities

$$\alpha \sum_{n=1}^{\infty} nA_n e = \int_0^{\infty} H(u)dK(u), \tag{2.4.12}$$

$$\alpha \sum_{n=1}^{\infty} n\tilde{A}_n e = \int_0^{\infty} H(u) \frac{1 - K(u)}{\kappa_1'} du, \tag{2.4.13}$$

$$\pi \sum_{n=1}^{\infty} nA_n e = \mu_1'^{-1}\kappa_1', \tag{2.4.14}$$

$$\pi \sum_{n=1}^{\infty} n\tilde{A}_n e = \mu_1'^{-1}(2\kappa_1')^{-1}\kappa_2', \tag{2.4.15}$$

where κ_2' is the second moment of $K(\cdot)$, hold. Formulas (2.4.12) and (2.4.13) are stated only for the case where $\alpha_{m+1} = 0$ and $K(0+) = 0$.

The vector $\beta^* = \Sigma_{n=1}^{\infty} nA_n e$, which will arise in a number of formulas, is, by virtue of formula (2.4.7), explicitly given by

$$\beta^* = \mu_1'^{-1}\kappa_1' e + \mu_1'^{-1}(I - A)T^{-1}e. \tag{2.4.16}$$

2.5. TWO APPLICATIONS OF PH-DISTRIBUTIONS

The greatest appeal of PH-distributions lies in the fact that in many situations they allow cumbersome numerical integrations to be replaced by matrix calculations. In this section, we discuss two instances of independent interest. Similar simplifications are available for other Stieltjes integrals, in which the integration is with respect to a PH-distribution. The matrix analytic calculations do require some skill in manipulating the formalism of phase type distributions. Some concern for computational organization is also needed in dealing with matrices of (seemingly) very high dimensions.

We first discuss the computation of the matrices A_n and $\tilde{A}_n$, $n \geq 0$, defined in (2.4.10), for the case where $K(\cdot)$ is a PH-distribution with (irreducible) representation (β, S) of order ν. For simplicity, we consider only the case where $\alpha_{m+1} = \beta_{\nu+1} = 0$. The following theorem, then, is a generalization of theorem 2.2.8.

Theorem 2.5.1. If the probability distribution $K(\cdot)$ is of phase type with representation (β, S) of order ν, the matrices A_n, $n \geq 0$, are given by

$$A_n = U_n \cdot (I \otimes S^o), \quad \text{for} \quad n \geq 0. \tag{2.5.1}$$

The matrices U_n of dimensions $m \times m\nu$ are given by

$$U_0 = -(I \otimes \beta)(T \otimes I + I \otimes S)^{-1},$$
$$U_n = -U_{n-1}(T^o A^o \otimes I)(T \otimes I + I \otimes S)^{-1}, \quad \text{for} \quad n \geq 1. \tag{2.5.2}$$

Proof. By using standard properties of the Kronecker product, we may write

$$A_n = \int_0^\infty P(n, t)\beta \exp(St) S^o dt$$
$$= \int_0^\infty P(n, t) \otimes \beta \exp(St) dt \cdot (I \otimes S^o) = U_n \cdot (I \otimes S^o),$$

for $n \geq 0$. By partial integration, we obtain

$$U_n = -\delta_{n0}(I \otimes \beta S^{-1}) - \int_0^\infty P'(n, t) \otimes \beta \exp(St) S^{-1} dt.$$

Now using the differential equations (2.4.9) for $P(n, t)$, it follows that for $n \geq 1$,

$$U_n = -U_n(T \otimes S^{-1}) - U_{n-1}(T^o A^o \otimes S^{-1}),$$

or

$$U_n = -U_{n-1}(T^o A^o \otimes I)(T \otimes I + I \otimes S)^{-1}.$$

For $n = 0$, we have

$$U_0 = -(I \otimes \beta)(I \otimes S^{-1}) - U_0(T \otimes S^{-1}),$$

or

$$U_0 = -(I \otimes \beta)(T \otimes I + I \otimes S)^{-1}.$$

Corollary 2.5.1. The matrices $\tilde{A}_n$, $n \geq 0$, are given by

$$\tilde{A}_n = U_n \cdot (I \otimes \kappa_1'^{-1} e), \quad \text{for} \quad n \geq 0. \tag{2.5.3}$$

Proof. By using the fact that $\kappa_1'^{-1}[1 - K(x)] = \kappa_1'^{-1}\beta \exp(Sx) e$, and applying the same argument.

Remarks. We may also evaluate the matrices $\tilde{A}_n$, $n \geq 0$, by replacing in (2.5.2) the vector β by θ, the unique vector satisfying $\theta(S + S^o B^o) = \mathbf{0}$, $\theta e = 1$, but this would obviously be computationally inefficient. Theorem 2.5.1 was first proved in M. F. Neuts [227], but with unnecessary concern over the nonsingularity of the matrix $T \otimes I + I \otimes S$, which was established in theorem 2.2.9.

Comments on Numerical Computation. At first glance the formulas (2.5.1), (2.5.2), and (2.5.3) may appear to be of little practical utility as they involve the inversion of a matrix of order $m\nu$ and manipulations with other matrices of comparable size. This is not the case. By an efficient use of the special structure of these equations, the matrices A_n, $n \geq 0$, can be computed by using only matrices of much smaller dimension than $m\nu$. Computer memory may also be economically used.

The matrix $T \otimes I + I \otimes S$ has the block-partitioned structure

$$\begin{vmatrix} T_{11}I + S & T_{12}I & T_{13}I & \cdots & T_{1m}I \\ T_{21}I & T_{22}I + S & T_{23}I & \cdots & T_{2m}I \\ T_{31}I & T_{32}I & T_{33}I + S & \cdots & T_{3m}I \\ \vdots & \vdots & \vdots & & \\ T_{m1}I & T_{m2}I & T_{m3}I & \cdots & T_{mm}I + S \end{vmatrix}, \tag{2.5.4}$$

where the blocks are square matrices of order ν.

Applying the row operations of the Gauss elimination procedure to the block structure in (2.5.4), the matrix may be reduced to block upper triangular form. This requires, in general, the inversion of m matrices of order ν. These matrices are computed once and are stored for future use.

It is now convenient to write the $m \times m\nu$ matrices U_n as $[V_1(n), \ldots, V_m(n)]$, for $n \geq 0$. The m matrices $V_i(n)$, $1 \leq i \leq m$, are of dimensions $m \times \nu$. The equations (2.5.1) and (2.5.2) may then be written as

$$A_n = [V_1(n)S^o, V_2(n)S^o, \ldots, V_m(n)S^o], \quad \text{for} \quad n \geq 0, \tag{2.5.5}$$

and

$$\sum_{i=1}^{m} V_i(0)T_{ij} + V_j(0)S = -\hat{B}_j,$$

$$\sum_{i=1}^{m} V_i(n)T_{ij} + V_j(n)S = -\sum_{i=1}^{m} V_i(n-1)T_i^{\,o}\alpha_j, \tag{2.5.6}$$

for $1 \leq j \leq m$, $n \geq 1$. The matrix $\hat{B}_j$ is an $m \times \nu$ matrix. Its j-th row is given by $\boldsymbol{\beta}$, and all its other rows are zero.

We now solve the first equation in (2.5.6) for the matrices $V_j(0)$, $1 \leq j \leq m$, by Gauss elimination. The matrix A_0 and $m \times \nu$ matrix $\sum_{i=1}^{m} T_i^{\,o} V_i(0)$ are computed and stored. The matrices $V_j(0)$, $1 \leq j \leq m$, have now served their purpose. The memory locations used to store them may now be reused. The matrices $V_j(1)$, $1 \leq j \leq m$, are computed next, and the matrices A_1 and $\sum_{i=1}^{m} T_i^{\,o} V_i(1)$ are evaluated. The latter may be written in the locations occupied by $\sum_{i=1}^{m} T_i^{\,o} V_i(0)$. The same procedure is continued for successive values of n, until an adequate number of the matrices A_n have been obtained.

The matrix $U = \sum_{n=0}^{\infty} U_n$ is obtained in a similar manner. Writing U as $[V_1, V_2, \ldots, V_n]$, we obtain upon summing the equations (2.5.2) on n that

$$\sum_{i=1}^{m} V_i(T_{ij} + T_i^o \alpha_j) + V_j S = -\hat{B}_j, \quad \text{for} \quad 1 \leq j \leq m. \tag{2.5.7}$$

It is now clear that

$$U(Q^* \otimes I + I \otimes S) = -I \otimes \beta, \tag{2.5.8}$$

and hence that

$$\begin{aligned} A &= -(I \otimes \beta)(Q^* \otimes I + I \otimes S)^{-1}(I \otimes S^o) \\ &= [V_1 S^o, V_2 S^o, \ldots, V_m S^o]. \end{aligned} \tag{2.5.9}$$

The system (2.5.7) is again solved by block Gauss elimination. The vector β^* of formula (2.4.16) is now computed without much additional effort.

The power of the procedure just discussed is clear when implemented, for example, to compute a variety of quantities for the $PH/PH/1$, which is discussed in detail in section 3.7. As we shall see, for this model, a large number of computationally explicit formulas are available.

Note that up to this point no special features of the representations (α, T) and (β, S) were used. If we are willing to deal only with distributions $F(\cdot)$ and $K(\cdot)$, which are, for example, mixtures of (generalized) Erlang distributions, we can very drastically reduce the computation times by exploiting the simple representations for these distributions. The matrix $T \otimes I + I \otimes S$ is then itself block upper triangular, and the computation of the matrices $V_j(n)$, $1 \leq j \leq m$, is greatly simplified. The extensively discussed queueing models with Erlang or hyperexponential service and interarrival times clearly become particular cases that are of high computational tractability.

The second application deals with the efficient computation of certain integrals arising in the study of the $GI/M/c$ queue. This queue has an embedded Markov chain P, whose columns with index $j \geq c$ have a simple special structure that needs not concern us here. The elements P_{ij}, $i \geq 0$, $0 \leq j \leq c - 1$, of the first c columns of P, are much more involved and are given by

$$\begin{aligned} P_{ij} &= 0, && \text{for} \quad i + 1 < j < c, \\ &= \int_0^\infty \binom{i+1}{j} e^{-\mu j t}(1 - e^{-\mu t})^{i-j+1} dF(t), && \text{for} \quad j \leq i + 1 \leq c, \\ &= \int_0^\infty dF(t) \int_0^t e^{-c\mu\tau} \frac{(c\mu\tau)^{i-c}}{(i-c)!} \binom{c}{j} e^{-\mu j(t-\tau)}[1 - e^{-\mu(t-\tau)}]^{c-j}\, c\mu\, d\tau, \\ & && \text{for} \quad i \geq c,\ 0 \leq j \leq c - 1. \end{aligned} \tag{2.5.10}$$

The simple probability argument leading to these expressions is given, for example, in D. Gross and C. M. Harris [115, p. 280]. In section 4.2 we shall give a general purpose algorithm to evaluate these quantities. Here we consider the case where $F(\cdot)$ is a PH-distribution with representation $(\boldsymbol{\alpha}, T)$ of order m with $\alpha_{m+1} = 0$.

Let us define the column vectors $\boldsymbol{h}_j(r)$, for $0 \le j \le r$, $1 \le r \le c$, by

$$\boldsymbol{h}_r(r) = (r\mu I - T)^{-1}\boldsymbol{T}^o, \qquad \text{for} \quad 1 \le r \le c,$$

$$\boldsymbol{h}_j(r) = (j+1)\mu(j\mu I - T)^{-1}\boldsymbol{h}_{j+1}(r), \quad \text{for} \quad 0 \le j \le r-1, \qquad (2.5.11)$$

and the row vectors $\boldsymbol{g}_\nu$, $\nu \ge 0$, by

$$\boldsymbol{g}_0 = c\mu\boldsymbol{\alpha}(c\mu I - T)^{-1},$$

$$\boldsymbol{g}_{\nu+1} = \boldsymbol{g}_\nu c\mu(c\mu I - T)^{-1}, \quad \text{for} \quad \nu \ge 0. \qquad (2.5.12)$$

Theorem 2.5.2. The quantities P_{ij}, $0 \le j \le c-1$, $i \ge 0$, are given by

$$\begin{aligned} P_{ij} &= 0, && \text{for} \quad i+1 < j \le c, \\ &= \boldsymbol{\alpha}\boldsymbol{h}_j(i+1), && \text{for} \quad j \le i+1 \le c, \\ &= \boldsymbol{g}_{i-c}\boldsymbol{h}_j(c), && \text{for} \quad i \ge c, \quad 0 \le j \le c-1. \end{aligned} \qquad (2.5.13)$$

Proof. We first treat the case where $i \ge c$. Replacing $dF(t)$ by $\boldsymbol{\alpha}\exp(Tt)\boldsymbol{T}^o dt$ in the double integral in (2.5.10), we obtain

$$\boldsymbol{\alpha}\int_0^\infty dt \int_0^t e^{-c\mu\tau}\frac{(c\mu\tau)^{i-c}}{(i-c)!}c\mu\binom{c}{j}e^{-\mu j(t-\tau)} \cdot [1 - e^{-\mu(t-\tau)}]^{c-j}\exp(Tt)d\tau\, \boldsymbol{T}^o,$$

which upon interchanging the integrals and setting $t - \tau = v$, becomes

$$\boldsymbol{\alpha}\int_0^\infty e^{-c\mu\tau}\frac{(c\mu\tau)^{i-c}}{(i-c)!}c\mu\exp(T\tau)d\tau \cdot \int_0^\infty \binom{c}{j}e^{-\mu j v} \cdot (1 - e^{-\mu v})^{c-j}\exp(Tv)dv\, \boldsymbol{T}^o = \boldsymbol{g}_{i-c}\boldsymbol{h}_j(c),$$

where

$$\boldsymbol{g}_\nu = \boldsymbol{\alpha}\int_0^\infty \exp[-(c\mu I - T)\tau]\frac{(c\mu\tau)^\nu}{\nu!}c\mu d\tau, \quad \text{for} \quad \nu \ge 0. \qquad (2.5.14)$$

The vectors $\boldsymbol{h}_j(r)$, $0 \le j \le r$, $1 \le r \le c$, are defined by

$$\boldsymbol{h}_j(r) = \int_0^\infty \binom{r}{j}e^{\mu(r-j)v}(1 - e^{-\mu v})^{r-j}\exp[-(r\mu I - T)v]dv \cdot \boldsymbol{T}^o. \qquad (2.5.15)$$

The second integral in the expression preceding (2.5.14) is then clearly equal to $h_j(c)$.

The recursion formula (2.5.12) for g_ν, $\nu \geq 0$, is readily verified. Partial integration in (2.5.14) yields

$$g_\nu = c\mu\alpha(c\mu I - T)^{-1} \int_0^\infty \exp[-(c\mu I - T)\tau] \frac{(c\mu\tau)^{\nu-1}}{(\nu-1)!} c\mu d\tau$$

$$= g_{\nu-1} \cdot c\mu(c\mu I - T)^{-1}, \quad \text{for} \quad \nu \geq 1.$$

Verification of the recurrence (2.5.11) for the $h_j(r)$ is only a little more involved. Clearly for $j = r$, we obtain that $h_r(r) = (r\mu I - T)^{-1} T^o$, for $1 \leq r \leq c$. For $0 \leq j \leq r - 1$, we have

$$h_j(r) = \binom{r}{j} \int_0^\infty e^{\mu(r-j)\nu}(1 - e^{-\mu\nu})^{r-j} \exp[-(r\mu I - T)\nu] d\nu \, T^o$$

$$= \mu(r-j)\binom{r}{j}(r\mu I - T)^{-1}\left\{\int_0^\infty \exp(T\nu)e^{-\mu j\nu}(1 - e^{-\mu\nu})^{r-j} d\nu \, T^o\right.$$

$$\left. + \int_0^\infty \exp(T\nu)e^{-\mu(j+1)\nu}(1 - e^{-\mu\nu})^{r-j-1} d\nu \, T^o\right\},$$

by partial integration. The integrals inside the braces are seen to be equal to $\binom{r}{j}^{-1} h_j(r)$ and $\binom{r}{j+1}^{-1} h_{j+1}(r)$ respectively.

Upon substitution and elementary manipulations, we obtain (2.5.11). It is finally clear from (2.5.10) that for $j \leq i + 1 \leq c$, we have $P_{ij} = \alpha h_j(i + 1)$, so that the proof is completed.

In computing the probability density of the steady-state queue length *at an arbitrary time t* for the $PH/M/c$ queue, we obtain quantities $\tilde{P}_{ij}$, which are defined exactly as the P_{ij} in formula (2.5.10), except that the differential $dF(t)$ is replaced by $\lambda_1'^{-1}[1 - F(t)]dt$, where λ_1' denotes the mean interarrival time. The formulas for the computation of the $\tilde{P}_{ij}$ are entirely similar to those for the P_{ij} and are given in the following.

Corollary 2.5.2. The probabilities $\tilde{P}_{ij}$ are given by

$$\begin{aligned} \tilde{P}_{ij} &= 0, && \text{for} \quad i + 1 < j < c, \\ &= \alpha \tilde{h}_j(i + 1), && \text{for} \quad j \leq i + 1 \leq c, \\ &= g_{i-c}\tilde{h}_j(c), && \text{for} \quad i \geq c, \quad 0 \leq j \leq c - 1. \end{aligned} \tag{2.5.16}$$

The vectors g_ν are defined as in formula (2.5.12) and

$$\begin{aligned} \tilde{h}_r(r) &= \lambda_1'^{-1}(r\mu I - T)^{-1}e, \quad \text{for} \quad 1 \leq r \leq c, \\ \tilde{h}_j(r) &= (j + 1)\mu(j\mu I - T)^{-1}\tilde{h}_{j+1}(r), \\ &\qquad \text{for} \quad 0 \leq j \leq r - 1, \quad 1 \leq r \leq c. \end{aligned} \tag{2.5.17}$$

Proof. It suffices to note that $\lambda_1'^{-1}[1 - F(t)] = \lambda_1'^{-1}\boldsymbol{\alpha} \exp(Tt)\boldsymbol{e}$.

Remarks

a. Theorem 2.5.2 and corollary 2.5.2, which were first proved and computationally implemented in D. E. Baily and M. F. Neuts [10], provide highly efficient algorithms for the computation of the first c columns of the matrices P and $\tilde{P}$, which arise in the $PH/M/c$ queue.

The inverses of the c matrices $(j\mu I - T)$, $1 \leq j \leq c$, need to be evaluated. For greatest efficiency, we compute the vectors $\boldsymbol{h}_j(r)$—and simultaneously the $\tilde{\boldsymbol{h}}_j(r)$—in the order

$$\begin{array}{llll}
\boldsymbol{h}_c(c) & & & \\
\boldsymbol{h}_c(c-1), & \boldsymbol{h}_{c-1}(c-1), & & \\
\boldsymbol{h}_c(c-2), & \boldsymbol{h}_{c-1}(c-2), & \boldsymbol{h}_{c-2}(c-2), & \\
\vdots & \vdots & \vdots & \\
\boldsymbol{h}_c(1), & \boldsymbol{h}_{c-1}(1), & \boldsymbol{h}_{c-2}(1), & \cdots \quad \boldsymbol{h}_1(1).
\end{array}$$

The vectors $\boldsymbol{h}_c(j)$ and $\tilde{\boldsymbol{h}}_c(j)$, for $0 \leq j \leq c - 1$, are retained in memory for use in the computation of the P_{ij} and $\tilde{P}_{ij}$ with $i \geq c$, $0 \leq j \leq c - 1$.

b. As the basis for a truncation procedure or as an accuracy check, we may use the following equalities, which are easily derived from the definitions of the P_{ij} and $\tilde{P}_{ij}$:

$$\sum_{i=c-1}^{\infty} \sum_{j=0}^{c-1} P_{ij} = c\mu\lambda_1',$$

$$\sum_{i=c-1}^{\infty} \sum_{j=0}^{c-1} \tilde{P}_{ij} = c\mu(2\lambda_1')^{-1}\lambda_2', \tag{2.5.18}$$

where λ_2' is the second moment of $F(\cdot)$.

c. The stable $GI/M/c$ queue has an extensive and classical literature. See, for example, J. W. Cohen [61], D. Gross and C. M. Harris [115], or T. L. Saaty [272]. It provides a classical example of a Markov chain with a modified geometric invariant vector $\boldsymbol{x}$, with components satisfying $\boldsymbol{x}_k = \boldsymbol{x}_{c-1}\xi^{k-c+1}$, for $k \geq c - 1$. The quantity ξ is the unique solution in (0, 1) of the equation

$$\xi = f[c\mu(1 - \xi)], \tag{2.5.19}$$

where $f(\cdot)$ denotes the Laplace-Stieltjes transform of $F(\cdot)$.

In order to compute the vector $\boldsymbol{x}$, we do not need the quantities P_{ij} for $i \geq c$, $0 \leq j \leq c - 1$. They do, however, play a significant role in the study of the *GI/M/c with group arrivals,* discussed in section 4.2. The numerical solution of the simple equation (2.5.19) is straightforward. It was discussed in M. L. Chaudhry and J. G. C. Templeton [50] and I. Sahin [273]. For related material, see also F. A. Al-Khayyal and D. Gross [3].

d. Integrals of the type shown in (2.5.10) were evaluated in R. E. Nance and U. N. Bhat [212] by expanding the binomial terms appearing in the integrands. The integrals of interest are then expressed as finite or infinite sums of somewhat simpler integrals. These sums have terms with alternating signs. Because of the likely loss of significance, this approach cannot be generally recommended.

e. Some variants of the *GI/M/c* queue differ from that classical model only in the analytic expressions for the elements of the first c columns of the transition probability matrix P. One such instance is the case where the service rate depends on the number of active servers. This model was discussed for $c = 2$ in U. N. Bhat [21]. It is possible to compute the quantities P_{ij}, $i \geq 0$, $0 \leq j \leq c - 1$, for this case also by appropriately modifying the derivations given above. Similar comments apply to the queue with heterogeneous exponential servers, treated by V. P. Singh [285].

2.6. CONCLUDING REMARKS

The use of special PH-distributions in the study of stochastic models is so widespread and varied that any survey of the literature must almost surely be incomplete. The term "distributions of phase type" is also frequently used for the particular case of generalized Erlang distributions. There is also, particularly in the earlier literature, a profusion of terminology, which may lead to confusion and occasional duplication of results. The classical Erlang distributions are particular cases of the gamma distributions and of χ^2-distributions. They arise from the former for integral values of the shape parameter and from the latter when the number of degrees of freedom is even. See, for example, L. Takács [292] and D. M. G. Wishart [305, 306].

The (generalized) Erlang distributions have a coefficient of variation less than one. Erlang renewal processes exhibit, particularly for higher orders, a very "regular" behavior. To offer an alternative, P. M. Morse [210] proposed the use of hyperexponential distributions, whose coefficient of variation always exceeds one. See, for example, J. W. Cohen [61, p. 138]. Without the formalism discussed in this chapter, the treatment, even of simple models involving hyperexponential distributions, becomes very laborious.

The classical Erlang distribution has a property that in section 2.1 we referred to as "disturbingly particular." It allows us, for example, in a single server queue with Erlang service times to consider that each customer requires m phases of service. Each phase then has an exponential distribution. We may now replace each customer upon arrival by m pseudo-customers and study, for example, the $GI/E_m/1$ queue as a $GI^{(m)}/M/1$ queue. This idea was suggested in a different context by D. G. Kendall [165] and has been widely applied.

Its appeal lies in the fact that, in particular cases, we may recover the number of customers in the queue from the number of pseudo-customers or phases that are present. For example, if in the $GI/E_m/1$ queue there are $i \geq 1$ phases present at any time, then the corresponding queue length is given by

$$i_1 = \left[\frac{i}{m}\right],$$

and the service in course is in its phase $i - mi_1$. For multi-server queues or for models involving mixtures of Erlang distributions, this recovery is no longer possible. The argument is limited to very special models. Even for these, it does not lead to algorithmically appealing expressions.

A versatile class of probability distributions on $[0, \infty)$ was introduced by G. Luchak [196–98]. See also D. M. G. Wishart [306]. These are mixtures of Erlang distributions of the form

$$\sum_{k=1}^{\infty} p_k E_k(\lambda; x), \tag{2.6.1}$$

where $p_k \geq 0$, $\sum_{k=1}^{\infty} p_k = 1$. These distributions are not necessarily of phase type, and their Laplace-Stieltjes transform need not be rational. If the mixing density $\{p_k\}$ is of phase type, and in particular when it has finite support, then the probability distributions in (2.6.1) are indeed PH-distributions, by theorem 2.2.5. This will always be the case for mixtures, obtained by fitting a finite sum of the type (2.6.1) to empirical data or by truncating an infinite sum for algorithmic use. The drawback of the form (2.6.1), at least for the matrix methods developed in this book, is that it limits us to PH-distributions with a very special representation. The matrix T is the same as for the Erlang distribution, and the vector $\boldsymbol{\alpha}$ is given by $(p_m, p_{m-1}, \ldots, p_1)$. It appears likely that the order of the matrix T needed to represent a given distribution $F(\cdot)$ will be very high. This is confirmed by work of W. Bux and U. Herzog [42, 43], who developed an algorithm for fitting distributions of the type (2.6.1) to empirical data.

This brings us to the largely unexplored area of statistical fitting and approximation by PH-distributions. It is elementary to prove that the sets of Cox distributions (with rational Laplace-Stieltjes transforms), the probability distributions of the form (2.6.1), and the PH-distributions are all three dense in the set of all probability distributions on $[0, \infty)$. This is occasionally stated as a loose justification for limiting a given analysis to any of these classes of distributions. The value of this simple theorem as an approximation theorem is largely illusory. No general approximation results are, in fact, known. Although the density plots of PH-distributions, even with m as small as five to ten, can be highly versatile, there are interesting distributions, for example, with steeply increasing or decreasing densities or with intervals of constancy, which are difficult to approximate. Foremost among these are "delayed distributions," for which $F(x) = 0$, for $0 \leq x \leq a$, for some $a > 0$. Such distributions are of interest to many applications, but even the simple delayed exponential distribution is difficult to approximate by phase type distributions.

One might suggest considering a delayed distribution as the convolution of a distribution, degenerate at $a > 0$, and a second distribution which is well approximated by a PH-distribution. The degenerate distribution at a is the limit of a sequence of Erlang distributions of increasing order. Although for some applications this suggestion has merit, the order of the Erlang distribution, needed for a close approximation, is generally so high as to render this approach impractical. The systems of differential equations arising from such an approximation will typically also be stiff, due to the high values of some of the entries in the coefficient matrices. A workable alternative, specifically to models involving delayed distributions, is to use a discretized model and to exploit the preservation under right shifts of the discrete PH-distributions, discussed in section 2.2. For some specific models that we have studied on an exploratory basis, this approach has given entirely satisfactory results. A similar concern apparently underlies the use of certain particular discrete distributions of phase type in reliability applications, as discussed, for example, in J. L. Bogdanoff [29–31] and V. N. Saksena [275].

In our experience, PH-distributions readily lend themselves to the representation of certain qualitative features of data, such as bimodality or increased tail probabilities. Their ability to serve as convenient numerical approximations to other useful families of probability distributions requires much further study. At the very least, they widely extend *in a computationally convenient manner* the family of probability distributions available for use in stochastic models.

A promising tool for the study and the fitting of PH-distributions is *computer graphics*. Their relationship to differential equations makes them natural objects for study by the *analog computer* (see R. L. Gue, [116]). A specific, but rudimentary, application, using generalized Erlang distributions of order three, was reported in G. Rossa [270]. Our own experiments on a digital computer with a graphics terminal have been highly encouraging, though as yet limited by time constraints. Various ideas on approximations deserve to be explored in this manner and could lead to worthwhile theoretical conjectures. We refer to U. Grenander [113], where this methodological approach is cogently advocated.

Chapter 3
Quasi-Birth-and-Death Processes

3.1. INTRODUCTION

In this chapter we consider the particular case of the Markov chains $\tilde{P}$ or processes $\tilde{Q}$, treated in chapter 1, where the matrix $\tilde{P}$ or $\tilde{Q}$ is *block tridiagonal.* This arises in many specific applications, a number of which will be discussed in this and subsequent chapters. The paper by R. V. Evans [89] and the Ph.D. thesis of V. Wallace [302] initiated the study of such Markov processes and emphasized the algorithmic significance of matrix-geometric solutions. V. Wallace also coined the term *quasi-birth-and-death processes* (QBD processes) and developed a computer routine for their analysis.

We now recognize that the block-tridiagonal case lies at the intersection of the classes of Markov chains of the types $GI/M/1$ and $M/G/1$. As such, it may be considered from two different viewpoints, and doing so leads to very detailed results. We shall, in the course of this chapter, present the key properties of Markov chains of $M/G/1$ type as they appear for the tridiagonal case. These properties will lead to results on the analogue of the *busy period* for queues described by QBD processes. It is well known that for queues of the $GI/M/1$ type the distribution of the busy period† is nearly intractable by classical methods. See, for

†As this book was nearing completion, V. Ramaswami [263] announced a derivation of the distribution of the busy period for the $GI/PH/1$ queue which is entirely free of transform methods. The distribution is explicitly given in terms of a matrix of mass-functions, which is itself the solution to a nonlinear matrix integral equation of Volterra type. These results also provide additional probabilistic interpretations for several constants that appear in the discussion of the $GI/PH/1$ in chapter 4.

example, B. W. Conolly [66, 67]. The structure of QBD processes is also important to the development of algorithms and analytic results for *waiting times.*

In addition to general results, this chapter also contains a number of special and surprisingly explicit results on some elementary queues that involve distributions of phase type. These results, presented here for the first time, are useful in the study of more complex models and, by themselves, have definite didactic interest.

The applications that require the material in this chapter typically lead to *continuous parameter* processes rather than to discrete parameter chains. In chapter 1 we emphasized the terminology of Markov chains; here we shall prefer that of processes. As we pointed out in section 1.7, there are no important mathematical differences between the corresponding theorems for $\tilde{P}$ and those for the generator $\tilde{Q}$. Our choice of terminology also serves to state the general results in the form needed for most applications and to familiarize the reader, not already so acquainted, with the terms and properties of Markov processes.

A *quasi-birth-and-death process*—henceforth called a *QBD process* —is a Markov process on the state space $E = \{(i, j), i \geq 0, 1 \leq j \leq m\}$, with infinitesimal generator $\tilde{Q}$, given by

$$\tilde{Q} = \begin{vmatrix} B_0 & A_0 & & & & \\ B_1 & A_1 & A_0 & & & \\ & A_2 & A_1 & A_0 & & \\ & & A_2 & A_1 & A_0 & \cdots \\ & & & A_2 & A_1 & \cdots \\ & & & \vdots & \vdots & \end{vmatrix}, \tag{3.1.1}$$

where $B_0 e + A_0 e = B_1 e + A_1 e + A_0 e = (A_0 + A_1 + A_2)e = \mathbf{0}$. As in chapter 1, we shall call the form (3.1.1) *canonical.* The elementary modifications, needed to handle more complicated boundary behavior, were discussed in section 1.5 and will be illustrated by examples. The generator $\tilde{Q}$ is also assumed to be *irreducible.* The matrix $A = A_0 + A_1 + A_2$ is a finite generator. The following theorem, then, is a particular case of theorem 1.7.1.

Theorem 3.1.1. The process $\tilde{Q}$ is positive recurrent if and only if the minimal nonnegative solution R to the matrix-quadratic equation

$$R^2 A_2 + R A_1 + A_0 = 0 \tag{3.1.2}$$

has all its eigenvalues inside the unit disk and the finite system of equations

$$x_0(B_0 + RB_1) = \mathbf{0}$$
$$x_0(I - R)^{-1}e = 1 \tag{3.1.3}$$

has a (unique) positive solution x_0.

If the matrix A is irreducible, then $\mathrm{sp}(R) < 1$ if and only if

$$\pi A_2 e > \pi A_0 e, \tag{3.1.4}$$

where π is the stationary probability vector of A.

The stationary probability vector $x = [x_0, x_1, \ldots]$ of $\tilde{Q}$ is given by

$$x_i = x_0 R^i, \quad \text{for} \quad i \geq 0. \tag{3.1.5}$$

The (equivalent) equalities

$$RA_2 e - A_0 e = RB_1 e - B_0 e = \mathbf{0} \tag{3.1.6}$$

hold.

3.2. THE BOUNDED AND UNBOUNDED *M/PH/*1 AND *PH/M/c* QUEUES

The elementary queues *M/PH/*1 and *PH/M/c* are particular cases of QBD processes. Their discussion, although primarily of pedagogical interest, is also of interest for several other reasons. All references to complex analysis and roots of transcendental equations are avoided. To see what this implies, the reader may wish to compare the lucid and detailed presentation of the transform solution for the $M/E_m/1$ and $E_m/M/1$ queues in L. Kleinrock [172; 1:126–33] to the approach discussed here.

The present solution is fully explicit. Except for the case of the unbounded *GI/M/c* queue, the matrix R may be explicitly obtained. The derivations exploit the formal properties of PH-distributions and can be presented to a lay audience without relying on the mathematically more demanding results on matrix-geometric solutions.

The *finite* versions of these queues may also be explicitly solved. This is somewhat surprising. Finite queues rarely admit elegant, closed-form solutions.

The M/PH/*1 *Queue. The service time distribution $H(\cdot)$ is of phase type with representation (β, S) of order ν. The mean service time is $\mu_1' = -\beta S^{-1}e$, and the Poisson arrival rate is λ. In discussing the unbounded queue, we assume that $\rho = \lambda\mu_1' < 1$. For ease of notation, we let $\beta_{\nu+1} = 0$.

The $M/PH/1$ queue may be studied as a QBD process with the state space $E = \{0, (i, j); i \geq 1, 1 \leq j \leq \nu\}$. The state 0 corresponds to the empty queue; the state (i, j) to having $i \geq 1$ customers in the system and the service process in the phase j, $1 \leq j \leq \nu$. The generator $\tilde{Q}$ is given by

$$\tilde{Q} = \begin{array}{c|ccccc|}
0 & -\lambda & \lambda\boldsymbol{\beta} & \mathbf{0} & \mathbf{0} & \cdots \\
1 & S^o & S - \lambda I & \lambda I & 0 & \cdots \\
2 & \mathbf{0} & S^oB^o & S - \lambda I & \lambda I & \cdots \\
3 & \mathbf{0} & 0 & S^oB^o & S - \lambda I & \cdots \\
4 & \mathbf{0} & 0 & 0 & S^oB^o & \cdots \\
5 & \mathbf{0} & 0 & 0 & 0 & \cdots \\
\vdots & \vdots & \vdots & \vdots & \vdots &
\end{array}. \tag{3.2.1}$$

The stationary probability vector $\boldsymbol{x} = [x_0, \boldsymbol{x}_1, \boldsymbol{x}_2, \ldots]$ is given by the following theorem.

Theorem 3.2.1. We have

$$x_0 = 1 - \rho,$$
$$\boldsymbol{x}_i = (1 - \rho)\boldsymbol{\beta}R^i, \quad \text{for} \quad i \geq 1. \tag{3.2.2}$$

The matrix R is given by

$$R = \lambda(\lambda I - \lambda B^{oo} - S)^{-1}, \tag{3.2.3}$$

where $B^{oo} = \boldsymbol{e} \cdot \boldsymbol{\beta}$.

Proof. A direct argument, not based on theorem 3.1.1, goes as follows. The steady-state equations are given by

$$-\lambda x_0 + \boldsymbol{x}_1 S^o = 0,$$
$$\lambda x_0\boldsymbol{\beta} + \boldsymbol{x}_1(S - \lambda I) + \boldsymbol{x}_2 S^oB^o = \mathbf{0}, \tag{3.2.4}$$
$$\lambda \boldsymbol{x}_{i-1} + \boldsymbol{x}_i(S - \lambda I) + \boldsymbol{x}_{i+1}S^oB^o = \mathbf{0}, \quad \text{for} \quad i \geq 2.$$

Multiplying each equation but the first on the right by the column vector $\boldsymbol{e}$ leads to

$$\boldsymbol{x}_1 S^o = \lambda x_0,$$
$$\boldsymbol{x}_{i+1} S^o = \lambda \boldsymbol{x}_i \boldsymbol{e}, \quad \text{for} \quad i \geq 1. \tag{3.2.5}$$

Since now $x_{i+1}S^oB^o = \lambda x_i B^{oo}$, for $i \geq 1$, we obtain

$$x_i(\lambda I - \lambda B^{oo} - S) = \lambda x_{i-1}, \quad \text{for} \quad i \geq 2, \tag{3.2.6}$$

and similarly,

$$x_1(\lambda I - \lambda B^{oo} - S) = \lambda x_0 \beta. \tag{3.2.7}$$

It only remains to show that $\lambda I - \lambda B^{oo} - S$ is nonsingular and that $x_0 = 1 - \rho$. The nonsingularity is clear from the fact that $S + \lambda B^{oo} - \lambda I$ is a *stable* matrix, but an elementary direct argument proceeds as follows.

Let $\boldsymbol{u}$ be in the (left) null space of $\lambda I - \lambda B^{oo} - S$, then

$$\boldsymbol{u}S + \lambda(\boldsymbol{u}\boldsymbol{e})\boldsymbol{\beta} - \lambda \boldsymbol{u} = \mathbf{0}. \tag{3.2.8}$$

If $\boldsymbol{ue} = 0$, then $\boldsymbol{u}(\lambda I - S) = \mathbf{0}$, implies that $\boldsymbol{u} = \mathbf{0}$. If $\boldsymbol{ue} \neq 0$, we may normalize $\boldsymbol{u}$ by setting $\boldsymbol{ue} = 1$. Multiplying in (3.2.8) on the right by $\boldsymbol{e}$, we then obtain that $\boldsymbol{u}S^o = 0$. It further follows that $\boldsymbol{u} = \lambda\boldsymbol{\beta}(\lambda I - S)^{-1}$, and therefore

$$\boldsymbol{u}S^o = \lambda\boldsymbol{\beta}(\lambda I - S)^{-1}S^o = 0. \tag{3.2.9}$$

By formula (2.2.6), $\boldsymbol{\beta}(\lambda I - S)^{-1}S^o$ is the Laplace-Stieltjes transform at $s = \lambda > 0$, of the probability distribution $H(\cdot)$. The equality (3.2.9) therefore cannot hold. The vector $\boldsymbol{u}$ must be zero, and the matrix $\lambda I - \lambda B^{oo} - S$ is nonsingular.

The irreducibility of the representation $(\boldsymbol{\beta}, S)$ leads by a standard argument to the irreducibility of the stable matrix $S + \lambda B^{oo} - \lambda I$, so that the matrix R in (3.2.3) is *positive.*

It will be shown in theorem 3.2.2 that $\text{sp}(R) < 1$, if $\rho < 1$. The quantity x_0 is therefore given by the normalizing equation

$$x_0 + x_0\boldsymbol{\beta}R(I - R)^{-1}\boldsymbol{e} = 1. \tag{3.2.10}$$

Substitution for R leads to

$$x_0 - \lambda x_0\boldsymbol{\beta}(\lambda B^{oo} + S)^{-1}\boldsymbol{e} = 1.$$

The inverse of $\lambda B^{oo} + S$ is calculated as follows:

$$\begin{aligned}(\lambda B^{oo} + S)^{-1} &= S^{-1}(I + \lambda B^{oo}S^{-1})^{-1} \\ &= S^{-1} \sum_{\nu=0}^{\infty} (-1)^\nu \lambda^\nu (B^{oo}S^{-1})^\nu \\ &= S^{-1}\left[I - \lambda \sum_{\nu=0}^{\infty} \rho^\nu B^{oo}S^{-1}\right] \\ &= S^{-1}[I - \lambda(1 - \rho)^{-1}B^{oo}S^{-1}].\end{aligned}$$

Upon substitution, we obtain

$$x_0 - \lambda x_0 \beta S^{-1} e + \lambda^2 x_0 (1-\rho)^{-1} \beta S^{-1} B^{oo} S^{-1} e$$
$$= x_0 + \rho x_0 + x_0 \rho^2 (1-\rho)^{-1} = x_0 (1-\rho)^{-1} = 1,$$

so that $x_0 = 1 - \rho$.

Corollary 3.2.1. The matrix $\lambda(\lambda I - \lambda B^{oo} - S)^{-1}$ may also be written as

$$\lambda(\lambda I - \lambda B^{oo} - S)^{-1} = \lambda(\lambda I - S)^{-1} + \frac{1}{f(\lambda)} \lambda^2 (\lambda I - S)^{-1} e \cdot \beta(\lambda I - S)^{-1}, \quad (3.2.11)$$

where $f(\lambda) = \beta(\lambda I - S)^{-1} S^o$.

Proof. We write

$$\lambda(\lambda I - \lambda B^{oo} - S)^{-1} = \lambda(\lambda I - S)^{-1}[I - \lambda B^{oo}(\lambda I - S)^{-1}]^{-1}.$$

The rightmost inverse is given by

$$I + \sum_{\nu=1}^{\infty} [\lambda B^{oo}(\lambda I - S)^{-1}]^{\nu} = I + \lambda B^{oo}(\lambda I - S)^{-1} \sum_{\nu=0}^{\infty} [\lambda \beta(\lambda I - S)^{-1} e]^{\nu}.$$

Since $\lambda\beta(\lambda I - S)^{-1} e = 1 - f(\lambda)$, we obtain

$$[I - \lambda B^{oo}(\lambda I - S)^{-1}]^{-1} = I + \frac{1}{f(\lambda)} \lambda B^{oo}(\lambda I - S)^{-1},$$

and formula (3.2.11) follows upon substitution.

Since the vectors $(\lambda I - S)^{-1} e$ and $\beta(\lambda I - S)^{-1}$ are positive, it is also clear from (3.2.11) that the matrix $\lambda(\lambda I - \lambda B^{oo} - S)^{-1}$ is positive.

Theorem 3.2.2. The spectral radius η of the positive matrix R, given by formula (3.2.3), is equal to $\eta = \lambda(\lambda - c^*)^{-1}$, where c^* is the unique solution of the equation

$$\lambda\beta(cI - S)^{-1} e = 1, \quad (3.2.12)$$

in the interval $(-\tau, \lambda)$, where $-\tau$ is the eigenvalue of maximal real part of S.

Proof. Let u be the left eigenvector of R corresponding to η, normalized by $ue = 1$. The equation $uR = \eta u$ then leads to $\lambda u = \lambda\eta u - \lambda\eta\beta - \eta u S$. Postmultiplying by e leads to $\eta = \lambda/d$, where $d = uS^o > 0$. Upon substitution, we obtain $u = \lambda\beta[(\lambda - d)I - S]^{-1}$, and since $ue = 1$, (3.2.12) follows. The function $\phi(c) = \lambda\beta(cI - S)^{-1} e$ is strictly decreasing, positive and convex on $(-\tau, \infty)$, and has a pole at

$-\tau$. As c tends to infinity, $\phi(c)$ tends to zero. It follows that (3.2.12) has a unique solution. If $\rho < 1$, then $\phi(0) = \rho < 1$, and c^* is negative. This implies that $0 < \eta < 1$. If $\rho = 1$, $\phi(0) = 1$, and $\eta = 1$. If $\phi(0) > 1$, c^* is positive and $\rho > 1$. Since $\eta > 0$, it follows that $c^* < \lambda$.

The stationary density $\{y_k, k \geq 0\}$ of the queue length following departures is readily derived from the vector $\boldsymbol{x}$. We obtain

$$y_k = C\boldsymbol{x}_{k+1}\boldsymbol{S}^o = (1 - \rho)C\boldsymbol{\beta}R^{k+1}\boldsymbol{S}^o = \lambda(1 - \rho)C\boldsymbol{\beta}R^k\boldsymbol{e}, \qquad (3.2.13)$$

for $k \geq 0$. The constant C is given by

$$\lambda(1 - \rho)C\boldsymbol{\beta}(I - R)^{-1}\boldsymbol{e} = 1.$$

We have already verified in the course of the proof of theorem 3.2.1 that $(1 - \rho)\boldsymbol{\beta}R(I - R)^{-1}\boldsymbol{e} = \rho$. This readily implies that $C = \lambda^{-1}$.

Remark. Theorem 3.2.1 and formula (3.2.13) provide the most efficient computational procedure for evaluating the joint density of the queue length and the service phase for the $M/PH/1$ queue. In order to compute this density it is necessary to evaluate neither the service time distribution nor the discrete density of the number of arrivals during a service.

The joint density of the queue length and the service phase is an essential ingredient in the algorithms for the stationary waiting time densities under service in random order or under the last-come, first-served disciplines, discussed in M. F. Neuts [225].

The Finite M/PH/1 Queue. QBD processes on a finite state space do not, in general, have a stationary probability vector $\boldsymbol{x}$ of a particularly simple analytic form. The finite $M/PH/1$ queue is an exception. We consider the $M/PH/1/K + 1$ queue in which at most $K + 1$ customers can be present in the system. Any customers arriving while there are $K + 1$ customers in the system are lost.

The finite analogue of the matrix given in (3.2.1) is similar in form, except that the final row $K + 1$ is given by

$$\boldsymbol{0} \;\; 0 \;\; 0 \;\; \cdots \;\; 0 \;\; \boldsymbol{S}^o\boldsymbol{B}^o \;\; S.$$

The stationary vector $\boldsymbol{x} = [x_0, \boldsymbol{x}_1, \ldots, \boldsymbol{x}_{K+1}]$ satisfies the equations

$$\begin{aligned} -\lambda x_0 + \boldsymbol{x}_1\boldsymbol{S}^o &= 0, \\ \lambda x_0\boldsymbol{\beta} + \boldsymbol{x}_1(S - \lambda I) + \boldsymbol{x}_2\boldsymbol{S}^o\boldsymbol{B}^o &= \boldsymbol{0}, \\ \lambda \boldsymbol{x}_{i-1} + \boldsymbol{x}_i(S - \lambda I) + \boldsymbol{x}_{i+1}\boldsymbol{S}^o\boldsymbol{B}^o &= \boldsymbol{0}, \quad \text{for} \quad 2 \leq i \leq K, \qquad (3.2.14) \\ \lambda \boldsymbol{x}_K + \boldsymbol{x}_{K+1}S &= \boldsymbol{0} \\ x_0 + \sum_{i=1}^{K+1} \boldsymbol{x}_i\boldsymbol{e} &= 1. \end{aligned}$$

Exactly the same calculations as in the proof of theorem 3.2.1 lead to the explicit solution, given by

$$x_i = x_0 \beta R^i, \quad \text{for} \quad 1 \leq i \leq K,$$
$$x_{K+1} = x_0 \beta R^K(-\lambda S^{-1}), \tag{3.2.15}$$

where the matrix R is given by formula (3.2.3) and x_0 is given by

$$x_0 = \left\{\beta\left[\sum_{i=0}^{K} R^i - \lambda R^K S^{-1}\right]e\right\}^{-1}. \tag{3.2.16}$$

We note that the formulas (3.2.15) and (3.2.16) are in a convenient form to compute the vector x for successive increasing values of K. The finite sum in (3.2.16) is not written in closed form, because the matrix $I - R$ may now be singular.

Once the vector x is known, a variety of other features of the finite $M/PH/1$ may be routinely computed. In particular, the waiting time distributions at departures and at arbitrary times may be found by adapting the procedure discussed in section 3.9. The distribution of the busy period, which for the general finite $M/G/1$ queue is very complicated—see, for example, L. W. Miller [205] and S. I. Rosenlund [269]—becomes for the present case a simple absorption time in a finite Markov process and may be computed by solving a system of differential equations. The method is discussed in M. F. Neuts [225].

The *overflow process* of a finite $M/PH/1$ queue is interesting. It may be informally described as a Poisson process, which is turned on only when the corresponding Markov process $\tilde{Q}$ is in one of the states $(K + 1, j)$, $1 \leq j \leq \nu$, and is interrupted at all other times. This is a particular case of a *Markov modulated Poisson process*—see, for example, J. Grandell [111]. This is in turn a subcase of the versatile Markovian point process, discussed in M. F. Neuts [232]. We shall return to the problem of overflows in queueing models in section 6.4.

The PH/M/c Queue. We now consider the $GI/M/c$ queue in which the arrivals form a PH-renewal process with interarrival time distribution $F(\cdot)$ of phase type with representation (α, T) of order m and mean $\lambda_1{}' = -\alpha T^{-1}e$.

The service rate of each of the c servers is denoted by μ. The $PH/M/c$ queue is a QBD process on the state space $E = \{(i, j), i \geq 0, 1 \leq j \leq m\}$, where i denotes the number of customers in the system and j is the phase of the arrival process. Its generator $\tilde{Q}$ is given by

$$\tilde{Q} = \begin{vmatrix} T & T^oA^o & 0 & 0 & \cdots & & & \\ \mu I & T-\mu I & T^oA^o & 0 & \cdots & & & \\ 0 & 2\mu I & T-2\mu I & T^oA^o & \cdots & & & \\ \vdots & \vdots & \vdots & \vdots & & & & \\ 0 & 0 & 0 & 0 & \cdots & c\mu I & T-c\mu I & T^oA^o & \cdots \\ 0 & 0 & 0 & 0 & \cdots & 0 & c\mu I & T-c\mu I & \cdots \\ \vdots & \vdots & \vdots & \vdots & & \vdots & \vdots & \vdots & \end{vmatrix}$$

(3.2.17)

The QBD process $\tilde{Q}$ is positive recurrent if and only if

$$c\mu > \boldsymbol{\pi}(T^oA^o)\boldsymbol{e} = \boldsymbol{\pi}T^o = \lambda_1{'}^{-1},$$

where $\boldsymbol{\pi}(T + T^oA^o) = \boldsymbol{0}$, $\boldsymbol{\pi e} = 1$. This, of course, is the classical equilibrium condition for the $GI/M/c$ queue.

The invariant probability vector $\boldsymbol{x} = [\boldsymbol{x}_0, \boldsymbol{x}_1, \ldots]$ of $\tilde{Q}$ is given by

$$\boldsymbol{x}_i = \boldsymbol{x}_{c-1}R^{i-c+1}, \quad \text{for} \quad i \geq c - 1, \tag{3.2.18}$$

where the matrix R is the minimal nonnegative solution of the equation

$$c\mu R^2 + R(T - c\mu I) + T^oA^o = 0. \tag{3.2.19}$$

The equality $c\mu R\boldsymbol{e} = T^o$ holds. The vectors $\boldsymbol{x}_0, \ldots, \boldsymbol{x}_{c-1}$, satisfy

$$\begin{aligned} \boldsymbol{x}_0T + \mu\boldsymbol{x}_1 &= \boldsymbol{0} \\ \boldsymbol{x}_{i-1}T^oA^o + \boldsymbol{x}_i(T - i\mu I) + (i + 1)\mu\boldsymbol{x}_{i+1} &= \boldsymbol{0}, \quad \text{for} \quad 1 \leq i \leq c - 2, \\ \boldsymbol{x}_{c-2}T^oA^o + \boldsymbol{x}_{c-1}[T - (c - 1)\mu I + c\mu R] &= \boldsymbol{0}, \\ \sum_{i=0}^{c-2} \boldsymbol{x}_i\boldsymbol{e} + \boldsymbol{x}_{c-1}(I - R)^{-1}\boldsymbol{e} &= 1. \end{aligned} \tag{3.2.20}$$

For $c = 1$, we obtain the equations

$$\boldsymbol{x}_0(T + \mu R) = \boldsymbol{0}, \qquad \boldsymbol{x}_0(I - R)^{-1}\boldsymbol{e} = 1.$$

For $c \geq 2$, multiplication of the first c equations in (3.2.20) by $\boldsymbol{e}$ leads to

$$\boldsymbol{x}_{i-1}T^o = i\mu\boldsymbol{x}_i\boldsymbol{e}, \quad \text{for} \quad 1 \leq i \leq c - 1. \tag{3.2.21}$$

We may therefore replace $\boldsymbol{x}_{i-1}T^oA^o$ by $i\mu\boldsymbol{x}_iA^{oo}$, and so obtain after elementary manipulations

$$x_0 = -\mu x_1 T^{-1},$$

$$x_i = (i+1)\mu x_{i+1}(i\mu I - i\mu A^{oo} - T)^{-1}, \quad \text{for} \quad 1 \le i \le c-2, \tag{3.2.22}$$

$$x_{c-1}[(c-1)\mu I - (c-1)\mu A^{oo} - T - c\mu R] = \mathbf{0}.$$

The last equation determines x_{c-1} up to a multiplicative constant. The other equations determine $x_0, \ldots, x_{c-2}$, up to that same constant, which is ultimately given by the last equation in (3.2.20). It is clear that $x_0, \ldots, x_{c-1}$, are positive vectors. The existence of the matrix inverses was proved in theorem 3.2.1.

The matrix R has positive rows only for those indices j for which $T_j^o > 0$. The other rows are zero. This enables us to compute the vector x for interarrival time distributions of special types even though their representation may require high values of m. For example, if $F(\cdot)$ is a mixture of r (generalized) Erlang distributions involving a total of m exponential components, the computation of the matrix R is essentially that of an $r \times m$, not an $m \times m$ matrix. The point of performing such a computation could be to study the fluctuations of the queue length in cases where the interarrival times are highly variable. See, for example, H. Heffes and J. M. Holtzman [135] for similar questions arising in telephone engineering.

It is entirely straightforward to carry through the matrix-geometric solution to the *PH/M/c* queue with batch service. A number of variants, obtained by using different assumptions on the sizes of the batches that may be allowed to enter service, lead to generators with the structure, discussed in general in chapter 1. The case of Erlang arrivals and service in groups of fixed size was treated by R. F. Love [193] using the procedure of counting the stages in the system. The assumption of Erlang arrivals leads in our approach to major simplifications, which we shall briefly discuss for the case of single services. The corresponding results for variants with batch service will be left to the initiative of the reader.

The $E_m/M/c$ is a highly simplified particular case. Only the m-th row of R is positive. The equation (3.2.19) may then be replaced by

$$c\mu u_m u_1 - (\lambda + c\mu)u_1 + \lambda = 0,$$

$$c\mu u_m u_i - (\lambda + c\mu)u_i + \lambda u_{i-1} = 0, \quad \text{for} \quad 2 \le i \le m, \tag{3.2.23}$$

where $u_j = R_{mj}$, $1 \le j \le m$.

These equations lead to

$$u_i = \frac{\lambda}{\lambda + c\mu(1-u_m)} u_{i-1}, \quad \text{for} \quad 2 \le i \le m,$$

$$u_1 = \frac{\lambda}{\lambda + c\mu(1-u_m)},$$

so that $u_i = u_1{}^i$, for $1 \leq i \leq m$, and hence

$$c\mu u_1{}^{m+1} - (\lambda + c\mu)u_1 + \lambda = 0. \tag{3.2.24}$$

It is easily verified that $u_m = u_1{}^m$ is the unique root in (0, 1) of the equation

$$z = \left(\frac{\lambda}{\lambda + c\mu - c\mu z}\right)^m. \tag{3.2.25}$$

The numerical solution of the $E_m/M/c$ queue (in steady state) requires the computation of *one* constant $u_m = R_{mm}$ by any appropriate root-finding technique applied to equation (3.2.25). All other quantities are derived from it by elementary operations.

Once the matrix R is known, the stationary distribution $\tilde{W}(\cdot)$ of the virtual waiting time for the $PH/M/c$ queue is readily obtained.

Theorem 3.2.3. The stationary distribution $\tilde{W}(\cdot)$ of the virtual waiting time is given by

$$\tilde{W}(x) = 1 - x_{c-1}R(I - R)^{-1}\exp[-c\mu x(I - R)]e, \tag{3.2.26}$$

for $x \geq 0$.

Proof. By direct calculation.

The following theorem is a "duality" result between the $PH/M/1$ and the $M/PH/1$ queues. We recall that the stationary density of the queue length prior to arrivals in the stable $GI/M/1$ queue is given by $z_k = (1 - \xi)\xi^k$, for $k \geq 0$, where ξ is the unique solution in (0, 1) of the equation

$$\xi = \int_0^\infty e^{-\mu(1-\xi)x}\,dF(x). \tag{3.2.27}$$

Theorem 3.2.4. The quantity $c = \mu(1 - \xi)$ is the unique positive solution of the equation

$$\mu\alpha(cI - T)^{-1}e = 1. \tag{3.2.28}$$

Proof. For the $PH/M/1$ queue, the equation (3.2.27) becomes

$$\xi = \alpha[\mu(1 - \xi)I - T]^{-1}T^o.$$

Setting $\mu(1 - \xi) = c$, we obtain

$$\mu^{-1} = c^{-1}[1 - \alpha(cI - T)^{-1}T^o] = \int_0^\infty [1 - F(x)]e^{-cx}\,dx$$

$$= \int_0^\infty e^{-cx}\alpha\exp(Tx)e\,dx = \alpha(cI - T)^{-1}e.$$

The same argument as in the proof of theorem 3.2.2 shows that if $\lambda_1'\mu > 1$, the equation (3.2.28) has a unique positive solution. Equation (3.2.28) is, of course, the "dual" equation of (3.2.12).

The Finite PH/M/c Queue. Substantially the same device that led to the explicit formulas for the stationary probability vector $\boldsymbol{x}$ for the finite $M/PH/1$ queue also leads to an explicit solution for the finite $PH/M/c$ queue. The solution appears, however, in a surprising "reversed" matrix-geometric form.

We shall assume that at most $K + c$, $K \geq 0$, customers can be in the system at any time. The finite $PH/M/c$ queue may then be studied in terms of a Markov process with state space $E = \{(i, j), 0 \leq i \leq K + c, 1 \leq j \leq m\}$ and generator $\tilde{Q}$, given by

$$\tilde{Q} = \begin{array}{r|cccccccccc|} 0 & T & T^oA^o & 0 & \cdots & 0 & 0 & 0 & \cdots & 0 & 0 \\ 1 & \mu I & T-\mu I & T^oA^o & \cdots & 0 & 0 & 0 & \cdots & 0 & 0 \\ 2 & 0 & 2\mu I & T-2\mu I & \cdots & 0 & 0 & 0 & \cdots & 0 & 0 \\ \vdots & \vdots & \vdots & \vdots & & \vdots & \vdots & \vdots & & \vdots & \vdots \\ c & 0 & 0 & 0 & \cdots & c\mu I & T-c\mu I & T^oA^o & \cdots & 0 & 0 \\ \vdots & \vdots & \vdots & \vdots & & \vdots & \vdots & \vdots & & \vdots & \vdots \\ K+c-1 & 0 & 0 & 0 & \cdots & 0 & 0 & 0 & \cdots & T-c\mu I & T^oA^o \\ K+c & 0 & 0 & 0 & \cdots & 0 & 0 & 0 & \cdots & c\mu I & T+T^oA^o-c\mu I \end{array}$$

(3.2.29)

The steady-state equations are given by

$$\begin{aligned} \boldsymbol{x}_0 T + \mu \boldsymbol{x}_1 &= \boldsymbol{0}, \\ \boldsymbol{x}_{i-1} T^o A^o + \boldsymbol{x}_i (T - i\mu I) + (i+1)\mu \boldsymbol{x}_{i+1} &= \boldsymbol{0}, \quad \text{for} \quad 1 \leq i \leq c-1, \\ \boldsymbol{x}_{i-1} T^o A^o + \boldsymbol{x}_i (T - c\mu I) + c\mu \boldsymbol{x}_{i+1} &= \boldsymbol{0}, \quad \text{for} \quad c \leq i \leq K+c-1, \\ \boldsymbol{x}_{K+c-1} T^o A^o + \boldsymbol{x}_{K+c}(T + T^o A^o - c\mu I) &= \boldsymbol{0}, \\ \sum_{i=0}^{K+c} \boldsymbol{x}_i \boldsymbol{e} &= 1. \end{aligned} \tag{3.2.30}$$

Multiplying all but the last equation by the column vector $\boldsymbol{e}$ leads to

$$\boldsymbol{x}_{i-1} T^o = \min(i, c)\mu \boldsymbol{x}_i \boldsymbol{e}, \quad \text{for} \quad 1 \leq i \leq K + c. \tag{3.2.31}$$

Multiplying the equations (3.2.31) on the right by the row vector $\boldsymbol{\alpha}$ and recalling that $T^o\boldsymbol{\alpha} = T^oA^o$, and $\boldsymbol{e} \cdot \boldsymbol{\alpha} = A^{oo}$, we obtain after routine substitutions

$$x_0 = -\mu x_1 T^{-1},$$
$$x_i = (i+1)\mu x_{i+1}(i\mu I - i\mu A^{oo} - T)^{-1}, \quad \text{for} \quad 1 \le i \le c-1,$$
$$x_i = c\mu x_{i+1}(c\mu I - c\mu A^{oo} - T)^{-1}, \quad \text{for} \quad c \le i \le K+c-1,$$
$$x_{K+c} = (x_{K+c} T^o)\alpha(c\mu I - c\mu A^{oo} - T)^{-1},$$
$$\sum_{i=0}^{K+c} x_i e = 1. \tag{3.2.32}$$

The penultimate equation in (3.2.32) gives x_{K+c} up to a multiplicative constant. Indeed, setting

$$\hat{R}(i) = i\mu(i\mu I - i\mu A^{oo} - T)^{-1}, \quad \text{for} \quad 1 \le i \le c,$$
$$\gamma = \frac{1}{c\mu} x_{K+c} T^o, \tag{3.2.33}$$

we obtain

$$x_0 = c\gamma\alpha\hat{R}^{K+1}(c) \prod_{\nu=1}^{c-1} \hat{R}(c-\nu)(-\mu T^{-1}),$$
$$x_i = \frac{c}{i}\gamma\alpha\hat{R}^{K+1}(c) \prod_{\nu=1}^{c-i} \hat{R}(c-\nu), \quad \text{for} \quad 1 \le i \le c-1, \tag{3.2.34}$$
$$x_i = \gamma\alpha\hat{R}^{K+c-i+1}(c), \quad \text{for} \quad c \le i \le K+c.$$

The constant γ is uniquely determined by the normalizing equation.

Remarks

a. The matrices $\hat{R}(i)$, $1 \le i \le c$, are positive. The matrix R, defined by equation (3.2.19), and the matrix $\hat{R}(c)$ are related by

$$R = R^2\hat{R}(c), \tag{3.2.35}$$

provided $\text{sp}(R) < 1$.

To see this, we write equation (3.2.19) as

$$c\mu R^2 + T^o A^o = R(c\mu I - c\mu A^{oo} - T) + c\mu Re \cdot \alpha.$$

When $\text{sp}(R) < 1$, then $c\mu Re = T^o$, so that the terms $T^o A^o$ and $c\mu Re \cdot \alpha$ cancel. Formula (3.2.35) is now immediate, but its probabilistic significance (if any) is not apparent.

b. The particular case $K = 0$ is of interest to telephone engineering—see, for example, A. Descloux [77]. It corresponds to a c-trunk *loss system* with exponential holding times and a PH-renewal process as arrival process.

Its explicit solution lends itself to a number of interesting interpretations. We first note that it readily follows from the steady-state equations that

$$x^* = \sum_{i=0}^{c} x_i = \pi. \tag{3.2.36}$$

The probability $\psi(c)$ that an arriving customer is lost is given by

$$\psi(c) = (x^* T^o)^{-1}(x_c T^o) = c\mu\lambda_1{}'\gamma, \tag{3.2.37}$$

since $x_c T^o = \gamma\alpha\hat{R}(c) T^o$, and $\hat{R}(c) T^o = c\mu e$.

It is useful to study the fluctuations of the number of busy channels for arrival streams which are more or less regular than a Poisson process, but which have the same mean interarrival times. For definiteness, we may consider, for example, the interarrival time distribution

$$F(x) = \sum_{\nu=1}^{m} \alpha_\nu E_\nu(\lambda, x), \tag{3.2.38}$$

where the parameters α_ν, $1 \le \nu \le m$, and λ are linked by the constraint

$$\frac{1}{\lambda} \sum_{\nu=1}^{m} \alpha_\nu \nu = \frac{1}{\lambda^*}. \tag{3.2.39}$$

For fixed values of λ^*, c, and μ, we may generate a highly versatile family of input processes with the same arrival rate. An easy way of studying the fluctuations of the number of busy channels is to consider the conditional densities

$$q_i(j) = \pi_j^{-1} x_{ij}, \quad \text{for} \quad 0 \le i \le c, \tag{3.2.40}$$

for various values of j, $1 \le j \le m'$, where $m' = \max\{\nu: \alpha_\nu > 0\}$. The quantity $q_i(j)$ is the stationary conditional probability that i channels are occupied, given that the system is in the arrival phase j.

Note that for matrices T of the form

$$\begin{vmatrix} -\lambda & \lambda & & & \\ & -\lambda & \lambda & & \\ & & \cdots & & \\ & & & -\lambda & \lambda \\ & & & & -\lambda \end{vmatrix},$$

it becomes very easy to solve a system of linear equations of the form

$$u(dI - dA^{oo} - T) = v, \qquad d \ge 0.$$

This is the main ingredient in the computation of the vector x. It can be written as an efficient subroutine. Computation of the quantities of interest with c as large as 300 is entirely straightforward. Even for m as large as 10, no excessive storage requirements or computation times are needed. It is particularly recommended to implement the algorithm in an interactive mode.

c. There does not appear to be an easy way to recover the simple solution, given by formulas (3.2.18) through (3.2.22), for the unbounded queue by letting K tend to infinity in the solution for the finite queue. For the stable $M/PH/1$ queue, such a passage to the limit is trivial.

d. The literature on the solution of the $M/PH/1$ and $PH/M/c$ queues, at least for the particular case of Erlang services or arrivals, is extensive. As was pointed out in chapter 2, the special properties of the Erlang distribution have mostly obscured the simple algebraic properties that lead to a general, explicit steady-state solution for these models and for their bounded versions. Alternate approaches to one or more of these models may be found in U. N. Bhat [19, 23], P. J. Brockwell [40], C. Burrows [41], J. W. Cohen [61], G. P. Cosmetatos [71], D. J. Daley [76], M. J. Fischer [98], D. P. Gaver [105], D. Gross and C. M. Harris [115], S. K. Gupta and J. K. Goyal [117, 123], S. K. Gupta [118–22], T. L. Healy [131], P. Hokstad [143], R. R. P. Jackson and D. J. Nickols [150], L. Kleinrock [172], D. König and D. Stoyan [175], R. F. Love [193], D. Z. Mittwoch [208], S. C. Moore and U. N. Bhat [209], P. M. Morse [210], M. F. Neuts [221], T. L. Saaty [272], D. N. Shanbhag [281], L. Takács [291], D. M. G. Wishart [305, 306], F. Wu [308], and G. Xu [309].

e. The $PH/M/c/c$ model may, because of its ease of computation, also be used to obtain approximate results for the $PH/M/\infty$ queue. The latter model is a particular case of the $PH/G/\infty$ queue treated in V. Ramaswami and M. F. Neuts [262]. The equations presented there lend themselves to convenient numerical computation. This is, in general, not the case for the available results on the $GI/M/\infty$ queue. These typically involve binomial generating functions, whose numerical inversion is highly unstable. See, for example, H. Heffes and J. M. Holtzman [135] and L. Takács [293].

3.3. THE BUSY PERIOD STRUCTURE

For the classical queueing models, the *busy period* is the length of time between the arrival of a customer to an empty queue and the first epoch thereafter that the queue becomes empty again. For more complex models, this is not the most useful concept, as its definition depends greatly on the boundary behavior of the queue.

For queues of the $M/G/1$ type, there is a more natural first passage time, for which G. Latouche [179] has proposed the term *fundamental period.* I have extensively studied this first passage time [219, 223, 228]. We consider a QBD process (or, more generally, a queue of the $M/G/1$ type), starting in the state $(i + 1, j)$, and study the first passage time to the set of states $\boldsymbol{i} = \{(i, 1), \ldots, (i, m)\}$, where i is sufficiently large so that the states (i, j'), $1 \leq j' \leq m$, are not boundary states.

We shall consider the length of time of that first passage as well as the number of left transitions during it. In queueing theory, the latter usually corresponds to the number of customers served during that first passage time. The following conditional probabilities are of basic importance.

By $G_{jj'}^{(\nu)}(k, x)$, we denote the conditional probability that a QBD process, starting in the state $(i + \nu, j)$ at time $t = 0$, reaches the set $\boldsymbol{i}$ for the first time no later than time x, after exactly k transitions to the left, and does so by entering the state (i, j'). This probability is so defined for $x \geq 0$, $\nu \geq 1$, $k \geq 1$, $1 \leq j \leq m$, $1 \leq j' \leq m$, and for any i such that the states (i, j'), $1 \leq j' \leq m$, are not boundary states.

It is clear from the structure of the matrix $\tilde{Q}$ in (3.1.1) that the probability $G_{jj'}^{(\nu)}(k, x)$ does not depend on i. The matrix with elements $G_{jj'}^{(\nu)}(k, x)$ will be denoted by $G^{(\nu)}(k, x)$. For ease of presentation, we also introduce the transform matrix

$$\hat{G}^{(\nu)}(z, s) = \sum_{k=1}^{\infty} z^k \int_0^{\infty} e^{-sx} dG^{(\nu)}(k, x), \tag{3.3.1}$$

for $|z| \leq 1$, $\operatorname{Re} s \geq 0$.

Lemma 3.3.1. The matrix $\hat{G}^{(\nu)}(z, s)$ satisfies

$$\hat{G}^{(\nu)}(z, s) = [\hat{G}^{(1)}(z, s)]^{\nu}, \quad \text{for} \quad \nu \geq 1. \tag{3.3.2}$$

Proof. The first passage time from $(i + \nu, j)$ to the set $\boldsymbol{i}$, reached in the state (i, j'), is the sum of ν first passage times, that is, from $(i + \nu, j)$ to some state $(i + \nu - 1, h_1)$ in the level $\boldsymbol{i} + \boldsymbol{\nu} - \mathbf{1}$; from there to some state $(i + \nu - 2, h_2)$ in the level $\boldsymbol{i} + \boldsymbol{\nu} - \mathbf{2}$; and so on. It readily follows from the Markov property that, given the indices $h_1, h_2, \ldots, h_{\nu-1}$, the pairs consisting of the times and the numbers of left transitions involved in each of those ν first passage times are *conditionally independent*.

By the law of total probability, we therefore obtain

$$G_{jj'}^{(\nu)}(k, x) = \sum_{h_1=1}^{m} \cdots \sum_{h_{\nu-1}=1}^{m} \sum_{k_1+\cdots+k_\nu=k} \int \cdots \int dG_{jh_1}^{(1)}(k_1, u_1) \cdot dG_{h_1 h_2}^{(1)}(k_2, u_2) \cdots dG_{h_{\nu-1} j'}^{(1)}(k_\nu, u_\nu), \tag{3.3.3}$$

where $k_1 \geq 1, \ldots, k_\nu \geq 1$, and the domain of integration is the region $\{u_1 \geq 0, \ldots, u_\nu \geq 0, u_1 + \cdots + u_\nu \leq x\}$. Upon taking the joint Laplace-Stieltjes transform and generating function of both sides, we obtain the stated matrix equality.

We may now denote $\hat{G}^{(1)}(z, s)$ by $\hat{G}(z, s)$ and define $\hat{G}^0(z, s)$ to be the identity matrix.

Lemma 3.3.2. The matrix $\hat{G}(z, s)$ satisfies the equation

$$\hat{G}(z, s) = z(sI - A_1)^{-1}A_2 + (sI - A_1)^{-1}A_0\hat{G}^2(z, s). \tag{3.3.4}$$

Proof. By conditioning on the time of and the state visited in the first transition away from the level $i + 1$.

The discussion of the nonnegative solutions (for $0 \leq z \leq 1, s \geq 0$) of the matrix equation (3.3.4) proceeds along very similar lines to that given for the equation for the rate matrix R in chapter 1. We shall only give the essential steps of the various proofs.

We introduce the matrices $C_0(s)$ and $C_2(s)$, defined by

$$C_0(s) = (sI - A_1)^{-1}A_2, \qquad C_2(s) = (sI - A_1)^{-1}A_0, \tag{3.3.5}$$

and we write C_0 and C_2 for the matrices obtained by setting $s = 0$.

Lemma 3.3.3. For $s \geq 0$, the matrix $C(s) = C_0(s) + C_2(s)$ is substochastic and has for $s > 0$ at least one row sum less than one.

If the generator A is irreducible, so is the *stochastic* matrix $C = C_0 + C_2$. The invariant probability vector γ of C is then given by

$$\gamma = (\pi A_1 e)^{-1}\pi A_1. \tag{3.3.6}$$

Proof. By direct calculation.

We now study the matrix-quadratic equation

$$X(z, s) = zC_0(s) + C_2(s)X^2(z, s), \tag{3.3.7}$$

for $0 \leq z \leq 1, s \geq 0$. Its minimal nonnegative solution $X^*(z, s)$ is that solution obtained by successive substitutions starting with the zero matrix. It is readily verified that the successive iterates are substochastic and entry-wise nondecreasing. It follows that $X^*(z, s)$ exists and is substochastic, and that $X^*(z, s) \leq X(z, s)$ for any other solution. In particular, we have that $X^*(z, s) \leq \hat{G}(z, s)$. The following theorem is of basic interest.

Theorem 3.3.1. The transform matrix $\hat{G}(z, s)$, $0 \leq z \leq 1$, $s \geq 0$, is equal to the minimal nonnegative solution $X^*(z, s)$ of equation (3.3.7).

Proof. Let $\{\hat{G}_n(z, s), n \geq 0\}$ be the sequence of matrices obtained by successive substitutions in equation (3.3.7), starting with $\hat{G}_0(z, s) = 0$. We shall prove by a combinatorial argument that

$$\lim_{n \to \infty} \hat{G}_n(z, s) = \hat{G}(z, s). \tag{3.3.8}$$

For every z and s satisfying $0 \leq z \leq 1$, $s \geq 0$, the minimal nonnegative solution $X^*(z, s)$ is the limit of the nondecreasing sequence of nonnegative matrices defined by

$$\begin{aligned} \hat{G}_1(z, s) &= zC_0(s), \\ \hat{G}_{n+1}(z, s) &= zC_0(s) + C_2(s)\hat{G}_n^2(z, s), \quad \text{for} \quad n \geq 1. \end{aligned} \tag{3.3.9}$$

We shall now identify sets of paths $T_n(j, j')$ leading from the state $(i + 1, j)$ to (i, j') without visiting the set i in between. The element $[\hat{G}_n(z, s)]_{jj'}$ of $\hat{G}_n(z, s)$ will then be seen to be the bivariate transform of the probability that the first passage from $(i + 1, j)$ to (i, j') takes place along a path in $T_n(j,j')$, involves exactly k left transitions, and is completed no later than time x.

The sets $T_n(j, j')$, $n \geq 1$, are recursively defined as follows. The set $T_1(j, j')$ contains all paths that go from $(i + 1, j)$ to (i, j') without visiting the set i in between *and involve exactly one left transition.* Paths in the set $T_{n+1}(j, j')$ are either in $T_1(j, j')$ or they move from $(i + 1, j)$ to some state $(i + 2, j'')$ at the first right transition. From the state $(i + 2, j'')$, these paths return to some state $(i + 1, j''')$ along a path that is in the set $T_n(j'', j''')$ (defined with respect to the level $i + 1$), and then go from $(i + 1, j''')$ to (i, j') along a path in $T_n(j''', j')$.

It is now clear from the definition of $T_n(j, j')$ and the second equation in (3.3.9) that $[\hat{G}_n(z, s)]_{jj'}$ has the stated interpretation.

We further consider $S_n(j, j')$, the set of paths leading from $(i + 1, j)$ to (i, j') without an intermediate visit to the set i *and involving at most n left transitions.* The set $S(j, j')$ is defined by $S(j, j') = \cup_n S_n(j, j')$.

Clearly $T_n(j, j') \subset S(j, j')$. The sets $T_n(j, j')$ form a monotone sequence, and $\lim_{n\to\infty} T_n(j, j') \subset S(j, j')$. By induction, we shall show that $S_n(j, j') \subset T_n(j, j')$, for all $n \geq 1$, and therefore

$$\lim_{n \to \infty} T_n(j, j') = S(j, j'). \tag{3.3.10}$$

Clearly $T_1(j, j') = S_1(j, j')$. Assuming that $S_n(j, j') \subset T_n(j, j')$, it follows from the monotonicity of the sets $T_n(j, j')$ that $S_n(j, j') \subset T_{n+1}(j, j')$. Therefore, it suffices to show that all paths from $(i + 1, j)$

to (i, j') with exactly $n + 1$ left transitions (which do not visit the set i in between) belong to $T_{n+1}(j, j')$. Any such path must visit the level $i + 2$ at its first right transition. We partition the remainder of such a path into the first return to $i + 1$ and thence the first passage from $i + 1$ to i, which is reached in the state (i, j'). Since the path involves exactly $n + 1$ left transitions, each of these first passage chains can involve at most n left transitions. From the recursive definition of the sets $T_n(j, j')$, it is now clear that this path belongs to the set $T_{n+1}(j, j')$. Therefore, $S_{n+1}(j, j') \subset T_{n+1}(j, j')$, and the induction argument is complete.

The element $[\hat{G}(z, s)]_{jj'}$ is clearly the joint transform of the probability that the first passage from $(i + 1, j)$ to the set i occurs along a path in $S(j, j')$, involves exactly k left transitions, and is completed before time x. By (3.3.10) and the continuity theorem of probability, the stated result follows.

It is now obvious that a necessary condition for the positive recurrence of the process $\tilde{Q}$ is that the matrix

$$G = \hat{G}(1-, 0+) \tag{3.3.11}$$

is *stochastic*. With a purely technical additional condition on the boundary matrices, we will see that this is also sufficient.

When the matrix A, and hence C, is irreducible, an elegant and familiar necessary and sufficient condition for the matrix G to be stochastic may be obtained. For the remainder of the discussion, we assume that A is *irreducible*.

Lemma 3.3.4. The vector $\delta^* = 2C_2e$ is given by

$$\delta^* = -2A_1{}^{-1}A_0e = 2e + 2A_1{}^{-1}A_2e. \tag{3.3.12}$$

The inequality $\gamma\delta^* < 1$ is equivalent to the inequality $\pi A_2 e > \pi A_0 e$.

Proof. The expressions for δ^* follow by direct calculation. Furthermore,

$$\gamma\delta^* = -2(\pi A_1 e)^{-1}(\pi A_0 e) = \frac{2\pi A_0 e}{\pi A_0 e + \pi A_2 e} < 1$$

is satisfied if and only if $\pi A_2 e > \pi A_0 e$.

The quantity ρ, defined by

$$\rho = \frac{\pi A_0 e}{\pi A_2 e}, \tag{3.3.13}$$

is called the *traffic intensity* of the QBD process. It is defined here by analogy with the corresponding term used in the theory of queues. Its probabilistic significance is discussed in what follows.

In order to discuss the condition under which the matrix G is stochastic, we consider the minimal nonnegative solution to the equation

$$G = C_0 + C_2 G^2. \tag{3.3.14}$$

Let η_1, $0 < \eta_1 \leq 1$, be the maximal eigenvalue of G, and let $\boldsymbol{u} \geq \mathbf{0}$ be a corresponding right eigenvector. We note that the same argument as in lemma 1.2.4 shows that $\eta_1 > 0$, whenever the QBD process $\tilde{Q}$ is irreducible.

Equation (3.3.14) implies that

$$\eta_1 \boldsymbol{u} = (C_0 + \eta_1{}^2 C_2)\boldsymbol{u}. \tag{3.3.15}$$

The irreducible nonnegative matrix $C_0 + \eta_1{}^2 C_2$, therefore, has the eigenvalue η_1, and the corresponding right eigenvector $\boldsymbol{u}$ is nonnegative. It follows that η_1 is the maximal eigenvalue of $C_0 + \eta_1{}^2 C_2$. The argument now proceeds exactly as in lemmas 1.3.2, 1.3.3, and 1.3.4. The quantity η_1 is seen to be the smallest solution in (0, 1] of the equation

$$z = \chi_1(z), \tag{3.3.16}$$

where $\chi_1(z)$ is the maximal eigenvalue of $C_0 + z^2 C_2$, $0 < z \leq 1$.

The spectral radius η_1 of G is equal to one if and only if

$$\chi_1{}'(1-) = \gamma\delta^* \leq 1. \tag{3.3.17}$$

Theorem 3.3.2. If $\gamma\delta^* \leq 1$, or equivalently $\boldsymbol{\pi} A_2 \boldsymbol{e} \geq \boldsymbol{\pi} A_0 \boldsymbol{e}$, then the matrix G is stochastic. It is the unique solution of the equation (3.3.14) in the set of stochastic matrices.

Proof. We normalize the positive right eigenvector $\boldsymbol{u}$ of G so that $\max_j u_j = 1$. Equation (3.3.14) leads to $\boldsymbol{u} = C\boldsymbol{u}$. Since C is stochastic, it follows that $\boldsymbol{u} = \boldsymbol{e}$, and in turn that G is stochastic.

Any stochastic solution G' of (3.3.14) must satisfy $G' \geq G$. Strict inequality of any pair of corresponding elements leads to a contradiction. G is therefore the *unique* stochastic solution.

Remarks.

a. The matrices $\hat{G}(1, s)$ and $\hat{G}(z, 0)$ are transforms of *semi-Markov* matrices, which we may denote by $G(x)$ and $\{G(k)\}$ respectively. $G_{jj'}(x)$ is the conditional probability that the QBD process, starting in the state $(i + 1, j)$, reaches the level i for the first time in the state (i, j'), no later than time x. The matrices $\{G(k)\}$ define a semi-Markov matrix *of lattice type.* $G_{jj'}(k)$ is the conditional probability that the QBD process, starting in the state $(i+1, j)$, reaches the level i in the state (i, j') at the k-th transition to the left. For the basic terminology and properties of semi-Markov processes and matrices, we refer to E. Çinlar [56].

b. From chapter 1 we know that the QBD process can be positive recurrent only if $\rho < 1$. One may show in general that when $\rho = 1$, the row sum means of the two semi-Markov matrices discussed in (*a*) are infinite. We shall only prove this for the (common) case where the matrix G is irreducible. This to avoid a lengthy discussion of the minor difficulties that arise when G is reducible.

c. Some QBD processes and a multitude of queueing models of $M/G/1$ type have been discussed in the literature by transform methods. These lead to the discussion of the roots in the unit disk of an equation similar to (1.6.5). It is then established that for the positive recurrent case, there are exactly m such zeros in the unit disk $\{z: |z| \leq 1\}$ and that the root of largest modulus occurs at $z = 1$. These m roots may be shown to be none other than the eigenvalues of the stochastic matrix G.

Moment Formulas

The row sum means of the semi-Markov matrices $G(\cdot)$ and $\{G(k)\}$ are the components of the column vectors $\tilde{\mu}_1$ and $\tilde{\mu}$, defined by

$$\tilde{\mu}_1 = \int_0^\infty x dG(x) e,$$

$$\tilde{\mu} = \sum_{k=1}^{\infty} kG(k) e. \tag{3.3.18}$$

These vectors play an important role in the sequel. Before we derive analytic expressions for them, we remind the reader that we limit this discussion to the case where the matrix G is irreducible and stochastic. The invariant probability vector of G is denoted by g. The matrix $\tilde{G} = e \cdot g$ has m identical rows equal to g. A classical result on finite Markov chains then states that the matrix $I - G + \tilde{G}$ is nonsingular. See, for example, J. Kemeny and J. L. Snell [164, p. 100].

Equation (3.3.4) may be written as

$$zA_2 - (sI - A_1)\hat{G}(z, s) + A_0\hat{G}^2(z, s) = 0. \tag{3.3.19}$$

Differentiation, either with respect to s or to z and setting $s = 0$, $z = 1$, leads to

$$M = C_0 + C_2(GM + MG),$$

$$M_1 = -A_1^{-1}G + C_2(GM_1 + M_1G), \tag{3.3.20}$$

where M and M_1 are defined by

$$M = \left(\frac{\partial}{\partial z}\hat{G}(z, s)\right)_{\substack{z=1\\s=0}}, \qquad M_1 \doteq \left(-\frac{\partial}{\partial s}\hat{G}(z, s)\right)_{\substack{z=1\\s=0}}. \tag{3.3.21}$$

Theorem 3.3.3. If $\rho = 1$, the matrices M and M_1 are infinite. If $\rho < 1$, the matrices M and M_1 are the unique solutions to the equations (3.3.20) and the vectors $\tilde{\mu}$ and $\tilde{\mu}_1$ are explicitly given by

$$\tilde{\mu} = Me = (I - G + \tilde{G})[\Delta(\tau)\tilde{G} - A]^{-1}A_2 e,$$

$$\tilde{\mu}_1 = M_1 e = (I - G + \tilde{G})[\Delta(\tau)\tilde{G} - A]^{-1}e, \tag{3.3.22}$$

where $\tau = A_2 e - A_0 e$, and $\Delta(\tau) = \text{diag}(\tau_1, \ldots, \tau_m)$.

Proof. We shall first prove the formulas (3.3.22). Postmultiplying by e in (3.3.20), we obtain

$$(A_0 + A_1 + A_0 G)\tilde{\mu} + A_2 e = \mathbf{0},$$

$$(A_0 + A_1 + A_0 G)\tilde{\mu}_1 + e = \mathbf{0}. \tag{3.3.23}$$

We note that

$$\begin{aligned}(A_0 + A_1 + A_1 G)(I - G + \tilde{G}) &= A_0 + A_1 - A_1 G - A_0 G^2 \\ &\quad + (2A_0 + A_1)\tilde{G} \\ &= A + (2A_0 + A_1)\tilde{G} \\ &= A - \Delta(\tau)\tilde{G},\end{aligned}$$

by using the equations (3.3.14) and $A_0 + A_1 + A_2 = A$.

If $\rho = 1$, the matrix $A - \Delta(\tau)\tilde{G}$, and hence also the matrix $A_0 + A_1 + A_0 G$, are *singular*, since $\pi[A - \Delta(\tau)\tilde{G}] = \mathbf{0}$. Next we show that for $\rho < 1$, the matrix $A - \Delta(\tau)\tilde{G}$ is nonsingular. Let u be a column vector in its null space, then

$$Au = \Delta(\tau)\tilde{G}u = (gu)(A_2 - A_0)e.$$

Premultiplying by π, we see that $gu = 0$, since $\pi A_2 e > \pi A_0 e$. This in turn implies that $Au = \mathbf{0}$, and therefore that $u = ke$. Since $gu = 0$, and $ge = 1$, it follows that $k = 0$.

For $\rho < 1$, we now readily obtain the formulas (3.3.22). Each of the equations (3.3.20) corresponds to a system of m^2 linear equations in m^2 unknowns. For $\rho = 1$, we have shown that the system is singular. For $\rho < 1$, we may show that each system has a nontrivial solution as follows. Performing successive substitutions in each of the equations (3.3.20), starting with $M = M_1 = 0$, we see that the successive iterates are element-wise nondecreasing and belong respectively to the compact sets of matrices $\{X: X \geq 0, Xe \leq \tilde{\mu}\}$ and $\{X: X \geq 0, Xe \leq \tilde{\mu}_1\}$. This proves the existence of a nonnegative solution.

The uniqueness is proved by showing that the coefficient matrix of the equations (3.3.20) may be written in the form $I - U$, where U is a nonnegative matrix of order m^2 with $\text{sp}(U) < 1$. For the details, we refer to M. F. Neuts [223].

Corollary 3.3.1. For $\rho < 1$, the equalities

$$g\tilde{\mu} = \frac{\pi A_2 e}{\pi A_2 e - \pi A_0 e} = \frac{1}{1-\rho},$$

$$g\tilde{\mu}_1 = \frac{1}{\pi A_2 e} \cdot \frac{1}{1-\rho}, \tag{3.3.24}$$

hold.

Proof. We have that

$$\pi[\Delta(\tau)\tilde{G} - A] = (\pi A_2 e - \pi A_0 e)g.$$

The stated formulas follow by direct calculations.

Remarks

a. The equalities (3.3.24) provide very useful internal accuracy checks. Their left-hand sides involve the matrices G and $\tilde{G}$ and the vector $\tilde{\mu}$ or $\tilde{\mu}_1$, each of which requires a fair amount of computation. The right-hand sides are directly expressed in terms of the data and of the easily computed vector π. For many applications, the right-hand sides are actually explicitly computable.

b. We may also obtain computationally useful formulas for the second moments of the duration of and the number of left transitions during the first passage studied here. These are not needed in the computation of the stationary vector x of $\tilde{Q}$, but play a useful role in the computation of the semi-Markov matrices $\{G(k)\}$ and $G(x)$. The matrices $M(2)$ and $M_1(2)$, defined by

$$M(2) = \left(\frac{\partial^2}{\partial z^2}\hat{G}(z, s)\right)_{\substack{z=1\\s=0}}, \qquad M_1(2) = \left(\frac{\partial^2}{\partial s^2}\hat{G}(z, s)\right)_{\substack{z=1\\s=0}}, \tag{3.3.25}$$

satisfy the equations

$$M(2) = 2C_2M^2 + C_2[M(2)G + GM(2)],$$

$$M_1(2) = 2C_2M_1{}^2 - 2A_1{}^{-1}M_1 + C_2[M_1(2)G + GM_1(2)]. \tag{3.3.26}$$

We see that these equations have the same coefficient matrix as in (3.3.20). For $\rho < 1$, both $M(2)$ and $M_1(2)$ are therefore uniquely determined nonnegative matrices. Since the constant terms in (3.3.26) involve the matrices M or M_1, their numerical solution requires a fair amount of effort and is again best carried out by iterative methods. As shown in M. F. Neuts [228], computationally useful explicit expressions may also be obtained for the vectors $M(2)\boldsymbol{e}$ and $M_1(2)\boldsymbol{e}$. Since this subject is marginal to our main discussion, we shall omit the details.

c. An appealing feature of algorithms based on purely probabilistic arguments is that most of the intermediate quantities computed during the implementation of the algorithm have interpretations of interest. These interpretations are not always obvious, so that their identification is often a pleasant mathematical pursuit. The following result is a case in point.

Corollary 3.3.2. The (j, k)-element of the matrix

$$D = (I - G + \tilde{G})[\Delta(\tau)\tilde{G} - A]^{-1}A_2 \tag{3.3.27}$$

is the conditional expected value of the number of times a left transition enters a state of the form (r, k), $r \geq i$, during a first passage from the state $(i + 1, j)$ to the level i.

Proof. Instead of counting only left transitions, we now need to consider m counting variables $N_1, \ldots, N_m$, where N_k denotes the number of left transitions into a state of the form (r, k), $r \geq i$, during a first passage from $(i + 1, j)$ to the level i. By exactly the same argument as used in the proofs of lemmas 3.3.1 and 3.3.2, we obtain the following multivariate analogue to equation (3.3.4) (with $s = 0$, for convenience),

$$A_2\Delta(\boldsymbol{z}) + A_1\hat{G}(\boldsymbol{z}, 0) + A_0\hat{G}^2(\boldsymbol{z}, 0) = 0, \tag{3.3.28}$$

where $\boldsymbol{z} = (z_1, \ldots, z_m)$ and $\Delta(\boldsymbol{z}) = \operatorname{diag}(z_1, \ldots, z_m)$.

Differentiation with respect to z_k and setting $\boldsymbol{z} = \boldsymbol{e}$, followed by calculations similar to those performed in the proof of theorem 3.3.3, leads to

$$\left(\frac{\partial}{\partial z_k}\hat{G}(\boldsymbol{z}, 0)\boldsymbol{e}\right)_{\boldsymbol{z}=\boldsymbol{e}} = (I - G + \tilde{G})[\Delta(\tau)\tilde{G} - A]^{-1}A_2\boldsymbol{e}_k, \tag{3.3.29}$$

where $\boldsymbol{e}_k$ is the unit vector with its k-th component equal to one. The right-hand side is clearly the k-th column of the matrix D, so that the (j, k)-element of the latter has the stated interpretation.

The Significance of the Traffic Intensity

The traffic intensity ρ, defined in (3.3.13), is clearly the ratio of the average transition rates $\boldsymbol{\pi} A_0 \mathbf{e}$ and $\boldsymbol{\pi} A_2 \mathbf{e}$, respectively to the right and to the left from any nonboundary level $\mathbf{i}$.

Another interpretation is obtained as follows. Consider the QBD process starting in the state $(i + 1, j)$. The probability that there are $k \geq 0$ right transitions before the first left transition is readily seen to be the j-th component of the vector $\boldsymbol{\nu}(k)$, given by

$$\boldsymbol{\nu}(k) = (-A_1{}^{-1}A_0)^k(-A_1{}^{-1}A_2)\mathbf{e}, \quad \text{for} \quad k \geq 0. \tag{3.3.30}$$

The mean vector $\tilde{\boldsymbol{\nu}} = \sum_{k=1}^{\infty} k\boldsymbol{\nu}(k)$ is then given by

$$\tilde{\boldsymbol{\nu}} = (A_1 + A_0)^{-1}A_0(A_1 + A_0)^{-1}A_2\mathbf{e} = -(A_1 + A_0)^{-1}A_0\mathbf{e}. \tag{3.3.31}$$

If we start the QBD process in one of the states $(i + 1, j)$ with the initial probability vector $(\boldsymbol{\pi} A_2 \mathbf{e})^{-1}\boldsymbol{\pi} A_2$, we see that the expected number of right transitions before the first left transition is given by

$$(\boldsymbol{\pi} A_2 \mathbf{e})^{-1}(\boldsymbol{\pi} A_2 \tilde{\boldsymbol{\nu}}) = (\boldsymbol{\pi} A_2 \mathbf{e})^{-1}(\boldsymbol{\pi} A_0 \mathbf{e}) = \rho. \tag{3.3.32}$$

A similar interpretation may be given for $1/\rho$ in which the roles of left and right transitions are interchanged. In queueing theory, the time until the first left transition, with the initial probability vector as given, is usually called the *average effective service time.* The traffic intensity then (usually) may be interpreted as the mean number of arrivals during a suitably defined average service time. For a specific example, see M. F. Neuts [229, 230] or section 6.2.

Formula (3.3.24) provides an interpretation for $(1 - \rho)^{-1}$. The inner product $\mathbf{g}\tilde{\boldsymbol{\mu}}$ is the average number of left transitions during a first passage from the level $\mathbf{i + 1}$ to the level $\mathbf{i}$, provided the initial state $(i + 1, j)$ is chosen with probability g_j, $1 \leq j \leq m$. In queueing theory, $(1 - \rho)^{-1}$ is (usually) the mean number of service completions during a suitably defined average fundamental period.

In some queueing models, such as the $GI/G/1$ queue, ρ is also the stationary probability that the server is busy. This interpretation is usually not preserved in queueing models described by QBD processes. See, for example, section 6.2.

The Busy Period and the Busy Cycle

By analogy with the usual definitions in the theory of queues, we may wish to consider for the QBD process defined in (3.1.1) the first passage time from the level **1** to the level **0** and the first return time of the level **0**. The former is the analogue of the *busy period*; the latter of the *busy cycle.* These notions clearly depend on the boundary states and on the specific structure of the matrix $\tilde{Q}$. They are therefore of considerably less significance than the fundamental period, which we have discussed up to now in this section.

We may define the matrices $\hat{G}^{(1,0)}(z, s)$ and $\hat{G}^{(0,0)}(z, s)$ in a manner entirely analogous to the matrix $\hat{G}(z, s)$. We consider a QBD process, starting at time $t = 0$ in the state $(1, j)$, and keep track of the time until the first visit to the level **0**, which occurs in some state $(0, j')$, as well as of the number of left transitions involved in this first passage time. The conditional probabilities $G_{jj'}^{(1,0)}(k, x)$ and the matrix of transforms $\hat{G}^{(1,0)}(z, s)$ are defined as before, but with obvious changes. The matrix $\hat{G}^{(0,0)}(z, s)$ is accordingly defined by considering the return times to the level **0**, *with at least one visit to a state outside the set* **0**.

A standard argument then leads to the matrix equations

$$\hat{G}^{(1,0)}(z, s) = z(sI - A_1)^{-1}B_1 + (sI - A_1)^{-1}A_0\hat{G}(z, s)\hat{G}^{(1,0)}(z, s),$$
$$\hat{G}^{(0,0)}(z, s) = (sI - B_0)^{-1}A_0\hat{G}^{(1,0)}(z, s), \tag{3.3.33}$$

which relate the matrices $\hat{G}^{(1,0)}(z, s)$ and $\hat{G}^{(0,0)}(z, s)$ to the matrix $\hat{G}(z, s)$. It is readily seen that when the matrix $G = \hat{G}(1, 0)$ is stochastic, so are the matrices $\hat{G}^{(1,0)}(1, 0)$ and $\hat{G}^{(0,0)}(1, 0)$.

By routine differentiations in the equations (3.3.33), we may relate the moments of the first passage times between **1** and **0**, as well as those of the return times of the level **0**, to the moments of the fundamental period. We cite only the following formulas among many, because of their useful interpretations in queueing theory.

$$\left(\frac{\partial}{\partial z}\hat{G}^{(1,0)}(z, s)\right)_{\substack{z=1\\s=0}} e = (-A_1{}^{-1}B_1)e + C_2(GM + MG)e$$
$$= C_0e + (M - C_0)e = \tilde{\mu}, \tag{3.3.34}$$

by the first formula in (3.3.20). Clearly also

$$\left(\frac{\partial}{\partial z}\hat{G}^{(0,0)}(z, s)\right)_{\substack{z=1\\s=0}} e = \tilde{\mu}. \tag{3.3.35}$$

A similar calculation as shown in (3.3.34) leads to the intuitive result

$$\left(-\frac{\partial}{\partial s}\hat{G}^{(1,0)}(z, s)\right)_{\substack{z=1\\s=0}} e = \tilde{\mu}_1, \tag{3.3.36}$$

but for the mean vector of the durations of the return times, we obtain the slightly more complicated expression

$$\left(-\frac{\partial}{\partial s}\hat{G}^{(0,0)}(z, s)\right)_{\substack{z=1\\s=0}} e = B_0{}^{-2}A_0 e + (-B_0{}^{-1}A_0)\tilde{\mu}_1$$

$$= (-B_0{}^{-1})e + (-B_0{}^{-1}A_0)\tilde{\mu}_1. \qquad (3.3.37)$$

The first term corresponds to the expected times to leave the set **0** from each of the states $(0, j)$. The matrix $(-B_0{}^{-1}A_0)$ in the second term has as its (j, j')-element, $1 \le j \le m$, $1 \le j' \le m$, the probability that, starting in the state $(0, j)$, the QBD process leaves the level **0** by moving to the state $(1, j')$.

The Matrix G for the $PH/M/1$ Queue

In the sequel, we shall obtain algorithmically useful formulas related to the busy period of fairly complex queues. Most of these formulas involve the matrix G, which is readily computable, but which is almost never available as an explicit function of the parameters. The $PH/M/1$ queue is an exception to this rule. By exploiting the matrix algebraic formalism of the PH-distributions, we may express the matrix G for the QBD process describing the $PH/M/1$ queue in a detailed, nearly explicit form.

The following results are of some interest in view of the known analytic complexity of the distribution of the busy period for a queue as simple as $GI/M/1$. See, for example, P. J. Brockwell [40], B. W. Conolly [66], and L. Takács [291]. Their lengthy and highly special derivation also stresses the merits of general structural results over the detailed but specialized formulas that occasionally may be obtained for specific stochastic models.

In discussing the $PH/M/1$ queue, we shall follow the same notation as in section 3.2. It is readily seen from equation (3.3.14) that the matrix G satisfies the equation

$$T^o A^o G^2 + (T - \mu I)G + \mu I = 0, \qquad (3.3.38)$$

and that ρ is given by

$$\rho = \frac{1}{\mu \lambda_1{}'}. \qquad (3.3.39)$$

Provided $\rho \le 1$, the matrix equation (3.3.38) has a unique stochastic solution G, which is strictly positive whenever the given representation $(\boldsymbol{\alpha}, T)$ of the interarrival time distribution $F(\cdot)$ is irreducible.

Theorem 3.3.4. For $\rho \leq 1$, the matrix H defined by

$$H = \mu(\mu I - T - \rho^{-1}T^o\Pi^o)^{-1} \tag{3.3.40}$$

is nonsingular and positive. The vector g is the left eigenvector of the matrix H, corresponding to its maximal eigenvalue, normalized by $ge = 1$. The maximal eigenvalue $\hat{\eta}$ of H satisfies

$$\hat{\eta} = 1 + \frac{g(\mu I - T)^{-1}T^o}{\alpha(\mu I - T)^{-1}T^o} \cdot (\rho^{-1} - 1). \tag{3.3.41}$$

The matrix G is given by

$$G = H + (1 - \rho^{-1})\frac{1}{f(\mu)}(\mu I - T)^{-1}T^o g, \tag{3.3.42}$$

where $f(\mu) = \alpha(\mu I - T)^{-1}T^o$. The vector g is the left invariant probability vector of G.

Proof. Premultiplying in (3.3.38) by π and recalling that $\pi T^o = \lambda_1'^{-1}$, we obtain

$$\left[\mu\pi - \frac{1}{\lambda_1'}\alpha G\right](I - G) = \mathbf{0}. \tag{3.3.43}$$

The irreducibility of the matrix G may be ascertained by considering the successive iterates obtained from equation (3.3.8) or from an argument by contradiction. Postmultiplying in (3.3.43) by the matrix $(I - G + \tilde{G})^{-1}$ and noting that

$$(I - G)(I - G + \tilde{G})^{-1} = I - \tilde{G},$$

equation (3.3.43) leads to

$$\alpha G = \rho^{-1}\pi + (1 - \rho^{-1})g, \tag{3.3.44}$$

and therefore

$$T^oA^oG^2 = \rho^{-1}T^o\Pi^oG + (1 - \rho^{-1})T^o \cdot g. \tag{3.3.45}$$

Upon substitution into (3.3.38), we obtain

$$(\mu I - T - \rho^{-1}T^o\Pi^o)G = \mu I + (1 - \rho^{-1})T^o g. \tag{3.3.46}$$

For $\rho = 1$, the matrix G is explicitly given by

$$G = H = \mu(\mu I - T - T^o\Pi^o)^{-1} \tag{3.3.47}$$

Since $T + T^o\Pi^o$ is an irreducible, semi-stable matrix, the inverse exists and is positive.

In general, for $\rho \leq 1$, we may write

$$\mu I - T - \rho^{-1}T^o\Pi^o = [I - \rho^{-1}T^o\Pi^o(\mu I - T)^{-1}](\mu I - T). \quad (3.3.48)$$

The matrix inside the square brackets is nonsingular, and its inverse, obtained by series expansion as in corollary 3.2.1, is explicitly given by

$$[I - \rho^{-1}T^o\Pi^o(\mu I - T)^{-1}]^{-1} = I + \frac{1}{\rho f(\mu)} T^o\Pi^o(\mu I - T)^{-1}. \quad (3.3.49)$$

It follows that the matrix H is also given by

$$H = \mu(\mu I - T)^{-1}\left(I + \frac{1}{\rho f(\mu)} T^o\Pi^o(\mu I - T)^{-1}\right)$$

$$= \mu(\mu I - T)^{-1} + \frac{\mu}{\rho f(\mu)} (\mu I - T)^{-1}T^o \cdot \pi(\mu I - T)^{-1}. \quad (3.3.50)$$

Since both the vector $(\mu I - T)^{-1}T^o$ and the vector $\pi(\mu I - T)^{-1}$ are positive (by the irreducibility of the representation and the positivity of π), it follows that H is a positive, nonsingular matrix.

Equation (3.3.46) now yields

$$G = H + \frac{1}{\mu}(1 - \rho^{-1})(HT^o) \cdot g. \quad (3.3.51)$$

The vector HT^o is evaluated by using formula (3.3.50) and is given by

$$HT^o = \frac{\mu}{f(\mu)} (\mu I - T)^{-1}T^o. \quad (3.3.52)$$

This expression is now substituted into (3.3.51). Premultiplying the resulting equation by the invariant probability vector g of G leads to

$$gH = \left[1 + (\rho^{-1} - 1)\frac{1}{f(\mu)} g(\mu I - T)^{-1}T^o\right]g, \quad (3.3.53)$$

which shows that g is the left eigenvector of H, corresponding to the positive eigenvalue $\hat{\eta} \geq 1$, which is given by the expression inside the square brackets.

It remains to be verified that G is positive and satisfies $Ge = e$. This is straightforward for $\rho = 1$, since then $G = H$. For $\rho < 1$, the proof requires some effort, since formula (3.3.51) then expresses G as the difference of two nonnegative matrices.

We first show that $Ge = e$. Since $\hat{\eta} > 1$, the matrix $I - H$ is nonsingular, and a direct calculation yields that

$$H(I - H)^{-1} = -\mu(T + \rho^{-1}T^o\Pi^o)^{-1}. \quad (3.3.54)$$

The equality $Ge = e$ is by formula (3.3.51) equivalent to

$$(I - H)^{-1}HT^o = \mu(1 - \rho^{-1})^{-1}e$$

or

$$(T + \rho^{-1}T^o\Pi^o)^{-1}T^o = -(1 - \rho^{-1})^{-1}e$$

or

$$T^o = -(1 - \rho^{-1})^{-1}(Te + \rho^{-1}T^o),$$

but this follows trivially from $Te + T^o = \mathbf{0}$.

In order to prove that the solution obtained in (3.3.51) is an irreducible nonnegative matrix, we proceed as follows. The matrix given in (3.3.51) satisfies the equation (3.3.38), and since

$$\alpha H = \mu\alpha(\mu I - T - \rho^{-1}T^o\Pi^o)^{-1} = \rho^{-1}\pi,$$

it satisfies equation (3.3.44) and hence also (3.3.46).

By theorem 3.3.2 we know that there exists an irreducible, stochastic matrix G, which satisfies the equation (3.3.38) and hence also (3.3.46). Denoting for a moment the right-hand side of (3.3.51) by G', we see that (3.3.46) leads to

$$(\mu I - T - \rho^{-1}T^o\Pi^o)(G - G') = 0.$$

Since the matrix H is nonsingular, it follows that $G' = G$. The solution in (3.3.51) is therefore irreducible and stochastic.

We may finally write the equation (3.3.38) in the equivalent form

$$\begin{aligned} G &= \mu(\mu I - T)^{-1}[I - (\mu I - T)^{-1}T^oA^oG]^{-1} \\ &= \mu(\mu I - T)^{-1}\left[I + \frac{1}{1 - \alpha G(\mu I - T)^{-1}T^o}(\mu I - T)^{-1}T^oA^oG\right]. \end{aligned} \tag{3.3.55}$$

Since $\alpha G(\mu I - T)^{-1}T^o$ is the Laplace-Stieltjes transform evaluated at $s = \mu > 0$ of a PH-distribution with representation $(\alpha G, T)$, the scalar factor inside the square brackets is positive. If we replace G in the right-hand side by $\mu(\mu I - T)^{-1}$, we obtain a strictly positive matrix, which is element-wise smaller than G. The matrix G is therefore strictly positive. This completes the proof of theorem 3.3.4.

By particularizing the formulas (3.3.22) to the present case, we may obtain explicit expressions for the mean duration of the busy period and the busy cycle for the $PH/M/1$ queue. Since $A_2e = \mu e$, we have the intuitively obvious relation

$$\tilde{\mu}_1 = \frac{1}{\mu}\tilde{\mu}. \tag{3.3.56}$$

The vector τ is given by $\tau = \mu e - T^o$, and therefore

$$\Delta(\tau)\tilde{G} - A = \mu e \cdot g - T^o \cdot g - T - T^o \cdot \alpha. \tag{3.3.57}$$

The vector $\nu = [\Delta(\tau)\tilde{G} - A]^{-1}e$ may be calculated explicitly by solving the system of equations

$$\mu(g\nu)e - (g\nu)T^o - (\alpha\nu)T^o - T\nu = e. \tag{3.3.58}$$

We obtain

$$\nu = [\mu(g\nu) - 1]T^{-1}e + [(g\nu) + (\alpha\nu)]e$$

and hence

$$g\nu = \frac{1}{\mu}\frac{1}{1-\rho}, \qquad \alpha\nu = \frac{\rho}{1-\rho}\Psi_1',$$

where Ψ_1' is the mean of the PH-distribution with representation (g, T).

The inner product $\pi\nu$ is given by

$$\pi\nu = [\mu(g\nu) - 1]\pi T^{-1}e + g\nu + \alpha\nu$$
$$= \frac{1}{1-\rho}\left(\frac{1}{\mu} + \rho\Psi_1' - \rho\frac{\lambda_2'}{2\lambda_1'}\right). \tag{3.3.59}$$

The vector $\alpha(I - G + \tilde{G})$ is given by

$$\alpha(I - G + \tilde{G}) = \alpha - \alpha G + g = \alpha + \rho^{-1}(g - \pi), \tag{3.3.60}$$

by virtue of formula (3.3.44).

The mean duration of the busy period is now clearly given by $\alpha\tilde{\mu}_1$. A customer arriving to an empty queue starts the QBD process in one of the states $(1, j)$, $1 \le j \le m$, according to the initial probability vector α, and the busy period is then the time until the level $\mathbf{0}$ is reached for the first time. We obtain explicitly

$$\alpha\tilde{\mu}_1 = \alpha\nu + \rho^{-1}g\nu - \rho^{-1}\pi\nu = \frac{1}{1-\rho}\frac{\lambda_2'}{2\lambda_1'} - \Psi_1'. \tag{3.3.61}$$

For the particular case of the $M/M/1$ queue, where $\lambda_1' = \Psi_1' = \lambda^{-1}$, and $\lambda_2' = 2\lambda^{-2}$, formula (3.3.61) reduces to the familiar formula

$$\alpha\tilde{\mu}_1 = \frac{1}{\mu}\left(1 - \frac{\lambda}{\mu}\right)^{-1}.$$

The expected duration of the busy cycle is given by $\alpha G\tilde{\mu}_1$, since at the end of the previous busy period the probability that the arrival phase is j is given by $(\alpha G)_j$, for $1 \le j \le m$. We may also obtain that mean directly from formula (3.3.37), as follows:

$$\alpha G\tilde{\mu}_1 = \alpha G(-T^{-1}e) + \alpha G(-T^{-1}T^oA^o)\tilde{\mu}_1 = -\alpha GT^{-1}e + \alpha\tilde{\mu}_1. \tag{3.3.62}$$

The first term is the mean of a PH-distribution with representation $(\alpha G, T)$. Such, indeed, is the probability distribution of the *idle period* for the $PH/M/1$ queue. The second term is the mean busy period, given by formula (3.3.61).

Remark. We see that even the mean busy period and the busy cycle involve the matrix G through the vectors αG and g. The characterization of these vectors by theorem 3.3.4 is specific, but not fully explicit. The dependence on the matrix G accounts for the difficulty in obtaining even these means by transform methods. Alternate expressions for these means will be obtained in chapter 4.

3.4. THE COMPUTATION OF THE MATRIX $G(x)$

The computation of the semi-Markov matrix $G(\cdot)$ with the transform matrix $\hat{G}(1, s)$ is a major task. It would be an essential step, however, in the numerical computation of the busy period distribution of, for example, the $PH/PH/1$ queue.

The equation (3.3.4) with $z = 1$ is equivalent to the integro-differential equation

$$G'(x) = A_2 + A_1 G(x) + A_0 \int_0^x G(x-u)G'(u)du, \quad \text{for} \quad x \geq 0, \tag{3.4.1}$$

with initial condition $G(0) = 0$. The solution of this equation by direct numerical integration is far from elementary. As we have not attempted to do so, we cannot report on our numerical experience. This may be an interesting subject for further investigation, and it may suggest an alternative to the elementary, but rather belabored, method discussed next.

If we consider the QBD process with generator $\tilde{Q}_1$, given by

$$\tilde{Q}_1 = \begin{vmatrix} 0 & 0 & 0 & 0 & 0 & \cdots \\ A_2 & A_1 & A_0 & 0 & 0 & \cdots \\ 0 & A_2 & A_1 & A_0 & 0 & \cdots \\ 0 & 0 & A_2 & A_1 & A_0 & \cdots \\ 0 & 0 & 0 & A_2 & A_1 & \cdots \\ \vdots & \vdots & \vdots & \vdots & \vdots & \end{vmatrix}, \tag{3.4.2}$$

in which the m states $(0, j)$, $1 \leq j \leq m$, of the level $\mathbf{0}$ are absorbing, then we may exploit the fact that $G_{jj'}(x)$ is the probability that, given that the process starts in the state $(1, j)$, absorption occurs no later than time x in the state $(0, j')$.

The j-th row of $G(x)$ may therefore be computed by solving a (suitably truncated) infinite system of differential equations, given by

$$
\begin{aligned}
y_0'(x) &= y_1(x)A_2, \\
y_1'(x) &= y_1(x)A_1 + y_2(x)A_2, \\
y_i'(x) &= y_{i-1}(x)A_0 + y_i(x)A_1 + y_{i+1}(x)A_2, \quad \text{for} \quad i \geq 2,
\end{aligned}
\tag{3.4.3}
$$

with initial conditions $y_{ik}(0) = \delta_{i1}\delta_{kj}$.

It is readily seen that $y_{ik}(x)$, $i \geq 0$, $1 \leq k \leq m$, is the conditional probability that, given that the process starts in the state $(1, j)$, it is in the state (i, k) at time x. For a given value of j, $G_{jj'}(x) = y_{0j'}(x)$.

The evaluation of $G(\cdot)$ by this procedure is elementary but requires a substantial amount of computation. The system (3.4.3) needs to be numerically integrated m times, once for each set of initial conditions.

There remains the problem of truncating the system (3.4.3). It is possible to compute for each j, $1 \leq j \leq m$, an index K_j, such that upon truncating the system (3.4.3) at the index K_j and setting $y_{K_j+1}(x) \equiv 0$, a probability mass at most ϵ is lost in each row of $G(\cdot)$. Specifically, the computed quantities $\bar{G}_{jj'}(x)$ satisfy

$$
\sum_{j'=1}^{m} G_{jj'}(x) - \sum_{j'=1}^{m} \bar{G}_{jj'}(x) < \epsilon, \quad \text{for} \quad x \geq 0. \tag{3.4.4}
$$

Most of the loss is concentrated in the tails, since a truncation of the system (3.4.3) corresponds to neglecting paths that wander far from the level **1** before their eventual absorption into the level **0**.

Note that if we set $y_{K_j+1}(x) = y_{K_j}(x)$, we obtain computed $\bar{G}_{jj'}(x)$, which sum to a probability distribution. This corresponds to making the level $K_j + 1$ a *reflecting barrier*, and may for some purposes be preferable to setting $y_{K_j+1}(x) = 0$.

The choice of the index K_j requires itself a nonnegligible amount of computation. Let $z_j(k)$ be the conditional probability that, given that the process $\tilde{Q}_1$ starts in the state $(1, j)$, it reaches the absorbing states of level **0** without visiting the level $k + 1$ in between.

If the probabilities $z_j(k)$, $k \geq 1$, are known, it suffices to choose the index K_j by

$$
K_j = \min\{k\colon z_j(k) \geq 1 - \epsilon\}. \tag{3.4.5}
$$

Since the sequence $\{z_j(k)\}$ is nondecreasing in k and tends to one, such a choice is always possible.

The probabilities $z_j(k)$, $k \geq 1$, may be computed by considering the more general quantities $u_{ir}(k)$, where for $1 \leq i \leq k$, $1 \leq r \leq m$, $u_{ir}(k)$ is the conditional probability that, starting in the state (i, r), the process $\tilde{Q}_1$ reaches absorption in $\mathbf{0}$ without visiting the level $k + 1$ in between. The vectors $\mathbf{u}_i(k)$ with components $u_{ir}(k)$ are readily seen to satisfy the system of linear equations

$$A_0\mathbf{e} + A_1\mathbf{u}_1(k) + A_2\mathbf{u}_2(k) = \mathbf{0},$$
$$A_0\mathbf{u}_{i-1}(k) + A_1\mathbf{u}_i(k) + A_2\mathbf{u}_{i+1}(k) = \mathbf{0}, \qquad 2 \le i \le k-1, \tag{3.4.6}$$
$$A_0\mathbf{u}_{k-1}(k) + A_1\mathbf{u}_k(k) = \mathbf{0}.$$

The quantity $z_j(k)$ is then given by the j-th component $u_{1j}(k)$ of $\mathbf{u}_1(k)$. The system of equations (3.4.6) may usually be solved without any difficulty, either directly or by an iterative procedure such as Gauss-Seidel. It is well suited for the latter, in particular because the solution for the index $k - 1$ may be modified trivially to provide a good starting solution for the system with index k.

A nice feature is that one computation for successively larger values of k provides all the truncation indices K_j. The principal drawback of this entire approach to the computation of $G(\cdot)$ lies in the formidable growth of the processing times for matrices A_0, A_1, and A_2 of higher order. Particularly when ρ is close to one, some or all of the indices K_j become large, and computation of the matrix $G(x)$ for large values of x becomes quite expensive in computer time. A particular version of this algorithm for the busy period of the $M/PH/1$ queue is discussed in M. F. Neuts [225].

3.5. THE EMBEDDED MARKOV CHAIN AT LEFT TRANSITIONS

We now consider the embedded Markov chain, obtained by observing the state of the QBD process with generator $\tilde{Q}$ of formula (3.1.1) immediately following the successive left transitions. This Markov chain has the state space $E = \{(i, j), i \ge 0, 1 \le j \le m\}$ and the transition probability matrix $\tilde{P}_L$, given by

$$\tilde{P}_L = \begin{vmatrix} \tilde{B}_0 & \tilde{B}_1 & \tilde{B}_2 & \tilde{B}_3 & \cdots \\ \tilde{C}_0 & \tilde{A}_1 & \tilde{A}_2 & \tilde{A}_3 & \cdots \\ 0 & \tilde{A}_0 & \tilde{A}_1 & \tilde{A}_2 & \cdots \\ 0 & 0 & \tilde{A}_0 & \tilde{A}_1 & \cdots \\ 0 & 0 & 0 & \tilde{A}_0 & \cdots \\ \vdots & \vdots & \vdots & \vdots & \end{vmatrix}, \tag{3.5.1}$$

where all the entries of $\tilde{P}_L$ are $m \times m$ matrices, given by

$$\begin{aligned}
\tilde{A}_k &= (-A_1{}^{-1}A_0)^k(-A_1{}^{-1}A_2), \quad \text{for} \quad k \geq 0, \\
\tilde{B}_k &= (-B_0{}^{-1}A_0)\tilde{A}_k, \quad \text{for} \quad k \geq 1, \\
\tilde{C}_0 &= (-A_1{}^{-1}B_1), \\
\tilde{B}_0 &= (-B_0{}^{-1}A_0)\tilde{C}_0.
\end{aligned} \tag{3.5.2}$$

We see that $\tilde{P}_L$ is a stochastic matrix of the $M/G/1$ type. As we have assumed throughout that $\tilde{Q}$ is irreducible, B_0 and A_1 are nonsingular, and the blocks given in (3.5.2) are therefore well defined.

In a number of applications, the matrices A_2 or B_1 have columns that are identically zero. States $(0, j)$ for which the j-th column of B_1 is zero are obviously unattainable by the chain $\tilde{P}_L$. The same applies to all states (i, j), $i \geq 1$, for which the j-th column of A_2 vanishes. In discussing, for example, the stationary probability vector of the Markov chain $\tilde{P}_L$, we should first delete all such ephemeral states and reduce the blocks defined in (3.5.2) accordingly.

This, however, results in clumsy notation, which detracts from the main argument. We shall instead use the notation as given and make the convention that all statements regarding irreducibility and positive recurrence apply to the smaller state space, obtained by deleting all such states. We shall verify in the following that all formulas for stationary probabilities correctly assign the value zero to these states.

As the matrix $\tilde{P}_L$ is of the $M/G/1$ type, its stationary probability vector and a variety of other quantities of interest may be obtained from the general theory of chains of the $M/G/1$ type. Because of the relationship of $\tilde{P}_L$ to the process $\tilde{Q}$, most of these quantities can be obtained in a computationally far more appealing form from the results for QBD processes. The results on Markov chains of $M/G/1$ type do, however, yield a relation involving the matrices R and G, obtained respectively from theorems 3.1.1 and 3.3.2. In addition to its theoretical interest, this result provides an additional accuracy check for those cases where both R and G are computed.

Theorem 3.5.1. The stationary probability vector

$$y = [y_0, y_1, y_2, \ldots]$$

of $\tilde{P}_L$ is given by

$$\begin{aligned}
y_0 &= cx_1B_1 = cx_0RB_1, \\
y_i &= cx_{i+1}A_2 = cx_0R^{i+1}A_2, \quad \text{for} \quad i \geq 1,
\end{aligned} \tag{3.5.3}$$

where

$$c = [x_0(I - R)^{-1}A_0e]^{-1}, \tag{3.5.4}$$

and the vectors x_i, $i \geq 0$, are as given by theorem 3.1.1.

Proof. Clearly the formulas (3.5.3) assign probability zero to states excluded because of zero columns in B_1 or A_2. The formulas (3.5.3) and (3.5.4) may be obtained by a heuristic argument as follows. The conditional probability that the QBD process $\tilde{Q}$ is in the state (i, j) at time $t + dt$, given that in the interval $(t, t + dt)$ a left transition has occurred, is given for the stationary version of the process $\tilde{Q}$ by the j-th component of the vectors

$$\left(x_1B_1e + \sum_{\nu=2}^{\infty} x_\nu A_2e\right)^{-1} x_1B_1, \quad \text{for} \quad i = 0,$$

and

$$\left(x_1B_1e + \sum_{\nu=2}^{\infty} x_\nu A_2e\right)^{-1} x_{i+1}A_2, \quad \text{for} \quad i \geq 1.$$

This is ascertained by considering the elementary probabilities of a left transition in $(t, t + dt)$ and of a left transition into the state (i, j) in that same interval. The term in parentheses may be written as

$$c^{-1} = x_1B_1e + \sum_{\nu=2}^{\infty} x_\nu A_2e = \sum_{\nu=1}^{\infty} x_\nu A_2e = x_0(I - R)^{-1}RA_2e$$
$$= x_0(I - R)^{-1}A_0e$$

by using the formulas (3.1.6).

This suggests the stated form for the vector y. A formal verification that y is indeed the stationary probability vector of $\tilde{P}_L$ is equivalent to checking that the equations

$$x_0RB_1\tilde{B}_0 + x_0R^2A_2\tilde{C}_0 = x_1B_1,$$

$$x_0RB_1\tilde{B}_k + \sum_{\nu=0}^{k} x_0R^{k-\nu+2}A_2\tilde{A}_\nu = x_{k+1}A_2, \quad \text{for} \quad k \geq 1,$$

are satisfied.

The first equation yields

$$x_0RB_1(-B_0{}^{-1}A_0)(-A_1{}^{-1}B_1) + x_0R^2A_0(-A_1{}^{-1}B_1)$$
$$= (x_0A_0 + x_2A_2)(-A_1{}^{-1}B_1) = x_1B_1,$$

since by virtue of the steady-state equations of the process $\tilde{Q}$, we have $x_0 = -x_1 B_1 B_0^{-1}$, and $x_0 A_0 + x_2 A_2 = -x_1 A_1$.

The verification of the equations for $k \geq 1$ proceeds along similar lines. The essential steps are as follows. The first term may be written as

$$x_0 R B_1 (-B_0^{-1} A_0)(-A_1^{-1} A_0)^k (-A_1^{-1} A_2) = x_0 A_0 (-A_1^{-1} A_0)^k (-A_1^{-1} A_2).$$

Next we write the equation $R^2 A_2 + R A_1 + A_0 = 0$, $k + 1$ times. We premultiply in the equation with index ν, $0 \leq \nu \leq k$, by $R^{k-\nu}$, and postmultiply by the matrix $(-A_1^{-1} A_0)^\nu (-A_1^{-1} A_2)$ and sum over ν. This yields

$$\sum_{\nu=0}^{k} R^{k-\nu+2} A_2 (-A_1^{-1} A_0)^\nu (-A_1^{-1} A_2)$$

$$= -\sum_{\nu=0}^{k} [R^{k-\nu+1} A_1 + R^{k-\nu} A_0](-A_1^{-1} A_0)^\nu (-A_1^{-1} A_2)$$

$$= -A_0 (-A_1^{-1} A_0)^k (-A_1^{-1} A_2) + R^{k+1} A_2.$$

Upon premultiplication of this equality by x_0, the second equation is obtained.

We now proceed to an argument that is essential to the discussion of the matrix $\tilde{P}_L$, viewed as a matrix of $M/G/1$ type. Let $G_{jj'}(k)$ be the conditional probability that the chain $\tilde{P}_L$ reaches the set i for the first time after exactly k transitions by hitting the state (i, j'), given that it starts in the state $(i + 1, j)$, $i \geq 1$, $1 \leq j \leq m$, $1 \leq j' \leq m$. The matrix with elements $G_{jj'}(k)$ is denoted by $G(k)$ and we introduce the matrix-generating function

$$\tilde{G}(z) = \sum_{k=1}^{\infty} G(k) z^k, \quad \text{for} \quad 0 \leq z \leq 1. \tag{3.5.5}$$

By an argument similar to that of theorem 3.3.1, we may prove that $\tilde{G}(z)$ is the minimal nonnegative solution for $0 \leq z \leq 1$ of the equation

$$\tilde{G}(z) = \sum_{n=0}^{\infty} z \tilde{A}_n \tilde{G}^n(z). \tag{3.5.6}$$

We have, in fact, the following intuitive result.

Lemma 3.5.1. The matrix $\tilde{G}(z)$ is equal to the matrix $\hat{G}(z, 0)$ discussed in section 3.3.

Proof. Equation (3.5.6) leads to

$$A_1\tilde{G}(z) = -zA_2 - A_0 \cdot \sum_{n=0}^{\infty} z\tilde{A}_n\tilde{G}^n(z) \cdot \tilde{G}(z) = -zA_2 - A_0\tilde{G}^2(z).$$

Similarly, successive substitutions in (3.3.4) with s set equal to zero leads to

$$\tilde{G}(z, 0) = \sum_{n=0}^{\infty} z\tilde{A}_n\hat{G}^n(z, 0),$$

so that the equations (3.3.4) and (3.5.6) are equivalent.

In order to limit our discussion to the most salient ideas, we limit ourselves to the case where the matrix G and hence also $\tilde{G}(z)$, for $0 < z \leq 1$, are irreducible. We also assume that $\rho < 1$, so that G is stochastic.

We now set out to study the first return time distributions of the level $\mathbf{0}$. Specifically, let $K_{jj'}(z)$, $1 \leq j \leq m$, $1 \leq j' \leq m$, be the probability generating function on ν of the conditional probability that a chain returns to the set $\mathbf{0}$ for the first time after exactly $\nu \geq 1$ steps by hitting the state $(0, j')$, given that it starts in the state $(0, j)$. The matrix $K(z)$ has elements $K_{jj'}(z)$. We similarly define the matrix $L(z)$, which corresponds to the first passage from the level $\mathbf{1}$ to the level $\mathbf{0}$.

Routine first passage arguments lead to the equations

$$L(z) = z\tilde{C}_0 + \sum_{n=1}^{\infty} z\tilde{A}_n\tilde{G}^{n-1}(z)L(z),$$

$$K(z) = z\tilde{B}_0 + \sum_{n=1}^{\infty} z\tilde{B}_n\tilde{G}^{n-1}(z)L(z). \tag{3.5.7}$$

By elementary calculations similar to those in the proof of lemma 3.5.1, we may verify that

$$L(z) = \hat{G}^{(1,0)}(z, 0), \qquad K(z) = \hat{G}^{(0,0)}(z, 0), \tag{3.5.8}$$

in terms of the matrices given by the equations (3.3.33).

Lemma 3.5.2. If G is a stochastic matrix, so are the matrices $K(1)$ and $L(1)$.

Proof. It is trivial to verify that $K(1)$ is stochastic if $L(1)$ is. Equation (3.3.33) leads to

$$L(1) = [I - (-A_1{}^{-1}A_0)G]^{-1}(-A_1{}^{-1}B_1). \tag{3.5.9}$$

The existence of the inverse is readily shown by an argument *a contrario*.

Since clearly

$$[I - (-A_1^{-1}A_0)G]e = (-A_1^{-1}B_1)e,$$

it follows that the matrix $L(1)$ is stochastic.

The vector κ is now defined to be the invariant probability vector of the matrix $K(1)$, and κ^* is defined by

$$\kappa^* = \left(\frac{d}{dz} K(z)e\right)_{z=1}. \tag{3.5.10}$$

Lemma 3.5.3. The vector κ^* is given by

$$\kappa^* = B_0^{-1}A_0(A_0G + A_1)^{-1}(A_0\tilde{\mu} + B_1e), \tag{3.5.11}$$

where the vector $\tilde{\mu}$ is given by formula (3.3.22).

Proof. The equations (3.3.33) lead to

$$\kappa^* = K'(1)e = (-B_0^{-1}A_0)L'(1)e$$

and

$$L'(1) = -A_1^{-1}B_1 + (-A_1^{-1}A_0)[ML(1) + GL'(1)].$$

The latter equation may be rewritten as

$$A_0ML(1) + A_0GL'(1) + A_1L'(1) + B_1 = 0,$$

which leads to

$$L'(1)e = -(A_0G + A_1)^{-1}(A_0\tilde{\mu} + B_1e).$$

The stated formula is now evident.

Theorem 3.5.2. The vector y_0, obtained in formula (3.5.3), is also given by

$$y_0 = (\kappa\kappa^*)^{-1}\kappa. \tag{3.5.12}$$

Proof. The proof of (3.5.12) is based on a neat argument, based on the theory of Markov renewal processes. The quantity $(y_{0j})^{-1}$ is, by a classical property of Markov chains, the mean recurrence time of the state $(0, j)$ in the Markov chain $\tilde{P}_L$.

If we now consider the chain $\tilde{P}_L$ only at its visits to the set **0** and record the indices of the states visited, as well as the numbers of transitions in $\tilde{P}_L$ between consecutive visits to **0**, we obtain an irreducible m-state Markov renewal process. Its sojourn times are integer valued, and the probability generating function of its transition probability matrix is none other than the matrix $K(z)$, obtained in formula (3.5.7). A classical property of Markov renewal processes, given, for example, in E. Çinlar [56, p. 155] or in J. J. Hunter [147, p. 196], expresses the mean recurrence time of the state $(0, j)$ in terms of the invariant probability vector $\boldsymbol{\kappa}$ of $K(1)$ and the vector $\boldsymbol{\kappa}^* = K'(1)\mathbf{e}$, of the row-sum means of $K(z)$.

Specifically, we have

$$(y_{0j})^{-1} = (\boldsymbol{\kappa}\boldsymbol{\kappa}^*)\kappa_j^{-1}, \quad \text{for} \quad 1 \leq j \leq m,$$

since the recurrence time in the Markov renewal process is clearly the same as that in the Markov chain $\tilde{P}_L$. The stated formula is now obvious.

Remarks. The relation

$$(\boldsymbol{\kappa}\boldsymbol{\kappa}^*)^{-1}\boldsymbol{\kappa} = [\mathbf{x}_0(I - R)^{-1}A_0\mathbf{e}]\mathbf{x}_0RB_1 \tag{3.5.13}$$

is an interesting one. In addition to given coefficient matrices, its left-hand side involves the matrix G, and its right-hand side the matrix R. If we have occasion to compute both the matrix R and the matrix G for the QBD process, formula (3.5.13) provides us with a powerful and far from obvious accuracy check.

There are other relations between the matrices R and G, which may also be used as accuracy checks. The pretty symmetric formula

$$R(A_0G - RA_2)G = A_0G - RA_2 \tag{3.5.14}$$

is readily obtained from the equations $R^2A_2 + RA_1 + A_0 = 0$ and $A_0G^2 + A_1G + A_2 = 0$, but its probabilistic significance is obscure.

3.6. THE BUSY PERIOD OF THE *PH/G/*1 QUEUE

The discussion in this section is a digression from the main subject of this book. The $PH/G/1$ queue is, in fact, a prime example of a queueing model with an embedded Markov chain of $M/G/1$ type. As such, it is a particular case of a class of queues, having as its input process the versatile Markovian process introduced in M. F. Neuts [232], which was studied in detail in V. Ramaswami [261]. Its inclusion here will lead to algorithmically useful information on the $PH/PH/1$ queue, discussed in section 3.7, where in particular a computationally explicit formula for the mean duration of the busy period will be obtained. Only the main steps of the proofs will be presented.

Arrivals are according to a PH-renewal process with underlying distribution $F(\cdot)$ with a given irreducible representation $(\boldsymbol{\alpha}, T)$ of order m, with $\alpha_{m+1} = 0$, and mean λ_1'. The service time distribution $K(\cdot)$ is general, but has finite mean μ_1'. In cases where $K(\cdot)$ is also assumed to be of phase type, its representation will be $(\boldsymbol{\beta}, S)$ of order ν with $\beta_{\nu+1} = 0$.

Considered at points of departure, the $PH/G/1$ queue is seen to have an *embedded Markov renewal process* on the state space $E = \{(i, j), i \geq 0, 1 \leq j \leq m\}$, where the index i denotes the queue length and the index j the phase of arrival, immediately following the epoch of departure. The transition probability matrix $\tilde{P}(\cdot)$ of the embedded Markov renewal process is given by

$$\tilde{P}(x) = \begin{vmatrix} B_0(x) & B_1(x) & B_2(x) & B_3(x) & \cdots \\ A_0(x) & A_1(x) & A_2(x) & A_3(x) & \cdots \\ 0 & A_0(x) & A_1(x) & A_2(x) & \cdots \\ 0 & 0 & A_0(x) & A_1(x) & \cdots \\ \vdots & \vdots & \vdots & \vdots & \end{vmatrix}, \tag{3.6.1}$$

where

$$\begin{aligned} A_n(x) &= \int_0^x P(n, u)dK(u), \\ B_n(x) &= \int_0^x \int_0^u P(0, u - v)T^o A^o P(n, v)dK(v)du, \\ &\qquad\qquad \text{for} \quad n \geq 0, \qquad x \geq 0. \end{aligned} \tag{3.6.2}$$

The matrices $P(n, u)$, $n \geq 0$, $u \geq 0$, are defined in section 2.4 and satisfy the differential equations (2.4.9).

We readily see that the Laplace-Stieltjes transforms $B_n^*(s)$ of the matrices $B_n(\cdot)$, $n \geq 0$, are related to the corresponding transforms $A_n^*(s)$ of the matrices $A_n(\cdot)$ by

$$B_n^*(s) = (sI - T)^{-1}T^o A^o A_n^*(s), \quad \text{for} \quad n \geq 0. \tag{3.6.3}$$

From this, it readily follows that the matrices $A_n = A_n(\infty)$, $n \geq 0$, are as given in the formulas (2.4.10) and that $B_n = B_n(\infty) = A^{oo}A_n$, $n \geq 0$, where as usually $(A^{oo})_{jj'} = \alpha_{j'}$. It follows easily from theorem 2.4.2 that the Markov renewal process $\tilde{P}(\cdot)$ is irreducible.

As in section 3.3, we introduce the matrix $\hat{G}(z, s)$. Its element $\hat{G}_{jj'}(z, s)$ is the joint transform of the number of services during and the duration of a first passage from the state $(i + 1, j)$ to the set $\boldsymbol{i}$, which is hit in the state (i, j'), $i \geq 0$.

By arguments similar to those in section 3.3, we may show that $\hat{G}(z, s)$ is the minimal nonnegative solution of the matrix equation

$$\hat{G}(z, s) = \sum_{n=0}^{\infty} zA_n^*(s)\hat{G}^n(z, s), \qquad 0 \leq z \leq 1, \qquad s \geq 0. \tag{3.6.4}$$

We shall not enter into a general discussion of this equation here. Relevant references are [219, 223, 224, 228], although some of the arguments based, in the earlier papers, on methods of complex analysis may, as in section 3.3, be replaced by simpler, purely probabilistic arguments.

The stability of the $PH/G/1$ queue is again equivalent to the fact that the minimal, nonnegative solution to the matrix equation

$$G = \sum_{n=0}^{\infty} A_n G^n \tag{3.6.5}$$

is *stochastic.* This in turn is equivalent to the condition

$$\rho = \pi\beta^* \leq 1, \tag{3.6.6}$$

where π is the invariant probability vector of the positive stochastic matrix $A = \sum_{n=0}^{\infty} A_n$, and the vector β^* is defined by

$$\beta^* = \sum_{n=0}^{\infty} nA_n e. \tag{3.6.7}$$

By virtue of formula (2.4.16), β^* is explicitly given by

$$\beta^* = \lambda_1'^{-1}\mu_1' e + \lambda_1'^{-1}(I - A)T^{-1}e, \tag{3.6.8}$$

and the condition (3.6.6) reduces to the familiar equilibrium condition

$$\mu_1' \leq \lambda_1'. \tag{3.6.9}$$

The matrix A_0 does not have zero columns, and the matrices A_n, $n \geq 1$, are positive by theorem 2.4.2. The matrix G is therefore strictly positive. Limiting our attention now to the stable queue, we denote by g the invariant probability vector of G, and define $\tilde{G}$ by $\tilde{G} = e \cdot g$.

The vectors $\tilde{\mu}$ and $\tilde{\mu}_1$ are defined by

$$\tilde{\mu} = \left(\frac{\partial}{\partial z}\hat{G}(z, s)e\right)_{\substack{z=1\\s=0}}, \qquad \tilde{\mu}_1 = \left(-\frac{\partial}{\partial s}\hat{G}(z, s)e\right)_{\substack{z=1\\s=0}}, \tag{3.6.10}$$

and are given by the following theorem.

Theorem 3.6.1. If $\rho = 1$, the vectors $\tilde{\mu}$ and $\tilde{\mu}_1$ are infinite. If $\rho < 1$, they are given by

$$\tilde{\mu} = \mu_1'^{-1}\tilde{\mu}_1 = (I - G + \tilde{G})[I - A + \tilde{G} - \Delta(\beta^*)\tilde{G}]^{-1}e. \tag{3.6.11}$$

Proof. The proof is substantially the same as that of theorem 3.3.3. By differentiating with respect to z in (3.6.4), setting $z = 1$, $s = 0$, and postmultiplying by $\boldsymbol{e}$, we obtain

$$\left(I - \sum_{n=1}^{\infty} A_n \sum_{\nu=0}^{n-1} G^{\nu}\right)\tilde{\boldsymbol{\mu}} = \boldsymbol{e}. \tag{3.6.12}$$

For $\tilde{\boldsymbol{\mu}}_1$, we obtain the same equation with the right-hand side replaced by $\mu_1{}'\boldsymbol{e}$.

The coefficient matrix, as used in (3.6.11), may be obtained by noting that

$$\begin{aligned}\left(I - \sum_{n=1}^{\infty} A_n \sum_{\nu=0}^{n-1} G^{\nu}\right)(I - G + \tilde{G}) &= I - G + \tilde{G} - \sum_{n=1}^{\infty} A_n \\ &\quad + \sum_{n=1}^{\infty} A_n G^n - \sum_{n=1}^{\infty} nA_n\tilde{G} \\ &= I - A + \tilde{G} - \Delta(\boldsymbol{\beta}^*)\tilde{G},\end{aligned}$$

by virtue of the equation (3.6.5).

If $\rho = 1$, we see that the vector $\boldsymbol{\pi}$ is in the left null-space of the latter matrix. The coefficient matrix in (3.6.12) is then singular, and $\boldsymbol{\mu}$ is infinite.

If $\rho < 1$, let $\boldsymbol{u}$ be in the right null-space of the matrix $I - A + \tilde{G} - \Delta(\boldsymbol{\beta}^*)\tilde{G}$, then

$$(I - A)\boldsymbol{u} + (\boldsymbol{g}\boldsymbol{u})(\boldsymbol{e} - \boldsymbol{\beta}^*) = \boldsymbol{0},$$

and upon premultiplication by $\boldsymbol{\pi}$, we obtain $\boldsymbol{g}\boldsymbol{u} = 0$, so that $\boldsymbol{u} = k\boldsymbol{e}$, since $(I - A)\boldsymbol{u} = \boldsymbol{0}$. Since $\boldsymbol{g}\boldsymbol{u} = 0$, and $\boldsymbol{g}\boldsymbol{e} = 1$, we obtain $k = 0$, so that the coefficient matrix in (3.6.12) is nonsingular. The stated formula (3.6.11) is now evident.

Corollary 3.6.1. The equality

$$\boldsymbol{g}\tilde{\boldsymbol{\mu}} = (1 - \rho)^{-1} \tag{3.6.13}$$

holds.

Proof. By direct calculation we have

$$\boldsymbol{\pi}[I - A + \tilde{G} - \Delta(\boldsymbol{\beta}^*)\tilde{G}] = (1 - \rho)\boldsymbol{g},$$

from which (3.6.13) follows upon substitution in the expression for $\boldsymbol{g}\tilde{\boldsymbol{\mu}}$, with $\tilde{\boldsymbol{\mu}}$ as given in (3.6.11).

The formulas (3.6.11) and (3.6.13) were first proved in M. F. Neuts [223]. We may, however, express $\tilde{\boldsymbol{\mu}}$ also in an equivalent form, shown for the first time here, which clearly brings out the role played by the generalized inverses $(I - G + \tilde{G})^{-1}$ and $(I - A + \Pi)^{-1}$ of $I - G$ and $I - A$ respectively.

Corollary 3.6.2. The vector $\tilde{\mu}$ may be written as

$$\tilde{\mu} = (1 - \rho)^{-1}[1 - g(I - A + \Pi)^{-1}\beta^*]e + (1 - \rho)^{-1}(I - G + \tilde{G})(I - A + \Pi)^{-1}\beta^*. \qquad (3.6.14)$$

Proof. We denote the vector $[I - A + \tilde{G} - \Delta(\beta^*)\tilde{G}]^{-1}e$ by ν, and obtain

$$(I - A)\nu = (g\nu)\beta^* + [1 - (g\nu)]e.$$

Premultiplying by π, we obtain $g\nu = (1 - \rho)^{-1}$. Adding $\Pi\nu = (\pi\nu)e$ to both sides of the preceding equation and recalling that $I - A + \Pi$ is nonsingular, we obtain

$$\nu = (1 - \rho)^{-1}(I - A + \Pi)^{-1}\beta^* + [(\pi\nu) - \rho(1 - \rho)^{-1}]e, \qquad (3.6.15)$$

since trivially $(I - A + \Pi)^{-1}e = e$.

We cannot determine $\pi\nu$ from this equation alone. This reflects the nonuniqueness of the generalized inverse. However, by equation (3.6.13), we have

$$g(I - G + \tilde{G})\nu = g\nu = (1 - \rho)^{-1},$$

so that (3.6.15) leads to

$$\pi\nu - \rho(1 - \rho)^{-1} = (1 - \rho)^{-1}[1 - g(I - A + \Pi)^{-1}\beta^*].$$

Upon substitution, we obtain formula (3.6.14).

Corollary 3.6.3. The mean number of customers served during a busy period of the stable $PH/G/1$ queue is given by $\alpha\tilde{\mu}$, and the mean duration of the busy period is $\alpha\tilde{\mu}_1 = \mu_1'\alpha\tilde{\mu}$.

Remarks

a. The computation, even of the mean busy period for the stable $PH/G/1$ queue, requires a substantial amount of effort. The relatively minor simplifications that we were able to obtain for the particular case of the $PH/M/1$ queue suggest that this complexity is inherent in the busy period for queues with other than Poisson arrivals.

It is tempting to neglect the dependence on the elapsed interarrival time at the beginning of each service in the hope of obtaining what might be an approximation to the mean busy period of the $PH/G/1$, or even the $GI/G/1$ queue. This was suggested in [115]. The argument given there would for the $PH/G/1$ model offer the quantity $\mu_1'[1 - \int_0^\infty H(u)dK(u)]^{-1}$, where $H(u)$ is the renewal function of the PH-arrival process, as an approximation to the mean busy period. As was shown in V. Ramaswami and D. Lucantoni [259], this quantity may differ greatly from the true mean, even for arrival processes that are not far removed from the Poisson process. The availability of detailed algorithms for such queueing models as $GI/PH/1$ and $PH/G/1$ makes it possible to examine the practical merits of the large number of bounds and approximations for the $GI/G/1$ queue that have been proposed in the literature.

b. Finite versions of the $E_k/G/1$ queue were considered by A. L. Truslove [297, 298]. The same analysis may be carried over to the $PH/G/1$ queue of finite capacity.

3.7. THE *PH/PH/1* QUEUE AS A QBD PROCESS

The $GI/G/1$ queue in which the interarrival and service time distributions are both of phase type is called the *PH/PH/1 queue.* It is a particular case of both the $PH/G/1$ queue, which was discussed in section 3.6, and the $GI/PH/1$ queue, which will be treated in detail in chapter 4. It may also be treated directly as an example of a QBD process. There is merit in presenting these various approaches separately. Some items of interest are easier to compute in one setting than another. Particular features of the interarrival or service time distributions may be better exploited in treating one embedded Markov process than another. This can be well illustrated by discussing the $PH/PH/1$ queue in the framework of QBD processes. We shall show that matrix operations of seemingly impractically high dimension may, for most cases of interest, be greatly simplified.

We denote the interarrival time distribution by $F(\cdot)$; its representation of dimension m by $(\boldsymbol{\alpha}, T)$; and its mean by λ_1'. The service time distribution $K(\cdot)$ has mean μ_1' and representation $(\boldsymbol{\beta}, S)$ of dimension ν. Both representations are irreducible. The $PH/PH/1$ queue may be studied as a QBD process on the state space

$$E = \{(0, j), 1 \leq j \leq m\} \cup \{(i, j, k), i \geq 1, 1 \leq j \leq m, 1 \leq k \leq \nu\}.$$

The index $i \geq 1$ denotes the number of customers in the system; the index j, $1 \leq j \leq m$, represents the phase of the PH-renewal process of arrivals, while the index k, $1 \leq k \leq \nu$, indicates the phase of the service in course. The states are labeled in the lexicographic order, that is, $(0, 1), \ldots, (0, m), (1, 1, 1), \ldots, (1, 1, \nu), (1, 2, 1), \ldots, (1, 2, \nu), \ldots.$

The generator $\tilde{Q}$ is given by

$$\tilde{Q} = \begin{vmatrix} T & T^oA^o \otimes \beta & 0 & 0 & \cdots \\ I \otimes S^o & A_1 & A_0 & 0 & \cdots \\ 0 & A_2 & A_1 & A_0 & \cdots \\ 0 & 0 & A_2 & A_1 & \cdots \\ \vdots & \vdots & \vdots & \vdots & \end{vmatrix}, \tag{3.7.1}$$

where

$$A_0 = T^oA^o \otimes I,$$
$$A_1 = T \otimes I + I \otimes S, \tag{3.7.2}$$
$$A_2 = I \otimes S^oB^o.$$

The matrix A is given by

$$A = (T + T^oA^o) \otimes I + I \otimes (S + S^oB^o), \tag{3.7.3}$$

and has the stationary probability vector $\pi \otimes \theta$, where π and θ are the stationary probability vectors of $T + T^oA^o$ and $S + S^oB^o$ respectively. Since

$$(\pi \otimes \theta)A_0(e \otimes e) = \pi T^o = \lambda_1'^{-1},$$
$$(\pi \otimes \theta)A_2(e \otimes e) = \theta S^o = \mu_1'^{-1},$$

formula (3.1.4) readily reduces to the familiar condition $\lambda_1' > \mu_1'$.

The rate matrix R, which is of order $m\nu$, is the minimal solution of the matrix-quadratic equation

$$R^2(I \otimes S^oB^o) + R(T \otimes I + I \otimes S) + T^oA^o \otimes I = 0. \tag{3.7.4}$$

The stationary probability vector z of $\tilde{Q}$ is of the form $[z_0, z_1, z_1R, z_1R^2, \ldots]$. The vectors z_0 and z_1 are of dimensions m and $m\nu$ respectively. They are obtained by solving the equations

$$z_0T + z_1(I \otimes S^o) = 0,$$
$$z_0(T^oA^o \otimes \beta) + z_1[T \otimes I + I \otimes S + R(I \otimes S^oB^o)] = 0, \tag{3.7.5}$$
$$z_0e + z_1(I - R)^{-1}e = 1.$$

It is convenient to eliminate z_0 from the first two equations. We note that the first equation yields

$$z_0T^o = z_1(e \otimes S^o),$$

so that

$$z_0(T^oA^o \otimes \beta) = (z_0T^o)(\alpha \otimes \beta) = z_1(A^{oo} \otimes S^oB^o).$$

The vector z_1 is therefore determined up to a multiplicative constant by the equation

$$z_1[T \otimes I + I \otimes S + A^{oo} \otimes S^o B^o + R(I \otimes S^o B^o)] = \mathbf{0}. \tag{3.7.6}$$

Two observations are in order. In chapter 4 we shall solve the $GI/PH/1$ queue by performing operations on matrices of order ν only. In principle this approach requires less storage but more overhead computation, even for the particular case of the $PH/PH/1$ queue. This is a general feature of probability models that involve PH-distributions. We typically have the choice between two approaches: a straightforward analysis based on a high-dimensional description of the state-space versus a mathematically more laborious analysis based on embedded Markov chains with special structure.

The second observation is that the detailed description of the state-space is particularly well suited to exploit special features of the representations of the interarrival and service time distributions. In section 3.8 we illustrate this by carrying out a detailed analysis of a model of which many particular cases have been discussed in the literature. For the present, we give some general instances of such simplifying features.

Consider the matrix $T^o A^o \otimes I$ of order $m\nu$. We see that if $T_i^o = 0$, the row of blocks $T_i^o \alpha_1 I$, $T_i^o \alpha_2 I$, $\ldots$, $T_i^o \alpha_m I$, is identically zero. The corresponding ν rows of the matrix R also vanish. Let an index i for which $T_i^o > 0$ be called an *absorption index*. If the representation $(\boldsymbol{\alpha}, T)$ has k' absorption indices, $1 \le k' \le m$, then the computation of R reduces to that of a matrix of dimensions $k'\nu \times m\nu$.

Many useful PH-distributions have representations with only a small number of absorption indices. When this is the case, further simplifications are readily obtained by appropriately partitioning the matrix R. To illustrate this, we assume for definiteness that

$$T_1^o > 0, T_2^o > 0, T_i^o = 0, \quad \text{for} \quad 3 \le i \le m.$$

We now partition the matrix R into square blocks of order ν in the obvious manner. Only the $2m$ matrices, denoted by $R(1, j)$ and $R(2, j)$, $1 \le j \le m$, which correspond to the first 2ν rows of R, need to be computed. The equation (3.7.4) leads to

$$\begin{aligned} &R(1, 1)R(1,j)S^o B^o + R(1, 2)R(2,j)S^o B^o \\ &\qquad + R(1,j)S + \sum_{h=1}^{m} R(1, h)T_{hj} + T_1^o \alpha_j I = 0, \\ &R(2, 1)R(1,j)S^o B^o + R(2, 2)R(2,j)S^o B^o \\ &\qquad + R(2,j)S + \sum_{h=1}^{m} R(2, h)T_{hj} + T_2^o \alpha_j I = 0, \end{aligned} \tag{3.7.7}$$

for $1 \le j \le m$.

We compute the m matrices

$$L(j) = -(T_{jj}I + S)^{-1}, \quad \text{for} \quad 1 \leq j \leq m, \tag{3.7.8}$$

and the vectors $\beta L(j)$, $1 \leq j \leq m$, only once and rewrite the equations (3.7.7) as

$$\begin{aligned} R(1,j) &= [R(1,1)R(1,j)S^o + R(1,2)R(2,j)S^o] \cdot \beta L(j) \\ &\quad + [\sum_{h \neq j} T_{hj}R(1,h)]L(j) + T_1{}^o\alpha_j L(j), \\ R(2,j) &= [R(2,1)R(1,j)S^o + R(2,2)R(2,j)S^o] \cdot \beta L(j) \\ &\quad + [\sum_{h \neq j} T_{hj}R(2,h)]L(j) + T_2{}^o\alpha_j L(j), \end{aligned} \tag{3.7.9}$$

for $1 \leq j \leq m$. The form (3.7.9) is now very convenient for solution by successive substitutions. Clearly the $2m$ vectors $R(1, j)S^o$ and $R(2, j)S^o$ are evaluated only once for each iteration.

It is clear that sparsity of the matrix T plays a significant role in simplifying these equations further. In contrast, special features of the matrix S only affect the algorithm to a minor extent. In order to illustrate the use of particular structural features of T further, we shall consider in section 3.8 the case of Erlang interarrival times.

The Idle Period. We consider the stationary $PH/PH/1$ queue at a departure epoch in which the queue becomes empty. The time until the next arrival is termed the (stationary) *idle period.*

Theorem 3.7.1. Given that at time t the stationary $PH/PH/1$ queue becomes empty, the conditional distribution of the idle period is of phase type with representation (ψ, T), where

$$\psi = \frac{z_1(I \otimes S^o)}{z_1(e \otimes S^o)}. \tag{3.7.10}$$

Proof. It is readily seen that the positive vector ψ is the conditional probability vector of the arrival phase at a time when the queue becomes empty. The time until the next arrival is clearly the time until absorption in the Markov process with generator

$$\begin{vmatrix} T & T^o \\ \mathbf{0} & 0 \end{vmatrix},$$

and initial probability vector $[\psi, 0]$.

3.8. THE SIMPLICITY OF ERLANG ARRIVALS

The simplifications discussed in the preceding section were based only on the number of absorption indices of the representation (α, T). In this section we shall consider additional simplifications that are induced by the particular features of the Erlang distributions.

Let the interarrival time distribution $F(\cdot)$ be of the form

$$F(x) = \sum_{j=1}^{m} \alpha_j E_{m-j+1}(\lambda; x); \quad \text{for} \quad x \geq 0, \tag{3.8.1}$$

with $\alpha_1 > 0$. The representation of $F(\cdot)$ is then given by $\alpha = (\alpha_1, \ldots, \alpha_m)$ and

$$T = \begin{vmatrix} -\lambda & \lambda & & & \\ & -\lambda & \lambda & & \\ & & \cdots & & \\ & & & -\lambda & \lambda \\ & & & & -\lambda \end{vmatrix}.$$

The representation (α, T) is irreducible if and only if $\alpha_1 > 0$. The particular choice $\alpha_1 = 1$ yields the $E_m/PH/1$ model. We assume that the queue is *stable* or, equivalently, that

$$m + 1 - \sum_{j=1}^{m} j\alpha_j > \lambda\mu_1'. \tag{3.8.2}$$

Since the index m is the sole absorption index, only the last ν rows of the matrix R are nonzero. We partition the $\nu \times m\nu$ matrix consisting of these rows into m square matrices $R(j)$, $1 \leq j \leq m$, of order ν. The equation (3.7.4) now leads to

$$\begin{aligned} R(m)R(1)S^oB^o - R(1)(\lambda I - S) + \lambda\alpha_1 I &= 0, \\ R(m)R(j)S^oB^o - R(j)(\lambda I - S) + \lambda R(j-1) + \lambda\alpha_j I &= 0, \\ \text{for} \quad 2 \leq j \leq m. \end{aligned} \tag{3.8.3}$$

As shown in theorem 3.2.1, the case $m = 1$ may be solved explicitly. The present discussion therefore deals with the case $m \geq 2$.

If we set $V = \lambda(\lambda I - S)^{-1}$ and $\nu = (1/\lambda)\beta V = \beta(\lambda I - S)^{-1}$, the equations (3.8.3) may be equivalently written as

$$\begin{aligned} R(1) &= R(m)R(1)S^o \cdot \nu + \alpha_1 V, \\ R(j) &= R(m)R(j)S^o \cdot \nu + R(j-1)V + \alpha_j V, \quad \text{for} \quad 2 \leq j \leq m. \end{aligned} \tag{3.8.4}$$

This form is already well suited for numerical solution by successive substitutions. In view of the small amount of overhead computation that is required to evaluate V and ν, we in fact recommend this procedure for the determination of R. Either from (3.7.4) or directly from (3.8.3), we obtain the formula

$$\sum_{j=1}^{m} R(j)S^o = \lambda e, \tag{3.8.5}$$

which may serve either as an accuracy check or as a stopping criterion in the iterative computation of R.

It is tempting, particularly in systems with so transparent a structure as those in (3.8.4), to consider various algebraic manipulations that bring them closer to an "explicit" solution. For purposes of numerical solution, such transformations are not always desirable, as they may adversely affect the amount of required overhead computation or the numerical stability of the resulting algorithm. This is a common drawback of the (relatively rare) explicit formulas that have been obtained for stochastic models under particular distributional assumptions.

It may be of interest to illustrate this by considering some transformations of the equations (3.8.4) which immediately come to the attention of anyone steeped in classical analytic methods.

Upon multiplying the equations for $1 \leq j \leq r$ in (3.8.4) by V^{r-j} respectively and adding the resulting equalities, we obtain the equivalent equations

$$R(r) = \sum_{j=1}^{r} R(m)R(j)S^o \cdot \nu V^{r-j} + \sum_{j=1}^{r} \alpha_j V^{r-j+1}, \quad \text{for} \quad 1 \leq r \leq m. \tag{3.8.6}$$

We see that the matrices $R(r)$, $1 \leq r \leq m-1$, are now expressed in terms of the matrix $R(m)$ and the vectors $R(j)S^o$, $1 \leq j \leq m$. The vectors may themselves be expressed in terms of the matrix $R(m)$. To see this, we postmultiply each of the equations in (3.8.6) by S^o, to obtain

$$R(r)S^o = R(m) \sum_{j=1}^{r} \delta_{r-j} R(j)S^o + \sum_{j=1}^{r} \alpha_j V^{r-j+1} S^o, \quad \text{for} \quad 1 \leq r \leq m, \tag{3.8.7}$$

where $\delta_\nu = \nu V^\nu S^o = \lambda^\nu \beta(\lambda I - S)^{-(\nu+1)} S^o$, for $0 \leq \nu \leq m-1$. Since $0 < \delta_0 < 1$, the matrix $I - \delta_0 R(m)$ is nonsingular, so that the vectors $R(r)S^o$, $1 \leq r \leq m$, are recursively expressed in terms of the matrix $R(m)$ by

$$R(r)S^o = [I - \delta_0 R(m)]^{-1}\left\{R(m)\sum_{j=1}^{r-1}\delta_{r-j}R(j)S^o + \sum_{j=1}^{r}\alpha_j V^{r-j+1}S^o\right\}, \tag{3.8.8}$$

for $1 \leq r \leq m$.

By using the equations (3.8.6) and (3.8.8) together, we may clearly eliminate the matrices $R(j)$, $1 \leq j \leq m - 1$, and obtain an equation for $R(m)$ alone. This is of limited utility in view of the considerable increase in arithmetic effort in solving that equation for $R(m)$.

A more workable alternative is to proceed as follows. We initialize the iterative computation by setting $R(j)$ equal to the rightmost term in (3.8.6) and compute the vectors $R(j)S^o$, $1 \leq j \leq m$, by using the formula (3.8.8). These vectors are now entered into the right-hand side of (3.8.4), to yield the next values for $R(j)$, $1 \leq j \leq m$. The value for $R(m)$ is returned to (3.8.8), to yield the next values of $R(j)S^o$, and so on.

The particular structure of the matrix T is so utilized to expedite convergence. At each step, the inverse of the current matrix $I - \delta_0 R(m)$ needs to be evaluated, and the overhead computations are somewhat more extensive. It is therefore not a priori clear that the preceding transformations do result in a reduced computation time.

The computation of the vectors $\mathbf{z}_0$ and $\mathbf{z}_1$ proceeds in the present case as follows. We denote the components of $\mathbf{z}_0$ by $z_0(j)$, for $1 \leq j \leq m$, and partition the vector $\mathbf{z}_1$ into m ν-vectors $\mathbf{z}_1(j)$, $1 \leq j \leq m$. The first equation in (3.7.5) now readily yields

$$z_0(j) = \frac{1}{\lambda}\sum_{r=1}^{j}\mathbf{z}_1(r)S^o, \quad \text{for} \quad 1 \leq j \leq m, \tag{3.8.9}$$

and hence

$$\mathbf{z}_0\mathbf{e} = \frac{1}{\lambda}\sum_{j=1}^{m}(m - j + 1)\mathbf{z}_1(j)S^o. \tag{3.8.10}$$

The normalizing equation yields, because of the special structure of the matrix R, that

$$\mathbf{z}_0\mathbf{e} + \sum_{j=1}^{m}\mathbf{z}_1(j)\mathbf{e} + \mathbf{z}_1(m)[I - R(m)]^{-1}\sum_{j=1}^{m}R(j)\mathbf{e} = 1. \tag{3.8.11}$$

The second equation in formula (3.7.5) leads to the equations

$$\begin{aligned} \lambda\alpha_1 z_0(m)\boldsymbol{\beta} + \mathbf{z}_1(1)(S - \lambda I) + \mathbf{z}_1(m)R(1)S^oB^o &= \mathbf{0}, \\ \lambda\alpha_j z_0(m)\boldsymbol{\beta} + \mathbf{z}_1(j)(S - \lambda I) + \lambda\mathbf{z}_1(j-1) + \mathbf{z}_1(m)R(j)S^oB^o &= \mathbf{0}, \\ \text{for} \quad 2 \leq j \leq m. \end{aligned} \tag{3.8.12}$$

These may be rewritten into the convenient form

$$z_1(1) = z_0(m)\alpha_1\boldsymbol{\beta} V + z_1(m)R(1)S^o \cdot \frac{1}{\lambda}\boldsymbol{\beta} V,$$

$$z_1(j) = z_1(j-1)V + z_0(m)\alpha_j\boldsymbol{\beta} V + z_1(m)R(j)S^o \cdot \frac{1}{\lambda}\boldsymbol{\beta} V, \quad \text{for} \quad 2 \leq j \leq m. \tag{3.8.13}$$

These equations may again be transformed into a number of equivalent forms, but the easiest way to proceed appears to be the following. We see that $z_0(m)$ appears as a common factor of the constant terms in the linear equations (3.8.13). Fixing $z_0(m)$ at an estimated numerical value of the right magnitude, such as, for example, $(1 - \rho)/m$, we see that the resulting system (3.8.13) is ideally suited for Gauss-Seidel iteration. Solving that system iteratively, we obtain the vector $\mathbf{z}_1$ up to a multiplicative constant. The latter and the vector $\mathbf{z}_0$ may now readily be obtained from the equations (3.8.9), (3.8.10), and (3.8.11). No new overhead computations are required.

Remark. References on $GI/G/1$ queues in which the interarrival and the service time distributions are both of phase type (or more particularly Erlang) include C. C. Carson [47], T. Kawamura [161], T. C. T. Kotiah, J. W. Thompson, and W. A. O'N. Waugh [176], and M. F. Neuts [227].

3.9. WAITING TIME DISTRIBUTIONS IN THE QBD PROCESS

Many waiting time distributions in queues and related models may be conveniently computed as the distributions of certain times until absorption in QBD processes. The relationship of QBD processes to systems of linear differential equations with constant coefficients is used to construct efficient, if occasionally time-consuming, algorithms. Highly tractable formulas for the lower order moments of waiting times may also be obtained.

In discussing the absorption time distributions of interest, we first consider the simplest case of boundary behavior. Cases with complicated boundary behavior may usually be reduced to messy but finite-state absorption time problems. The treatment of the latter is usually easy and ad hoc. It will be illustrated for specific models in the subsequent chapters.

We consider the Markov process with the generator $\tilde{Q}^o$, given by

$$\tilde{Q}^o = \begin{vmatrix} 0 & 0 & 0 & 0 & \cdots \\ A_2 & D & 0 & 0 & \cdots \\ 0 & A_2 & D & 0 & \cdots \\ 0 & 0 & A_2 & D & \cdots \\ \vdots & \vdots & \vdots & \vdots & \end{vmatrix}, \tag{3.9.1}$$

in which all elements are square matrices of order m. The m states in the level $\mathbf{0}$ are absorbing. The matrix D has negative diagonal elements, nonnegative off-diagonal elements, and the inverse D^{-1} exists. The nonnegative matrix A_2 is such that $D\boldsymbol{e} + A_2\boldsymbol{e} = \mathbf{0}$. The matrix $K = -D^{-1}A_2$ is then clearly stochastic.

For use herein, we shall assume that there exists a matrix $\tilde{K}$ such that the matrix $I - K + \tilde{K}$ is nonsingular. In cases where K is irreducible, the matrix $\tilde{K}$ may be chosen as $\tilde{K} = \boldsymbol{e} \cdot \boldsymbol{\kappa}$, where $\boldsymbol{\kappa}$ is the invariant probability vector of K. This follows from the useful property in [164, p. 100], which we have already frequently used before.

The initial probability vector of the process $\tilde{Q}^o$ is denoted by $\boldsymbol{y}(0) = [\boldsymbol{y}_0(0), \boldsymbol{y}_1(0), \boldsymbol{y}_2(0), \ldots]$. We define $W_j(x)$, $x \geq 0$, $1 \leq j \leq m$, to be the probability that absorption occurs in the state $(0, j)$ no later than time x. The vector $\boldsymbol{W}(x)$ has the components $W_j(x)$, $1 \leq j \leq m$.

We now consider the system of differential equations

$$\boldsymbol{y}_0{}'(x) = \boldsymbol{y}_1(x)A_2,$$

$$\boldsymbol{y}_k{}'(x) = \boldsymbol{y}_k(x)D + \boldsymbol{y}_{k+1}(x)A_2, \quad \text{for} \quad k \geq 1, \qquad x \geq 0, \tag{3.9.2}$$

with the initial conditions $\boldsymbol{y}_k(0)$, for $k \geq 0$.

An elementary probability argument then shows that the quantity $y_{kj}(x)$ is the probability that at time x, the process $\tilde{Q}^o$ is in the state (k, j), $k \geq 0$, $1 \leq j \leq m$. The vector $\boldsymbol{W}(x)$ is therefore given by

$$\boldsymbol{W}(x) = \boldsymbol{y}_0(x), \quad \text{for} \quad x \geq 0, \tag{3.9.3}$$

and its density portion by

$$\boldsymbol{W}'(x) = \boldsymbol{y}_0{}'(x) = \boldsymbol{y}_1(x)A_2, \quad \text{for} \quad x > 0. \tag{3.9.4}$$

The vector $\boldsymbol{w}(s)$ of Laplace-Stieltjes transforms of $\boldsymbol{W}(x)$ is clearly given by

$$\boldsymbol{w}(s) = \sum_{k=0}^{\infty} \boldsymbol{y}_k(0)[(sI - D)^{-1}A_2]^k, \quad \text{for} \quad s \geq 0. \tag{3.9.5}$$

For $s = 0$, we obtain the vector

$$w(0) = \sum_{k=0}^{\infty} y_k(0)K^k, \tag{3.9.6}$$

whose components yield the probabilities of absorption into the states $(0, j)$ for $1 \le j \le m$. Since K is stochastic, we clearly have $w(0)e = 1$.

Theorem 3.9.1. The mean vector $-w'(0)$ is finite if and only if the vector $\sum_{k=1}^{\infty} ky_k(0)$ is finite. It is then given by

$$-w'(0) = \sum_{k=1}^{\infty} y_k(0) \sum_{\nu=0}^{k-1} K^\nu(-D^{-1})K^{k-\nu}. \tag{3.9.7}$$

If we set

$$Y = \sum_{k=0}^{\infty} y_k(0), \qquad Y^* = \sum_{k=1}^{\infty} y_k(0) \sum_{\nu=0}^{k-1} K^\nu, \tag{3.9.8}$$

then we have

$$-w'(0)e = Y^*(-D^{-1}e) = [Y - w(0) + Y^*\tilde{K}](I - K + \tilde{K})^{-1}(-D^{-1})e. \tag{3.9.9}$$

When the matrix K is *irreducible,* then

$$Y^*\tilde{K}(I - K + \tilde{K})^{-1} = Y^*\tilde{K} = (\nu e)\kappa, \tag{3.9.10}$$

where $\nu = \sum_{k=1}^{\infty} ky_k(0)$.

Proof. Upon differentiation in formula (3.9.5) and setting $s = 0$, we obtain

$$-w'(0) = \sum_{k=1}^{\infty} y_k(0) \sum_{\nu=0}^{k-1} K^\nu[(-D)^{-2}A_2]K^{k-\nu-1},$$

which is clearly equivalent to the stated formula (3.9.7). Upon post-multiplication by e, we obtain

$$-w'(0)e = Y^*(-D^{-1}e).$$

We now have

$$\begin{aligned} Y^*(I - K + \tilde{K}) &= \sum_{k=1}^{\infty} y_k(0) \sum_{\nu=0}^{k-1} K^\nu(I - K + \tilde{K}) \\ &= \sum_{k=1}^{\infty} y_k(0) - \sum_{k=1}^{\infty} y_k(0)K^k + Y^*\tilde{K} \\ &= Y - w(0) + Y^*\tilde{K}. \end{aligned}$$

When K is irreducible, $K\tilde{K} = \tilde{K}$, so that

$$Y^*\tilde{K} = \sum_{k=1}^{\infty} k y_k(0)\tilde{K} = (\nu e)\kappa.$$

Remarks

a. The second moment vector may be similarly computed, but the required algebraic manipulations are much more laborious. They are similar to the calculations described in M. F. Neuts [228] and will be omitted here.

b. In most applications, the quantity $-\boldsymbol{w}'(0)\boldsymbol{e}$ is of greatest interest. As nearly always K is irreducible, the major simplification given by formula (3.9.10) can be carried through. The computation of the mean $-\boldsymbol{w}'(0)\boldsymbol{e}$ is then straightforward.

c. If the vector $-\boldsymbol{w}'(0)$ itself is required, the following steps are advisable. One first computes $-\boldsymbol{w}'(0)\boldsymbol{e}$ directly by use of the simplified formulas. As the vector $-\boldsymbol{w}'(0)$ is positive, the known sum of its components may now be used to stop computation in the term-wise summation of the series in formula (3.9.7). The latter series is less forbidding than it looks. The matrices

$$X(k) = \sum_{\nu=0}^{k-1} K^{\nu}(-D^{-1})K^{k-\nu}, \quad \text{for} \quad k \geq 1,$$

may readily be computed by using the recursion

$$X(1) = (-D^{-1})K,$$

$$X(k+1) = (-D^{-1})K^{k+1} + KX(k), \quad \text{for} \quad k \geq 1,$$

which only requires the storage of two matrices of order m and the performance of two matrix multiplications per term.

d. In the rare cases where the matrix K is reducible, there usually are special simplifying features whose recognition results in major savings in computational effort. An example may be found in D. M. Lucantoni [195].

e. The numerical solution of the differential equations (3.9.2) needs little further discussion. It is clear how the system is to be truncated. The numerical integration may be carried out by any one of a number of classical methods, in which the very special nature of the coefficient matrix should be taken into account.

There is one important special case, particularly germane to the subject of this book, in which the computation of the vector $\boldsymbol{W}(x)$, $x \geq 0$, may be greatly simplified. The procedure, which we are about to discuss, has many important applications in queueing theory.

The Case Where $y(0)$ Is Matrix-Geometric

We consider the case where the vector $y(0)$ is of the form

$$y_k(0) = x_0 R^k V, \quad \text{for} \quad k \geq 0, \tag{3.9.11}$$

where x_0 is a nonnegative vector, R and V are nonnegative matrices, and $\text{sp}(R) = \eta < 1$. Furthermore, $x_0(I - R)^{-1}Ve = 1$.

For ease of presentation, we shall limit ourselves to the case where the matrices R and $D + A_2$ are *irreducible.* When this is not the case, the main results may still be established but require somewhat more detailed arguments. First a preliminary result.

We denote by u^o and u respectively positive right and left eigenvectors of R, corresponding to η, and normalized by $uu^o = 1$. Since $0 < \eta < 1$, the matrix $D + \eta A_2$ is stable. We denote its eigenvalue of maximum real part by $-\xi < 0$. We have $-\xi^o < -\xi < 0$, where $-\xi^o$ is the eigenvalue of maximum real part of the matrix D. By z^o and z, we denote positive right and left eigenvectors of the matrix $D + \eta A_2$, normalized by $zz^o = 1$.

Lemma 3.9.1. The eigenvalue of maximum real part of the matrix $I \otimes D + R^T \otimes A_2$ is $-\xi$. The corresponding left and right eigenvectors are $u^{oT} \otimes z$ and $u^T \otimes z^o$.

Proof. We have

$$(u^{oT} \otimes z)(I \otimes D + R^T \otimes A_2) = u^{oT} \otimes zD + \eta u^{oT} \otimes zA_2$$
$$= u^{oT} \otimes z(D + \eta A_2) = -\xi(u^{oT} \otimes z)$$

and

$$(I \otimes D + R^T \otimes A_2)(u^T \otimes z^o) = u^T \otimes Dz^o + \eta u^T \otimes A_2 z^o$$
$$= u^T \otimes (D + \eta A_2)z^o = -\xi(u^T \otimes z^o).$$

The matrix $I \otimes D + R^T \otimes A_2$ is not necessarily stable, but it has nonnegative off-diagonal elements. The matrix $I \otimes D + R^T \otimes A_2 - \lambda I \otimes I$, however, is stable for sufficiently large $\lambda > 0$. Since the eigenvectors of that matrix, corresponding to $-(\lambda + \xi)$, are positive, it follows that $-(\lambda + \xi)$ is the eigenvalue of maximum real part. The corresponding result for the matrix $I \otimes D + R^T \otimes A_2$ is now obvious.

Corollary 3.9.1. The matrix $I \otimes D + R^T \otimes A_2$ is nonsingular.

We now consider the Laplace-Stieltjes transform $w(s)$ of $W(\cdot)$, given by

$$w(s) = x_0 \sum_{k=0}^{\infty} R^k V[(sI - D)^{-1}A_2]^k = x_0 \Psi^*(s). \tag{3.9.12}$$

The matrix $\Psi^*(s)$ clearly satisfies the equation

$$\Psi^*(s) = V + R\Psi^*(s)(sI - D)^{-1}A_2, \tag{3.9.13}$$

which may be written into a more convenient form as follows. For a matrix C of dimensions $m_1 \times m_2$, we define $\tau(C)$ to be the m_1m_2-vector obtained by forming the *direct sum* of the rows of C.

If now

$$\phi^*(s) = \tau[\Psi^*(s)], \qquad \nu = \tau(V), \tag{3.9.14}$$

then the equation (3.9.13) may be written as

$$\phi^*(s) = \nu + \phi^*(s)[R^T \otimes (sI - D)^{-1}A_2]. \tag{3.9.15}$$

By using classical properties of Kronecker products, this equation may be transformed as follows

$$\phi^*(s)[I \otimes (sI - D)^{-1}][I \otimes (sI - D) - R^T \otimes A_2] = \nu,$$

and therefore

$$\phi^*(s) = \nu + \nu(sI \otimes I - I \otimes D - R^T \otimes A_2)^{-1}(R^T \otimes A_2). \tag{3.9.16}$$

Let $\Psi(\cdot)$ be the matrix of mass-functions with Laplace-Stieltjes transform $\Psi^*(s)$ and let $\phi(x) = \tau[\Psi(x)]$, for $x \geq 0$, then upon inversion of the transforms, the equation (3.9.16) yields

$$\begin{aligned}\phi(x) &= \nu + \nu \int_0^x \exp[(I \otimes D + R^T \otimes A_2)u]du(R^T \otimes A_2) \\ &= \nu + \nu(I \otimes D + R^T \otimes A_2)^{-1} \\ &\quad \cdot \{\exp[(I \otimes D + R^T \otimes A_2)x] - I \otimes I\}(R^T \otimes A_2),\end{aligned} \tag{3.9.17}$$

for $x \geq 0$.

Now let V^o be the square matrix of order m for which

$$\tau(V^o) = -\nu(I \otimes D + R^T \otimes A_2)^{-1} = \nu^o. \tag{3.9.18}$$

Furthermore, let $\theta(\cdot)$ be defined by

$$\theta(x) = \nu^o \exp[(I \otimes D + R^T \otimes A_2)x], \quad \text{for} \quad x \geq 0, \tag{3.9.19}$$

and let the $m \times m$ matrix $\Theta(x)$ be such that $\tau[\Theta(x)] = \theta(x)$, then it is clear from (3.9.19) that $\theta(\cdot)$ satisfies the differential equation

$$\theta'(x) = \theta(x)(I \otimes D + R^T \otimes A_2), \quad \text{for} \quad x \geq 0,$$

with $\theta(0) = \nu^o$. This is equivalent to the matrix-differential equation

$$\Theta'(x) = \Theta(x)D + R\Theta(x)A_2, \quad \text{for} \quad x \geq 0, \tag{3.9.20}$$

with initial conditions $\Theta(0) = V^o$.

In terms of the matrix $\Theta(\cdot)$, it follows from (3.9.17),

$$\Psi(x) = V + RV^oA_2 - R\Theta(x)A_2, \quad \text{for} \quad x \geq 0. \tag{3.9.21}$$

The vector $W(\cdot)$ is then clearly given by

$$W(x) = x_0V + x_0RV^oA_2 - x_0R\Theta(x)A_2, \quad \text{for} \quad x \geq 0. \tag{3.9.22}$$

We note that in the derivation of formula (3.9.22) only the nonsingularity of the matrix $I \otimes D + R^T \otimes A_2$ was used. This is valid under far less stringent conditions than those of lemma 3.9.1. For an example, we refer to D. M. Lucantoni [195], where the present procedure was computationally implemented. The preceding discussion is summarized in the following theorem, which in its statement emphasizes the simplicity of the required algorithmic steps.

Theorem 3.9.2. The vector $W(\cdot)$ with transform given in formula (3.9.12) may be found by implementing the following algorithmic steps.

Step 1. Evaluate the matrix V^o by solving the matrix equation

$$V^oD + RV^oA_2 = -V. \tag{3.9.23}$$

Step 2. Integrate the matrix-differential equation

$$\Theta'(x) = \Theta(x)D + R\Theta(x)A_2, \quad \text{for} \quad x \geq 0, \tag{3.9.24}$$

with initial conditions $\Theta(0) = V^o$.

Step 3. For each $x \geq 0$, evaluate

$$\begin{aligned} W(x) &= x_0V + x_0RV^oA_2 - x_0R\Theta(x)A_2 \\ &= -x_0V^oD - x_0R\Theta(x)A_2. \end{aligned} \tag{3.9.25}$$

Remarks

a. The computation of the vector $W(\cdot)$ is now reduced to the numerical solution of a highly structured system of m^2 linear equations and to the integration of a system of m^2 linear differential equations with constant coefficients. The equations (3.9.23) are well suited for numerical solution by Gauss-Seidel iteration, and the equation (3.9.24) may be integrated by a classical procedure such as the Runge-Kutta method. We note that the matrix $R\Theta(x)A_2$ also appears in the formula (3.9.25), so that by efficient programming a significant number of matrix multiplications can be saved.

b. In chapter 4 this procedure is applied to the computation of waiting time distributions in the *GI/PH/*1 queue. By way of a simple verification, we may consider here the particular case of the waiting time distribution at arrivals in the stationary *GI/M/*1 queue. The matrix R is there the scalar η, $0 < \eta < 1$, which is the unique solution in (0, 1) of the equation $\eta = f(\mu - \mu\eta)$, where $f(\cdot)$ is the Laplace-Stieltjes transform of the interarrival time distribution $F(\cdot)$. Furthermore $x_0 = 1 - \eta$, $V = 1$, $D = -\mu$, and $A_2 = \mu$. The scalar V^o is given by $V^o = [\mu(1 - \eta)]^{-1}$, so that

$$\Theta(x) = [\mu(1 - \eta)]^{-1} e^{-\mu(1-\eta)x}, \quad \text{for} \quad x \geq 0.$$

Finally,

$$\begin{aligned} W(x) &= 1 - \eta + (1 - \eta)\eta\mu[\mu(1 - \eta)]^{-1} \\ &\quad - (1 - \eta)\eta\mu[\mu(1 - \eta)]^{-1} e^{-\mu(1-\eta)x} \\ &= 1 - \eta + \eta[1 - e^{-\mu(1-\eta)x}], \quad \text{for} \quad x \geq 0. \end{aligned}$$

This is a well-known result.

c. Theorem 3.9.2 leads to interesting asymptotic results in which the irreducibility assumptions of lemma 3.9.1 play a significant role. By virtue of that lemma and a classical property of the matrix-exponential function, we have that

$$\exp[(I \otimes D + R^T \otimes A_2)x] = e^{-\xi x} L + o(e^{-\xi x}), \tag{3.9.26}$$

as $x \to \infty$. The matrix L is of order m^2 and is idempotent. It is explicitly given by

$$L = (\mathbf{u}^T \otimes \mathbf{z}^o)(\mathbf{u}^{oT} \otimes \mathbf{z}). \tag{3.9.27}$$

Lemma 3.9.2. The vector $-\nu(I \otimes D + R^T \otimes A_2)^{-1} L (R^T \otimes A_2)$ is the direct sum of the rows of the square matrix

$$\eta\xi^{-1}(\mathbf{u} \cdot V\mathbf{z}^o)(\mathbf{u}^{oT} \cdot \mathbf{z}A_2). \tag{3.9.28}$$

Proof. We have

$$\begin{aligned} &-\nu(I \otimes D + R^T \otimes A_2)^{-1} L (R^T \otimes A_2) \\ &\qquad = -\nu(I \otimes D + R^T \otimes A_2)^{-1}(\mathbf{u}^T \otimes \mathbf{z}^o) \cdot (\mathbf{u}^{oT} \otimes \mathbf{z})(R^T \otimes A_2) \\ &\qquad = \xi^{-1}\nu(\mathbf{u}^T \otimes \mathbf{z}^o) \cdot \eta(\mathbf{u}^{oT} \otimes \mathbf{z}A_2). \end{aligned}$$

The $[(i - 1)m + j]$-th component of this vector is given by

$$\eta\xi^{-1} \sum_{\nu=1}^{m} \sum_{r=1}^{m} V_{\nu r} u_\nu z_r^o u_i^o (\mathbf{z}A_2)_j = \eta\xi^{-1}(\mathbf{u} \cdot V\mathbf{z}^o)[\mathbf{u}^o \cdot \mathbf{z}A_2]_{ij},$$

for $1 \leq i \leq m$, $1 \leq j \leq m$.

Theorem 3.9.3. The vector $W(\cdot)$ satisfies

$$W(\infty) - W(x) = \eta\xi^{-1}(\mathbf{u} \cdot V\mathbf{z}^o)\mathbf{x}_0(\mathbf{u}^o \cdot \mathbf{z}A_2)e^{-\xi x} + o(e^{-\xi x}), \tag{3.9.29}$$

as $x \to \infty$.

Proof. It follows from formula (3.9.17) that

$$\phi(\infty) - \phi(x) = -\nu(I \otimes D + R^T \otimes A_2)^{-1} \cdot \exp[(I \otimes D + R^T \otimes A_2)x](R^T \otimes A_2),$$

for $x \geq 0$. By (3.9.26) and lemma 3.9.2, it follows that

$$\Psi(\infty) - \Psi(x) = \eta\xi^{-1}(\mathbf{u} \cdot V\mathbf{z}^o)(\mathbf{u}^o \cdot \mathbf{z}A_2)e^{-\xi x} + o(e^{-\xi x}), \quad (3.9.30)$$

as $x \to \infty$, which implies the stated result.

Remarks

a. Theorem 3.9.3 is useful in terminating the often lengthy computation of the tails of the waiting time distributions. The eigenvalues η and $-\xi$ may be efficiently computed by implementing Elsner's algorithm, described in section 1.9. This algorithm also produces a pair of corresponding left eigenvectors. The computation of the right eigenvectors is then routine.

b. The preceding discussion also suggests an alternate way to compute the mean vector $-\mathbf{w}'(0)$ for the case where the vector $\mathbf{y}(0)$ is matrix-geometric. From formula (3.9.16), we obtain

$$-\phi^{*\prime}(0) = \nu(I \otimes D + R^T \otimes A_2)^{-2}(R^T \otimes A_2)$$
$$= -\nu^o(I \otimes D + R^T \otimes A_2)^{-1}(R^T \otimes A_2).$$

It follows that

$$-\mathbf{w}'(0) = \int_0^\infty x\,dW(x) = \mathbf{x}_0 R V^1 A_2, \quad (3.9.31)$$

where the matrix V^1 is the unique solution to the equation

$$V^1 D + R V^1 A_2 = -V^o. \quad (3.9.32)$$

Waiting Times Under Other Queue Disciplines

The simple structure of the matrix $\tilde{Q}^o$, given in (3.9.1), typically arises in the computation of the distribution of the virtual waiting time or the waiting time at arrival epochs. In some queueing models it is possible to study the waiting time distributions under queue disciplines such as *last-come, first-served* or *service in random order* by identifying these waiting times with the times until absorption in suitable Markov processes. This approach has not yet been systematically developed, except for the $M/PH/1$ queue as discussed in M. F. Neuts [225].

For example, the waiting time under service in random order in the stationary $M/PH/1$ queue is identified with the absorption time in the process with generator $\tilde{Q}^o$, given by

$$\tilde{Q}^o = \begin{vmatrix} 0 & \mathbf{0} & \mathbf{0} & \mathbf{0} & \cdots \\ S^o & S - \lambda I & \lambda I & 0 & \cdots \\ \tfrac{1}{2}S^o & \tfrac{1}{2}S^o B^o & S - \lambda I & \lambda I & \cdots \\ \tfrac{1}{3}S^o & 0 & \tfrac{2}{3}S^o B^o & S - \lambda I & \cdots \\ \tfrac{1}{4}S^o & 0 & 0 & \tfrac{3}{4}S^o B^o & \cdots \\ \vdots & \vdots & \vdots & \vdots & \end{vmatrix}. \tag{3.9.33}$$

The initial probability vector is that given by theorem 3.2.1. These represent the situation of the stationary $M/PH/1$ queue as seen by a customer C^*, who arrives at time $t = 0$. The process $\tilde{Q}^o$ is in the state (i, j), $i \geq 1$, $1 \leq j \leq m$, if there are i customers *other than* C^* in the queue. We see that at the next departure the customer C^* will be chosen for service with probability $1/i$ and that he will continue waiting with probability $1 - 1/i$.

The paths of the process $\tilde{Q}^o$ are clearly no longer nonincreasing in i. This raises the problem of truncating the infinite system of differential equations for the absorption time distribution. In order to do so, it is necessary to study the distribution of the maximum excursion of the process $\tilde{Q}^o$ prior to absorption. This discussion proceeds along similar lines to that given in section 3.4. The substantial amount of overhead computation makes the evaluation of such waiting time distributions a major task. This and the consideration of waiting time distributions under a variety of priority rules are subjects worthy of further algorithmic investigations, since the comparison of queue disciplines on the basis of *mean* waiting times alone may lead to misleading conclusions.

Chapter 4
The $GI/PH/1$ Queue and Related Models

4.1. THE $GI/PH/1$ QUEUE

We consider the single server queue in which customers arrive one at a time, according to a renewal process with interarrival time distribution $F(\cdot)$ of finite mean λ_1', and satisfying $F(0+) = 0$. The successive service times are mutually independent and are also independent of the interarrival times. They have a common PH-distribution $K(\cdot)$ with irreducible representation $(\boldsymbol{\beta}, S)$ of order ν. The mean service time μ_1' is given by $\mu_1' = -\boldsymbol{\beta} S^{-1}\mathbf{e}$. For simplicity of exposition, we assume that $\beta_{\nu+1} = 0$. The case where $\beta_{\nu+1} > 0$ corresponds, in fact, to group service in which the allowable group sizes are independent and geometrically distributed. This and many other variants involving group service are easily studied by elementary modifications of the discussion for service of single customers.

We define the matrices $P(n, t)$, $n \geq 0$, $t \geq 0$, satisfying the recursive system of differential equations

$$P'(0, t) = P(0, t)S,$$

$$P'(n, t) = P(n, t)S + P(n - 1, t)S^{o}B^{o}, \quad \text{for} \quad n \geq 1, \tag{4.1.1}$$

with $P(n, 0) = \delta_{n0}I$, for $n \geq 0$. The system (4.1.1) was discussed in detail in section 2.4.

The $GI/PH/1$ queue is first considered at points of arrival. By I_r, we denote the queue length immediately *prior* to the r-th arrival; J_r is the phase state of the service in course, immediately *after* the r-th arrival, and τ_r stands for the interarrival time between the $(r - 1)$st and r-th arrivals. Conveniently choosing the time origin at an epoch of arrival and setting $\tau_0 = 0$, we readily see that the sequence of triplets $\{(I_r, J_r, \tau_r), r \geq 0\}$ is a *Markov renewal sequence* on the state space $E^o = \{(i, j, x), i \geq 0, 1 \leq j \leq \nu, x \geq 0\}$. Its transition probability matrix $\hat{P}(\cdot)$ is given by

$$\hat{P}(x) = \begin{vmatrix} B_0(x) & A_0(x) & 0 & \cdots \\ B_1(x) & A_1(x) & A_0(x) & \cdots \\ B_2(x) & A_2(x) & A_1(x) & \cdots \\ B_3(x) & A_3(x) & A_2(x) & \cdots \\ \vdots & \vdots & \vdots & \end{vmatrix}, \tag{4.1.2}$$

for $x \geq 0$, where

$$A_k(x) = \int_0^x P(k, t)dF(t),$$

$$B_k(x) = \sum_{r=k+1}^{\infty} \int_0^x P(r, t)dF(t)B^{oo}, \quad \text{for} \quad k \geq 0. \tag{4.1.3}$$

All but the first column of matrices in $\hat{P}(\cdot)$ are readily obtained by considering the possible changes in the queue length and the service phase from one arrival epoch to the next. The expression for the matrices $B_k(\cdot)$ requires some explanation. The (j, h)-element of $B_k(x)$ is given by

$$\left(\sum_{r=k+1}^{\infty} \int_0^x P(r, t)dF(t)e\right)_j \beta_h$$

$$= \left(\int_0^x \exp[(S + S^o B^o)t]dF(t)e - \sum_{r=0}^{k} \int_0^x P(r, t)dF(t)e\right)_j \beta_h$$

$$= \left(F(x)e - \sum_{r=0}^{k} \int_0^x P(r, t)dF(t)e\right)_j \beta_h.$$

The final expression in brackets is clearly the probability that, starting from an arrival and from the state (k, j), the queue empties out before the next arrival, which occurs not more than x time units later. The arriving customer will then start service in the phase h, $1 \leq h \leq \nu$, with probability β_h. It is to obtain this convenient form of the matrices $B_k(x)$ that above we defined J_r to be the service phase immediately *after* the r-th arrival epoch. If there are other customers in the queue, the service phases before and after arrivals are the same (with probability one).

The embedded Markov chain $\{(I_r, J_r), r \geq 0\}$ has the transition probability matrix $\tilde{P} = \hat{P}(\infty)$, which is clearly of the canonical form (1.1.5). The computation of the matrices $A_k = A_k(\infty)$ and $B_k = B_k(\infty) = B^{oo} - \Sigma_{r=0}^{k} A_r B^{oo}$, $k \geq 0$, was discussed in section 2.4. A comment on the matrices B_k is in order. Any column with index h for which $\beta_h = 0$ clearly vanishes in all the matrices B_k. In order to obtain an irreducible stochastic matrix $\tilde{P}$, all states $(0, h)$ for which $\beta_h = 0$ should be deleted, and the first row and column of blocks in $\tilde{P}$ ought to be appropriately redefined. It is notationally convenient not to do so. We shall see in the following that the computed invariant vector correctly assigns probability zero to such states. Whenever we say that $\tilde{P}$ is irreducible it will be understood that any ephemeral states $(0, h)$ with $\beta_h = 0$ have been excluded from the state space. Since by theorem 2.4.2 the matrices A_k, $k \geq 1$, are positive, and since A_0 does not have vanishing rows or columns, it is readily seen that $\tilde{P}$ is indeed *irreducible.*

The matrix $A = \Sigma_{k=0}^{\infty} A_k$ is strictly positive and stochastic. Its invariant probability vector $\boldsymbol{\theta}$ satisfies $\boldsymbol{\theta}(S + S^o B^o) = \mathbf{0}$, $\boldsymbol{\theta} \boldsymbol{e} = 1$. The vector $\boldsymbol{\beta}^* = \Sigma_{k=1}^{\infty} k A_k \boldsymbol{e}$ was computed earlier. By virtue of formula (2.4.16), we have

$$\boldsymbol{\beta}^* = \lambda_1' \mu_1'^{-1} \boldsymbol{e} + \mu_1'^{-1}(I - A)S^{-1}\boldsymbol{e}. \tag{4.1.4}$$

The inner product $\boldsymbol{\theta}\boldsymbol{\beta}^*$ reduces to $\lambda_1' \mu_1'^{-1}$, so that the Markov chain $\tilde{P}$ is positive recurrent if and only if the classical equilibrium condition $\lambda_1' > \mu_1'$ holds.

The Stationary Queue Length at Arrivals

Assuming henceforth that $\lambda_1' > \mu_1'$, and noting that the matrix A_0 does not have vanishing rows, we see that theorems 1.3.3 and 1.3.4 yield that the minimal solution R of $R = \Sigma_{k=0}^{\infty} R^k A_k$ is *strictly positive* and of spectral radius less than one.

The invariant probability vector $\boldsymbol{x} = [\boldsymbol{x}_0, \boldsymbol{x}_1, \boldsymbol{x}_2, \ldots]$ of $\tilde{P}$, partitioned into ν-vectors, is then given by the following theorem.

Theorem 4.1.1. The vector $\boldsymbol{x}$ is given by

$$\boldsymbol{x}_k = C\boldsymbol{\beta}R^k, \quad \text{for} \quad k \geq 0, \tag{4.1.5}$$

where

$$C = [\boldsymbol{\beta}(I - R)^{-1}\boldsymbol{e}]^{-1}. \tag{4.1.6}$$

The mean queue length L_1 prior to arrivals is given by

$$L_1 = C\boldsymbol{\beta}(I - R)^{-2}\boldsymbol{e} - 1. \tag{4.1.7}$$

Proof. The matrix $B[R]$ is given by

$$B[R] = \sum_{k=0}^{\infty} R^k \sum_{r=k+1}^{\infty} A_r B^{oo} = (I - R)^{-1} \sum_{r=1}^{\infty} (I - R^r) A_r B^{oo}$$

$$= (I - R)^{-1}(A - R)B^{oo} = B^{oo}.$$

The matrix B^{oo} is stochastic and has the invariant vector $\boldsymbol{\beta}$, whose components corresponding to the nonephemeral states are positive. Setting now $\mathbf{x}_0 = C\boldsymbol{\beta}$, it is readily verified that the invariant vector $\mathbf{x}$ of $\tilde{P}$ is given by (4.1.5) and (4.1.6).

Since $\sum_{k=1}^{\infty} kR^k = (I - R)^{-2}R = (I - R)^{-2} - (I - R)^{-1}$, the expression (4.1.7) for L_1 is readily obtained.

Remarks

a. The trivial final simplification of $B[R]$ to B^{oo} went unnoticed in theorem 4 of M. F. Neuts [227].

b. The quantity $C^{-1} = \boldsymbol{\beta}(I - R)^{-1}\mathbf{e}$ is the mean number of arrivals or, equivalently, customers served, during the busy period of the $GI/PH/1$ queue. This readily follows from the probabilistic significance of the matrix R. The j-th component of $(I + R + R^2 + \cdots)\mathbf{e} = (I - R)^{-1}\mathbf{e}$ is, as pointed out in the proof of theorem 1.2.1, the expected number of transitions before the first return to level $\mathbf{0}$, given that the chain starts in the state $(0, j)$. This initial state is chosen with probability β_j.

Other relations between the number of services during a busy period of $GI/PH/1$ and the taboo probabilities defined in section 1.2 were investigated by V. Ramaswami [263]. The resulting formulas are ideally suited for numerical computation.

The Stationary Queue Length at Time t

For queueing models that have an embedded Markov renewal process, there is a classical argument based on the key renewal theorem, which relates the stationary queue length density at an arbitrary time t to that at the epochs of transition of the embedded Markov renewal process. For the sake of completeness, we shall present the argument in detail for the present model. In the sequel, we shall state only the resulting expressions. The key renewal theorem leads to formally different expressions, depending on whether the underlying renewal process is lattice or nonlattice. In order to focus on the most useful of these two cases, we shall now impose the purely technical restriction that the interarrival time distribution $F(\cdot)$ is *nonlattice*. The same property is then inherited by the Markov renewal process $\hat{P}(\cdot)$.

A few preliminary properties of the Markov renewal process $\hat{P}(\cdot)$ are needed for use in the sequel. It is clear that all the row sum distributions of $\hat{P}(\cdot)$ are identical and equal to the interarrival time distribution $F(\cdot)$. All row sum means are therefore equal to λ_1'. The Markov renewal process $\hat{P}(\cdot)$ is *positive recurrent* if and only if $\lambda_1' > \mu_1'$. The inner product E^* of the invariant probability vector $\boldsymbol{x}$ of $\tilde{P}$ and the column vector $\lambda_1'\boldsymbol{e}$ of row sum means is an important quantity, called the *fundamental mean* of the Markov renewal process. For the present case, we clearly have

$$E^* = \lambda_1'\boldsymbol{x}\boldsymbol{e} = \lambda_1'. \tag{4.1.8}$$

A classical result on positive recurrent Markov renewal processes asserts that the *mean recurrence time* $\tau(k, j)$ of the state (k, j) of $\hat{P}(\cdot)$ is given by

$$\tau(k,j) = \frac{E^*}{x_{kj}} = \frac{\lambda_1'}{x_{kj}}, \quad \text{for} \quad k \geq 0, \qquad 1 \leq j \leq \nu. \tag{4.1.9}$$

The relevant theorem may be found, for example, in E. Çinlar [56, theorem 6.12, p. 155] or J. J. Hunter [147, theorem 2.11, p. 196].

We next consider the *matrix renewal function* $M(\cdot)$ of the Markov renewal process $\hat{P}(\cdot)$. Given the initial conditions $I_0 = k_0, J_0 = j_0$, the element $M_{k_0,j_0;k,j}(t)$ of the infinite matrix $M(t)$ is the conditional expected number of visits to the state (k, j) in $[0, t]$. The matrix $M(t)$ is then given by the convolution-power series

$$M(t) = \sum_{n=0}^{\infty} \hat{P}^{(n)}(t), \quad \text{for} \quad t \geq 0. \tag{4.1.10}$$

Let $I(t)$ denote the queue length at time t and, provided that $I(t) > 0$, let $J(t)$ be the phase of the service in course at time t. When $I(t) = 0$, the corresponding $J(t)$ is not defined. We are now interested in the continuous-parameter process $\{[I(t), J(t)], t \geq 0\}$. The time-dependent joint distribution of the queue length and the service phase is given by the conditional probabilities

$$\begin{aligned} Y(0; t) &= P\{I(t) = 0 \mid I_0 = k_0, J_0 = j_0\}, \\ Y(k, j; t) &= P\{I(t) = k, J(t) = j \mid I_0 = k_0, J_0 = j_0\}, \end{aligned} \tag{4.1.11}$$

for $k \geq 1$, $1 \leq j \leq \nu$, $t \geq 0$. These probabilities are simply related to the entries of the matrix $M(\cdot)$.

Lemma 4.1.1. The conditional probabilities, defined in (4.1.11), are given by

$$Y(0; t) = \sum_{k_1=0}^{\infty} \sum_{j_1=1}^{\nu} \int_0^t dM_{k_0 j_0; k_1 j_1}(u)$$

$$\cdot \left(\sum_{r=k_1+1}^{\infty} P(r, t-u)e \right)_{j_1} [1 - F(t-u)], \tag{4.1.12}$$

for $t \geq 0$, and

$$Y(k, j; t) = \sum_{k_1=k-1}^{\infty} \sum_{j_1=1}^{\nu} \int_0^t dM_{k_0 j_0; k_1 j_1}(u)$$

$$\cdot P_{j_1 j}(k_1 - k + 1, t - u)[1 - F(t-u)], \tag{4.1.13}$$

for $k \geq 1, 1 \leq j \leq \nu, t \geq 0$.

Proof. As the argument needed to prove (4.1.12) and (4.1.13) is classical, an informal discussion will suffice. It is readily seen from (4.1.10) that $dM_{k_0 j_0; k_1 j_1}(u)$ is the elementary conditional probability that in the interval $(u, u + du)$, the Markov renewal process $\hat{P}(\cdot)$ enters the state (k_1, j_1). In the $GI/PH/1$ queue, this corresponds to an arrival after which the queue length is $k_1 + 1$ and the service phase is j_1.

In order to prove (4.1.13), we note that the term $P_{j_1 j}(k_1 - k + 1, t - u)[1 - F(t - u)]$ is the conditional probability that during the interval (u, t), exactly $k_1 - k + 1$ departures occur, that the service phase changes from j_1 to j, and that no arrival occurs during that interval, given that $k_1 + 1$ customers are present at time u with the service in the phase j_1.

The corresponding term in (4.1.12) is similarly interpreted, except that during (u, t) all $k_1 + 1$ customers present at time u depart, so as to leave the queue empty at time t. The proof is completed by applying the law of total probability.

We shall now show that the limits

$$y_0 = \lim_{t \to \infty} Y(0; t),$$

$$y_{kj} = \lim_{t \to \infty} Y(k, j; t), \quad \text{for} \quad k \geq 1, \qquad 1 \leq j \leq \nu, \tag{4.1.14}$$

exist and are simply related to the components of the invariant probability vector x, given by theorem 4.1.1.

Theorem 4.1.2. The limits, defined in (4.1.14), exist and are given by

$$y_0 = \lambda_1'^{-1} \sum_{j_1=1}^{\nu} \sum_{k_1=0}^{\infty} x_{k_1 j_1} \int_0^{\infty} \sum_{r=k_1+1}^{\infty} [P(r, u)e]_{j_1} [1 - F(u)] du,$$

$$y_{kj} = \lambda_1'^{-1} \sum_{j_1=1}^{\nu} \sum_{k_1=k-1}^{\infty} x_{k_1 j_1} \int_0^{\infty} P_{j_1 j}(k_1 - k + 1, u)[1 - F(u)]du, \tag{4.1.15}$$

for $k \geq 1, 1 \leq j \leq \nu$.

Upon partitioning the probability vector y as $[y_0, y_1, y_2, \ldots]$, where the ν-vectors y_k, $k \geq 1$, have the components y_{kj}, $1 \leq j \leq \nu$, and upon defining the matrix $\Psi[R]$ by

$$\Psi[R] = \lambda_1'^{-1} \sum_{r=0}^{\infty} R^r \int_0^{\infty} P(r, u)[1 - F(u)]du, \tag{4.1.16}$$

we have

$$y_0 = 1 - C\beta(I - R)^{-1}\Psi[R]e,$$
$$y_k = C\beta R^{k-1}\Psi[R], \quad \text{for} \quad k \geq 1. \tag{4.1.17}$$

The scalar C is as given by formula (4.1.6).

Proof. By the key renewal theorem, the term

$$\int_0^t dM_{k_0, j_0; k_1, j_1}(u) P_{j_1 j}(k_1 - k + 1, t - u)[1 - F(t - u)],$$

tends to the limit

$$[\tau(k_1, j_1)]^{-1} \int_0^{\infty} P_{j_1 j}(k_1 - k + 1, u)[1 - F(u)]du.$$

The interchange of limits, needed to obtain (4.1.15), is justified by the Lebesgue-dominated convergence theorem. Upon writing the formulas (4.1.15) in matrix form and taking into account the results given in (4.1.15) and (4.1.5), we obtain the expressions stated in (4.1.17).

We shall now obtain a computationally useful relationship between the vector y and the vector x, by which the computation of the matrix $\Psi[R]$ may be avoided. To this effect, a number of preliminary results are needed.

Lemma 4.1.2. The matrices

$$\psi(k) = \lambda_1'^{-1} \int_0^{\infty} P(k, t)[1 - F(t)]dt, \qquad k \geq 0, \tag{4.1.18}$$

are related to the matrices A_k, $k \geq 0$, by

$$A_0 = I + \lambda_1'\psi(0)S,$$
$$A_k = \lambda_1'\psi(k)S + \lambda_1'\psi(k-1)S^oB^o, \quad \text{for} \quad k \geq 1. \tag{4.1.19}$$

The matrices R and $\Psi[R]$ are related by

$$R = I + \lambda_1{}'\Psi[R]S + \lambda_1{}'R\Psi[R]S^oB^o. \tag{4.1.20}$$

Proof. By partial integration and using the differential equations (4.1.1), we obtain

$$\begin{aligned} A_k &= \int_0^\infty P(k, t)dF(t) = -\int_0^\infty P(k, t)d[1 - F(t)] \\ &= \delta_{k0}I + \int_0^\infty P'(k, t)[1 - F(t)]dt \\ &= \delta_{k0}I + \lambda_1{}'\psi(k)S + (1 - \delta_{k0})\lambda_1{}'\psi(k-1)S^oB^o, \quad \text{for} \quad k \geq 0. \end{aligned}$$

Now premultiplying the equation of index k in (4.1.19) by R^k and summing, we obtain formula (4.1.20).

Theorem 4.1.3. The matrix $\Psi[R]$ is explicitly related to the matrix R by

$$\Psi[R] = \lambda_1{}'^{-1}(R - RB^{oo} - I)S^{-1}. \tag{4.1.21}$$

The vector y is given by

$$\begin{aligned} y_0 &= 1 - \lambda_1{}'^{-1}\mu_1{}' = 1 - \rho. \\ y_k &= C\lambda_1{}'^{-1}[\beta R^k - \beta R^{k-1} - (\beta R^k e)\beta]S^{-1}, \quad \text{for} \quad k \geq 1. \end{aligned} \tag{4.1.22}$$

Proof. Postmultiplying by e in (4.1.20), we obtain

$$(I - R)e = \lambda_1{}'(I - R)\Psi[R]S^o,$$

and since $I - R$ is nonsingular, $\lambda_1{}'\Psi[R]S^o = e$. This in turn may be written as $\lambda_1{}'\Psi[R]S^oB^o = B^{oo}$, which upon substitution in (4.1.20) yields

$$\lambda_1{}'\Psi[R]S = R - RB^{oo} - I.$$

Since S is nonsingular, formula (4.1.21) follows.

Substitution of the expression in (4.1.21) into the formulas (4.1.17) now yields

$$\begin{aligned} y_0 &= 1 - C\lambda_1{}'^{-1}\beta(I - R)^{-1}(R - I - RB^{oo})S^{-1}e \\ &= 1 + C\lambda_1{}'^{-1}\beta S^{-1}e - C\mu_1{}'\lambda_1{}'^{-1}\beta(I - R)^{-1}Re \\ &= 1 - \mu_1{}'\lambda_1{}'^{-1} = 1 - \rho, \end{aligned}$$

since $\beta S^{-1}e = -\mu_1{}'$, and C is as given by formula (4.1.6). The expression for y_k, $k \geq 1$, is immediate from (4.1.17) and (4.1.21).

Remark. The proof of theorem 4.1.3 is more direct than the proof of the formulas (4.1.22), which was given in M. F. Neuts [227], where the relation (4.1.21) had not yet been perceived.

Corollary 4.1.1. The marginal stationary queue length density at an arbitrary time is given by

$$y_0 = 1 - \rho,$$

$$y_k e = C\rho\beta R^k e + C\lambda_1{}'^{-1}\beta R^{k-1}(R - I)S^{-1}e, \quad \text{for} \quad k \geq 1. \tag{4.1.23}$$

The mean L_2 of that density is given by

$$\begin{aligned} L_2 &= \rho C\beta(I - R)^{-2}e - \rho - C\lambda_1{}'^{-1}\beta(I - R)^{-1}S^{-1}e \\ &= \rho L_1 - C\lambda_1{}'^{-1}\beta(I - R)^{-1}S^{-1}e, \end{aligned} \tag{4.1.24}$$

where L_1 is as in formula (4.1.7).

Proof. By direct calculation.

Remark. The second term in (4.1.24) may be written as $\lambda_1{}'^{-1}\psi_1{}'$, where $\psi_1{}'$ is the mean of a PH-distribution with representation (κ, S), where

$$\kappa = C\beta(I - R)^{-1}.$$

It is clear from (4.1.6) that κ is a probability vector. We see that $\psi_1{}'$ is the stationary mean residual service time of the customer *in service* at an arrival. Formula (4.1.24) may therefore be rewritten in the interesting form

$$L_2 = \rho\left(L_1 + \frac{\psi_1{}'}{\mu_1{}'}\right). \tag{4.1.25}$$

The Stationary Waiting Time Distributions

In discussing waiting times, we may consider the length of time an arriving customer waits *before entering service* under the first-come, first-served discipline. We may also consider the length of *time in the system,* which consists of the preceding waiting time plus the service time of the customer. Finally, we may consider the *virtual waiting time,* which is the length of time a virtual customer would have to wait if he joined the queue at some arbitrary time t. We denote the stationary distributions of these three waiting times by $W(\cdot)$, $W_1(\cdot)$, and $\hat{W}(\cdot)$ respectively, and distinguish their Laplace-Stieltjes transforms by the addition of asterisks.

Theorem 4.1.4. The Laplace-Stieltjes transforms of the waiting time distributions $W(\cdot)$, $W_1(\cdot)$, and $\hat{W}(\cdot)$ are given by

$$W^*(s) = C\beta \sum_{k=0}^{\infty} R^k[(sI - S)^{-1}S^oB^o]^k e, \tag{4.1.26}$$

$$W_1^*(s) = C\beta \sum_{k=0}^{\infty} R^k[(sI - S)^{-1}S^oB^o]^{k+1} e, \tag{4.1.27}$$

$$\hat{W}^*(s) = 1 - \rho + \lambda_1'^{-1}C\beta \sum_{k=1}^{\infty} R^{k-1}(R - RB^{oo} - I) \cdot S^{-1}[(sI - S)^{-1}S^oB^o]^k e, \tag{4.1.28}$$

for Re $s \geq 0$.

Proof. The same argument is applied in all three cases. If upon arrival a customer finds the queue in the state (k, j), $k \geq 1$, $1 \leq j \leq \nu$, then the conditional probability that his waiting time does not exceed x has the Laplace-Stieltjes transform, given by the j-th component of the vector

$$(sI - S)^{-1}S^oh^{k-1}(s) = (sI - S)^{-1}S^o[\beta(sI - S)^{-1}S^o]^{k-1} = [(sI - S)^{-1}S^oB^o]^k e,$$

where $h(s)$ is the Laplace-Stieltjes transform of the service time distribution. The three waiting time distributions differ only in the joint densities of the queue length and service phase at the arrival epoch considered. The stated formulas follow by applying the law of total probability.

Corollary 4.1.2. The mean waiting time $\overline{W}$ at arrivals is given by

$$\overline{W} = \lambda_1'L_2 - \mu_1'. \tag{4.1.29}$$

The mean time in the system is given by $\overline{W}_1 = \overline{W} + \mu_1' = \lambda_1'L_2$, and the mean virtual waiting time is given by

$$\overline{\hat{W}} = \rho\overline{W} + \frac{1}{2}\lambda_1'^{-1}\mu_2', \tag{4.1.30}$$

where μ_2' is the second moment of the service time.

Proof. We write $W^*(s)$ as

$$W^*(s) = C + C\beta \sum_{k=1}^{\infty} R^k(sI - S)^{-1}S^oh^{k-1}(s),$$

where $h(s) = \beta(sI - S)^{-1}S^o$. Upon differentiation and setting $s = 0$, we obtain

$$\overline{W} = -W^{*\prime}(0+) = \mu_1{}'C\beta \sum_{k=2}^{\infty} (k-1)R^k(-S^{-1})S^o$$

$$+ C\beta \sum_{k=1}^{\infty} R^k S^{-2}S^o$$

$$= \mu_1{}'C\beta(I-R)^{-2}R^2e - C\beta(I-R)^{-1}RS^{-1}e$$

$$= \mu_1{}'[C\beta(I-R)^{-2}e + C - 2] - [C\beta(I-R)^{-1}S^{-1}e + C\mu_1{}']$$

$$= \mu_1{}'[C\beta(I-R)^{-2}e - 2] - C\beta(I-R)^{-1}S^{-1}e$$

$$= \lambda_1{}'L_2 - \mu_1{}'.$$

The expression for $\overline{W}_1$ is obvious. For $\hat{\overline{W}}$, we have

$$\hat{\overline{W}} = -\hat{W}^{*\prime}(0+)$$

$$= -\rho C\beta \sum_{k=2}^{\infty} (k-1)R^{k-1}(R - I - RB^{oo})S^{-2}S^o$$

$$+ \lambda_1{}'^{-1}C\beta \sum_{k=1}^{\infty} R^{k-1}(R - I - RB^{oo})S^{-3}S^o$$

$$= -\rho C\beta(I-R)^{-1}RS^{-1}e + \rho\mu_1{}'C\beta(I-R)^{-2}R^2e$$

$$+ \lambda_1{}'^{-1}C\beta S^{-2}e + \lambda_1{}'^{-1}C\beta(I-R)^{-1}Re \cdot \beta S^{-2}e$$

$$= \rho\mu_1{}'C\beta(I-R)^{-2}e - 2\rho\mu_1{}' - \rho C\beta(I-R)^{-1}S^{-1}e + \frac{1}{2}\lambda_1{}'^{-1}\mu_2{}'$$

$$= \rho\overline{W} + \frac{1}{2}\lambda_1{}'^{-1}\mu_2{}'.$$

The last formula may also be written as

$$\hat{\overline{W}} = \rho\left(\overline{W} + \frac{\mu_2{}'}{2\mu_1{}'}\right), \tag{4.1.31}$$

and we recall that $\frac{1}{2}\mu_1{}'^{-1}\mu_2{}'$ is the mean of the probability distribution $K^*(x) = \int_0^x [1 - K(u)]\mu_1{}'^{-1}du$, the random modification of the service time distribution.

Remarks. It is interesting to note that all three mean waiting times depend on the matrix R only through L_2, the mean stationary queue length at an arbitrary time. The dependence on the matrix R is highly implicit, but for the *GI/PH/*1 the results of corollary 4.1.2 express the mean waiting times in a readily computable form. Similarly, much more involved expressions may be derived for the second and higher moments. This derivation is an exercise in massive matrix manipulations. It would be an interesting project to program a computer to perform such manipulations in symbolic form. There is clearly a point where the quotation by Leibniz used as the epigraph of this book also applies to analytic calculations.

The mean stationary virtual waiting time for the $PH/G/1$ queue may be computed by particularizing the formulas given in V. Ramaswami [261]. The complexity of the expressions for the mean waiting times in both the $PH/G/1$ and the $GI/PH/1$ queue, and in particular their dependence on the minimal solutions of nonlinear matrix equations, makes it highly unlikely that reasonably tractable formulas for these quantities in the general $GI/G/1$ queue will be obtainable. Some frustration over this state of affairs is voiced in L. Kleinrock [172]. It would be a worthwhile project to study the quality of known bounds and approximations to the mean waiting time in the $GI/G/1$ queue by exact calculations for various $GI/PH/1$ and $PH/G/1$ queues. Such a study should not be limited to the Erlang case. The "regularity" of Erlang arrivals or services may indeed place such approximations in too favorable a light.

Computation of the Waiting Time Distributions

The method, discussed in section 3.9, leads to very efficient algorithmic procedures to compute each of the probability distributions $W(\cdot)$, $W_1(\cdot)$, and $\hat{W}(\cdot)$. In each case the numerical computation will be reduced to the solution of a highly structured system of ν^2 linear differential equations with constant coefficients.

Each waiting time distribution may be viewed as the distribution of the time until absorption in the absorbing QBD process with generator $\tilde{Q}^o$, given by

$$\tilde{Q}^o = \begin{vmatrix} 0 & 0 & 0 & 0 & 0 & \cdots \\ S^oB^o & S & 0 & 0 & 0 & \cdots \\ 0 & S^oB^o & S & 0 & 0 & \cdots \\ 0 & 0 & S^oB^o & S & 0 & \cdots \\ 0 & 0 & 0 & S^oB^o & S & \cdots \\ \vdots & \vdots & \vdots & \vdots & \vdots & \end{vmatrix} \qquad (4.1.32)$$

with an appropriate initial probability vector. Since in all three cases the initial probability vector is of matrix-geometric form, the procedure of section 3.9 may be applied. Only the essential steps, adapted to the particular cases under discussion, will be presented.

For $W(\cdot)$ and $W_1(\cdot)$, we consider the matrix

$$\Psi^*(s) = \sum_{k=0}^{\infty} R^k[(sI - S)^{-1}S^oB^o]^k. \tag{4.1.33}$$

which, as in section 3.9, may be rewritten as

$$\phi^*(s) = \nu + \nu[sI \otimes I - I \otimes S - R^T \otimes S^oB^o]^{-1}(R^T \otimes S^oB^o), \tag{4.1.34}$$

where $\phi^*(s) = \tau[\Psi^*(s)]$, and $\nu = \tau(I)$, the ν^2-vectors obtained by forming the direct sums of the rows of the given matrices. We let $\Psi(\cdot)$ be the matrix of mass-functions with Laplace-Stieltjes transform $\Psi^*(s)$ and we set $\phi(x) = \tau[\Psi(x)]$, for $x \geq 0$. Upon inversion, equation (4.1.34) leads to

$$\begin{aligned}\phi(x) &= \nu + \nu \int_0^x \exp[(I \otimes S + R^T \otimes S^oB^o)u]du(R^T \otimes S^oB^o)\\ &= \nu + \nu(I \otimes S + R^T \otimes S^oB^o)^{-1}\\ &\quad \cdot \{\exp[(I \otimes S + R^T \otimes S^oB^o)x - I \otimes I\}(R^T \otimes S^oB^o),\end{aligned} \tag{4.1.35}$$

for $x \geq 0$.

Let V^o be the square matrix of order ν, such that $\tau(V^o) = \nu^o = -\nu(I \otimes S + R^T \otimes S^oB^o)^{-1}$, then V^o is the unique solution to the equation

$$V^oS + RV^oS^oB^o = -I. \tag{4.1.36}$$

In the present case, the matrix V^o may be explicitly expressed in terms of familiar matrices. Postmultiplication by $\boldsymbol{e}$ in (4.1.36) leads to $(I - R)V^oS^o = \boldsymbol{e}$, so that the matrix $V^oS^oB^o$ may be replaced by the known matrix $(I - R)^{-1}B^{oo}$. Doing so, we obtain the useful formula

$$V^o = (I - R)^{-1}(R - I - RB^{oo})S^{-1}. \tag{4.1.37}$$

Let now the ν^2-vector $\boldsymbol{\theta}(\cdot)$ be defined by

$$\boldsymbol{\theta}(x) = \nu^o \exp[(I \otimes S + R^T \otimes S^oB^o)x], \quad \text{for} \quad x \geq 0, \tag{4.1.38}$$

and let $\Theta(x)$ be the $\nu \times \nu$ matrix such that $\tau[\Theta(x)] = \boldsymbol{\theta}(x)$; it then follows that $\Theta(\cdot)$ satisfies the matrix differential equation

$$\Theta'(x) = \Theta(x)S + R\Theta(x)S^oB^o, \quad \text{for} \quad x \geq 0, \tag{4.1.39}$$

with initial condition $\Theta(0) = V^o$.

The matrix $\Psi(x)$ is then, by virtue of (4.1.35), given by

$$\Psi(x) = I + RV^oS^oB^o - R\Theta(x)S^oB^o$$
$$= -V^oS - R\Theta(x)S^oB^o, \quad \text{for} \quad x \geq 0. \tag{4.1.40}$$

The probability distribution $W(\cdot)$ is given by

$$W(x) = C\beta V^oS^o - C\beta R\Theta(x)S^o$$
$$= 1 - C\beta R\Theta(x)S^o, \quad \text{for} \quad x \geq 0. \tag{4.1.41}$$

As it should be, $W(0+)$ is equal to C, the stationary probability that an arriving customer finds the queue empty.

In order to compute $W_1(\cdot)$ we consider the matrix $\Psi_1^*(s)$, given by

$$\Psi_1^*(s) = \Psi^*(s)(sI - S)^{-1}. \tag{4.1.42}$$

The corresponding matrix $\Psi_1(\cdot)$ satisfies the matrix differential equation

$$\Psi_1'(x) = \Psi_1(x)S + \Psi(x), \quad \text{for} \quad x \geq 0, \tag{4.1.43}$$

with $\Psi_1(0) = 0$.

The probability distribution $W_1(\cdot)$ is then given by

$$W_1(x) = C\beta\Psi_1(x)S^o, \quad \text{for} \quad x \geq 0. \tag{4.1.44}$$

By efficient programming, the differential equation (4.1.43) may be solved as one computes the matrix $\Psi(x)$ from the formulas (4.1.39) and (4.1.40). The distribution $W_1(\cdot)$ of the time-in-system is then obtained with only a modicum of additional computation time.

The computation of the distribution $\hat{W}(\cdot)$ of the virtual waiting time proceeds along similar lines. We write the equation (4.1.28) as

$$\hat{W}^*(s) = 1 - \rho + \lambda_1'^{-1}C\beta\hat{\Psi}^*(s)(sI - S)^{-1}S^o, \tag{4.1.45}$$

where the matrix $\hat{\Psi}^*(s)$ satisfies

$$\hat{\Psi}^*(s) = (R - I - RB^{oo})S^{-1} + R\hat{\Psi}^*(s)(sI - S)^{-1}S^oB^o. \tag{4.1.46}$$

The first term on the right-hand side now plays the role of the matrix V, and the corresponding matrix $\hat{V}^o$ is the solution to the equation

$$\hat{V}^oS + R\hat{V}^oS^oB^o = -(R - I - RB^{oo})S^{-1}. \tag{4.1.47}$$

By the same derivation as before, we obtain $\hat{V}^o$ explicitly as

$$\hat{V}^o = -(R - I - RB^{oo})S^{-2}$$
$$- R(I - R)^{-1}(R - I - RB^{oo})S^{-1}B^{oo}S^{-1}$$
$$= -V^oS^{-1} + RV^o(I - B^{oo})S^{-1}, \tag{4.1.48}$$

where V^o is given by formula (4.1.37).

The matrix $\hat{\Theta}(\cdot)$ is correspondingly obtained by solving the matrix differential equation

$$\hat{\Theta}'(x) = \hat{\Theta}(x)S + R\hat{\Theta}(x)S^oB^o, \quad \text{for} \quad x \geq 0, \tag{4.1.49}$$

with

$$\hat{\Theta}(0) = \hat{V}^o. \tag{4.1.50}$$

The matrix $\hat{\Psi}(\cdot)$ of mass-functions with the transform matrix $\hat{\Psi}^*(s)$ is now expressed in terms of $\hat{\Theta}(\cdot)$ by

$$\begin{aligned}\hat{\Psi}(x) &= (R - I - RB^{oo})S^{-1} + R\hat{V}^oS^oB^o - R\hat{\Theta}(x)S^oB^o \\ &= -\hat{V}^oS - R\hat{\Theta}(x)S^oB^o, \quad \text{for} \quad x \geq 0.\end{aligned} \tag{4.1.51}$$

As we evaluate the matrix $\hat{\Psi}(x)$, we simultaneously also solve the differential equation

$$\hat{\Psi}_1'(x) = \hat{\Psi}_1(x)S + \hat{\Psi}(x), \quad \text{for} \quad x \geq 0, \tag{4.1.52}$$

with $\hat{\Psi}_1(0) = 0$. The distribution $\hat{W}(\cdot)$ is finally obtained from the formula

$$\hat{W}(x) = 1 - \rho + \lambda_1'^{-1}C\beta\hat{\Psi}_1(x)S^o, \quad \text{for} \quad x \geq 0. \tag{4.1.53}$$

Remarks

a. It is readily seen that the density portions on $(0, \infty)$ of each of the distributions $W(\cdot)$, $W_1(\cdot)$, and $\hat{W}(\cdot)$ may be obtained with little extra effort from the auxiliary quantities evaluated in the course of the preceding computations.

b. The literature on the stationary waiting time distributions of the $GI/G/1$ queue is extensive. In particular, the distribution $W(\cdot)$ has been investigated by a variety of analytic methods, mostly related to Lindley's equation [190]. The case where the service time distribution has a rational Laplace-Stieltjes transform leads to detailed analytic results—see, for example, J. W. Cohen [61]—but, because of the issues discussed in section 1.6, the computational implementation of these results is not always easy or reliable. For the $GI/PH/1$ queue, the service time distribution clearly has a rational Laplace-Stieltjes transform, but the relationship of the PH-distributions to Markov processes and to systems of linear differential equations with real, constant coefficients leads here to more explicit results. Formula (4.1.41) gives a fully explicit solution to Lindley's equation for the $GI/PH/1$ queue. It appears to be difficult to verify by direct substitution that the stated form of $W(\cdot)$ indeed satisfies that equation. This is not astonishing. Lindley's equation is a prime result of the combinatorial methods by which most known theorems on the $GI/G/1$ queue have been proved.

The results presented here utilize the recursive approaches common in the study of Markov processes. We conjecture that a technically arduous extension of the matrix-geometric solutions for Markov chains could be carried through to operator-geometric solutions for structured Markov processes on an uncountable state space. If this is the case, the *GI/G/1* queue and some of its variants would gain a new and enlightening methodological approach.

Asymptotic Behavior of the Waiting Time Distributions

The matrix R is positive and the generator $S + S^oB^o$ is irreducible. The results on the exponential tail behavior proved in section 3.9 are therefore applicable. We define the quantity $-\xi < 0$ to be the eigenvalue of maximum real part of the matrix $S + \eta S^oB^o$, where $\eta = \text{sp}(R) \in (0, 1)$.

The discussion in section 3.9 immediately yields that

$$\Psi(\infty) - \Psi(x) = Ke^{-\xi x} + o(e^{-\xi x}),$$
$$\hat{\Psi}(\infty) - \hat{\Psi}(x) = \hat{K}e^{-\xi x} + o(e^{-\xi x}), \tag{4.1.54}$$

as $x \to \infty$. The positive matrices K and $\hat{K}$ are defined below.

The matrices $\Psi_1(x)$ and $\hat{\Psi}_1(x)$ are related to the matrices $\Psi(x)$ and $\hat{\Psi}(x)$ by

$$\Psi_1(x) = \int_0^x \Psi(u)\exp[S(x-u)]du,$$
$$\hat{\Psi}_1(x) = \int_0^x \hat{\Psi}(u)\exp[S(x-u)]du, \quad \text{for} \quad x \geq 0, \tag{4.1.55}$$

so that

$$\Psi_1(\infty) - \Psi_1(x) = -K(\xi I + S)^{-1}e^{-\xi x} + o(e^{-\xi x}),$$
$$\hat{\Psi}_1(\infty) - \hat{\Psi}_1(x) = -\hat{K}(\xi I + S)^{-1}e^{-\xi x} + o(e^{-\xi x}), \tag{4.1.56}$$

as $x \to \infty$.

We note that the matrix $(\xi I + S)^{-1}$ is nonpositive and clearly does not have any zero columns. The coefficient matrices of $e^{-\xi x}$ in (4.1.56) are therefore positive. The following theorem now readily follows from theorem 3.9.3.

Theorem 4.1.5. The probability distributions $W(\cdot)$, $W_1(\cdot)$, and $\hat{W}(\cdot)$ of the waiting times in the stable *GI/PH/1* queue are asymptotically exponential with the same parameter ξ; that is,

$$1 - W(x) = ke^{-\xi x} + o(e^{-\xi x}),$$
$$1 - W_1(x) = k_1 e^{-\xi x} + o(e^{-\xi x}), \tag{4.1.57}$$
$$1 - \hat{W}(x) = \hat{k} e^{-\xi x} + o(e^{-\xi x}), \quad \text{as} \quad x \to \infty.$$

The constants are given by

$$k = C\eta(1-\eta)^{-1}(\beta u^o),$$
$$k_1 = -\beta(\xi I + S)^{-1} S^o \cdot k = \eta^{-1} k, \tag{4.1.58}$$
$$\hat{k} = -\lambda_1'^{-1} \xi^{-1}(1-\eta)\beta(\xi I + S)^{-1} S^o \cdot k$$
$$= \lambda_1'^{-1} \xi^{-1} C(\beta u^o),$$

where the vector u^o is a right eigenvector of R corresponding to η. A left eigenvector of R, corresponding to η, is u, and the pair is normalized by $ue = uu^o = 1$.

Proof. In view of theorem 3.9.3, it only remains to verify the stated expressions for the constants k, k_1, and $\hat{k}$. By applying formula (3.9.30), it follows that the matrices K and $\hat{K}$ are given by

$$K = \eta\xi^{-1}(uz^o)(u^o \cdot zS^oB^o) = \eta\xi^{-1}(uz^o)(zS^o)(u^o \cdot \beta),$$
$$\hat{K} = \eta\xi^{-1}[u \cdot (R - RB^{oo} - I)S^{-1}z^o](u^o \cdot zS^oB^o). \tag{4.1.59}$$

The inner product in the square brackets may be simplified as follows.

$$u(R - RB^{oo} - I)S^{-1}z^o = (\eta - 1)(uS^{-1}z^o) - \eta(\beta S^{-1}z^o),$$

since $uR = \eta u$ and $ue = 1$. From $(S + \eta S^oB^o)z^o = -\xi z^o$, we obtain

$$uz^o - \eta(\beta z^o) = -\xi uS^{-1}z^o,$$

and

$$\xi^{-1}(1-\eta)(\beta z^o) = -\beta S^{-1}z^o,$$

so that

$$u(R - RB^{oo} - I)S^{-1}z^o = (1-\eta)\xi^{-1}(uz^o).$$

The left eigenvector z of $S + \eta S^oB^o$, corresponding to $-\xi$ and normalized by $ze = zz^o = 1$, *is equal to the vector* u. To see this, we recall that

$$A^*(z) = \sum_{k=0}^{\infty} A_k z^k = \int_0^{\infty} \exp[(S + zS^oB^o)t]\,dF(t),$$

for $0 \leq z \leq 1$. Moreover, by equation (1.3.4), we have $uA^*(\eta) = \eta u$, but it also follows from the specific form of $A^*(\eta)$ and the definitions of

$-\xi$ and z that $zA^*(\eta) = z\int_0^\infty e^{-\xi t}dF(t) = f^*(\xi)z$. From the uniqueness of the maximal eigenvalue and $ue = ze = 1$, it follows that $u = z$, and $\eta = f^*(\xi)$.

Moreover $z(S + \eta S^o B^o) = -\xi z$ leads upon postmultiplication by e to $zS^o = \xi(1-\eta)^{-1}$. Finally, $uz^o = 1$, and

$$\beta(u^o \cdot \beta)(\xi I + S)^{-1}S^o B^o e = \beta(\xi I + S)^{-1}S^o \cdot (\beta u^o) = -\eta^{-1}(\beta u^o),$$

since $\beta(\xi I + S)^{-1}S^o = -\eta^{-1}$. The stated formulas now follow from (4.1.54) and (4.1.56) upon substitution into (4.1.41), (4.1.44), and (4.1.53), and by routine calculations using the preceding particular expressions.

4.2. GENERALIZATIONS AND VARIANTS OF THE *GI/PH/1* QUEUE

An interesting feature of queueing models is that relatively minor variations in the physical descriptions of the models may lead to entirely new and often difficult mathematical problems. In the extensive literature on queues a large number of variants of the classical queues have been discussed, mostly by ad hoc methods. These have frequently tended to obscure the unifying structural properties common to many such variants. The mathematical treatment of broad classes of related models becomes evident once their common structure is exhibited. This does not mean that all is then said. The more involved physical description of a model usually results in increased dimension of, for example, the blocks in the partitioned embedded Markov chain. If one's objective is to propose solutions that are computable for realistic parameter values, substantial problems of dimensionality remain to be resolved.

In this section, we discuss a number of variants and generalizations of the *GI/PH/1* queue, which are problems of practical significance. These models all have embedded Markov chains that, in the stable case, have a (modified) matrix-geometric invariant vector. The main theoretical results therefore follow quite readily from the theorems established earlier in this book. This will enable us to concentrate more specifically on the algorithmic issues raised by each model.

4.2a. The *SM/PH/1* Queue

In this model, the arrivals occur at the epochs of transition of a K-state Markov renewal process with transition probability matrix $F(\cdot)$. We assume that the stochastic matrix $F = F(\infty)$ is irreducible. Semi-

Markovian arrival processes arise very naturally in a variety of concrete situations. They deserve greater attention in the applied study of queues. In section 4.2b, we discuss a number of such applications.

The $SM/PH/1$ queue has an embedded Markov chain $\{(I_n, J_n, L_n), n \geq 0\}$, where $I_n \geq 0$ is the queue length prior to the n-th arrival. The random variable $J_n \in (1, \ldots, \nu)$ denotes the service phase immediately after the n-th arrival, and $L_n \in (1, \ldots, K)$ indicates the state of the Markov renewal process $F(\cdot)$ of arrivals immediately after the n-th arrival. The state space E consists of the triples (i, j, h), $i \geq 0$, $1 \leq j \leq \nu$, $1 \leq h \leq K$, written in lexicographic order. The service time distribution is as defined in section 4.1, and the same convention regarding ephemeral states is also made here.

The transition probability matrix $\tilde{P}$ is of the canonical form (1.1.5), with the matrices A_k and B_k, $k \geq 0$, given by

$$A_k = \int_0^\infty P(k, u) \otimes dF(u),$$

$$B_k = \sum_{r=k+1}^{\infty} \int_0^\infty P(r, u)B^{oo} \otimes dF(u) \tag{4.2.1}$$

$$= \sum_{r=k+1}^{\infty} A_r(B^{oo} \otimes I), \quad \text{for} \quad k \geq 0.$$

The formal analysis of the $SM/PH/1$ model proceeds along entirely similar lines to that of the $GI/PH/1$ queue. The matrix A is given by

$$A = \int_0^\infty \exp[(S + S^oB^o)u] \otimes dF(u), \tag{4.2.2}$$

and its invariant probability vector is given by $\pi \otimes \phi$, where π is the stationary probability vector of $S + S^oB^o$ and ϕ is the invariant probability vector of the irreducible stochastic matrix F. The equilibrium condition is routinely obtained from

$$(\pi \otimes \phi) \int_0^\infty \sum_{k=1}^{\infty} kP(k, u) \otimes dF(u) \cdot (e \otimes e)$$

$$= \mu_1'^{-1}\phi \int_0^\infty u dF(u)e = \mu_1'^{-1}(\phi\phi^*) = \mu_1'^{-1}\lambda_1' > 1. \tag{4.2.3}$$

The vector ϕ^* is the column vector of row sum means of the semi-Markov matrix $F(\cdot)$. The quantity $\lambda_1' = \phi\phi^*$ is called the *fundamental mean* of the Markov renewal process of arrivals.

The matrix R is defined and computed in the usual manner. It is now a strictly positive matrix of order νK. A calculation similar to that in the proof of theorem 4.1.1 leads to

$$B[R] = (I - R)^{-1}(A - R)(B^{oo} \otimes I). \tag{4.2.4}$$

The vector x_0 is no longer as explicit as in the case of the $GI/PH/1$ queue. It readily follows from (4.2.4), however, that the matrix $B[R]$ is of the form

$$\begin{vmatrix} \beta_1 D_1 & \beta_2 D_1 & \cdots & \beta_\nu D_1 \\ \beta_1 D_2 & \beta_2 D_2 & \cdots & \beta_\nu D_2 \\ \cdot & \cdot & & \cdot \\ \cdot & \cdot & & \cdot \\ \cdot & \cdot & & \cdot \\ \beta_1 D_\nu & \beta_2 D_\nu & \cdots & \beta_\nu D_\nu \end{vmatrix}, \tag{4.2.5}$$

where the $K \times K$ matrices $D_1, \ldots, D_\nu$ are positive and stochastic. This implies that x_0 is of the form

$$x_0 = \beta \otimes u, \tag{4.2.6}$$

where the vector u satisfies $u = u \sum_{j=1}^{\nu} \beta_j D_j$ and is uniquely determined by the normalizing condition $x_0(I - R)^{-1}e = 1$.

In order to identify and compute the matrices D_j, $1 \leq j \leq \nu$, we partition the matrix $(I - R)^{-1}(A - R)$ into square blocks of order K. It is then clear from formula (4.2.4) that D_j is the sum of those blocks in the j-th row of the partitioned matrix. Let D be the $K\nu \times K$ matrix

$$\begin{vmatrix} D_1 \\ \cdot \\ \cdot \\ \cdot \\ D_\nu \end{vmatrix},$$

we may then write

$$(I - R)^{-1}(A - R)(e \otimes I) = D, \tag{4.2.7}$$

or equivalently, since $A(e \otimes I) = e \otimes F$,

$$D = RD + e \otimes F - R(e \otimes I). \tag{4.2.8}$$

Let R be partitioned into the blocks $R(h, h')$, $1 \leq h, h' \leq \nu$, of order K, then (4.2.8) may be written as

$$D_j = \sum_{h=1}^{\nu} R(j, h)D_h + F - \sum_{h=1}^{\nu} R(j, h), \quad \text{for} \quad 1 \leq j \leq \nu. \tag{4.2.9}$$

This system may be recursively solved by a block Gauss-Seidel procedure, that is, by the recursive scheme

$$D_j^{(n+1)} = [I - R(j, j)]^{-1}\left\{\sum_{h<j} R(j, h)D_h^{(n+1)} + \sum_{h>j} R(j, h)D_h^{(n)} + F - \sum_{h=1}^{\nu} R(j, h)\right\}, \qquad (4.2.10)$$

for $n \geq 0$, $1 \leq j \leq \nu$, with $D_j^{(o)} = F - \sum_{h=1}^{\nu} R(j, h)$, for $1 \leq j \leq \nu$. A convenient stopping criterion is obtained from the fact that the matrices D_j are stochastic.

The computation of the vector $\boldsymbol{u}$ is then straightforward. The vector $\boldsymbol{\nu} = (I - R)^{-1}\boldsymbol{e}$ is also best computed by Gauss-Seidel iteration applied to the system of equations $\boldsymbol{\nu} = R\boldsymbol{\nu} + \boldsymbol{e}$.

Once the matrix R and the vector $\boldsymbol{u}$ have been computed, we may study other features of the stationary queue, such as the queue length density at an arbitrary time and the waiting time distributions. The required derivations are similar to those for the $GI/PH/1$ queue. The resulting expressions, along with a specific application and related numerical examples, may be found in M. F. Neuts and S. Chakravarthy [237].

When the service time distribution is exponential, we obtain the *SM/M/1 model*, studied by E. Çinlar [53] and M. F. Neuts [227]. For this model, very detailed explicit results may be obtained. With exponential service times, these results may easily be extended to the multi-server $SM/M/c$ queue.

Denoting the service rate by $\mu = \mu_1'^{-1}$, it follows from (4.2.3) that the $SM/M/1$ queue is *stable* if and only if

$$\mu(\boldsymbol{\phi\phi}^*) = \mu\lambda_1' > 1. \qquad (4.2.11)$$

The matrix A is now equal to F, and the matrix $B[R]$ of order K is given by

$$B[R] = (I - R)^{-1}(F - R). \qquad (4.2.12)$$

The classical equation for the matrix R may be written as

$$R = \int_0^{\infty} \exp[-\mu u(I - R)]dF(u). \qquad (4.2.13)$$

Theorem 4.2.1. The vector $\boldsymbol{x}_0 = \boldsymbol{\phi}(I - R)$ is strictly positive, and the invariant probability vector $\boldsymbol{x} = [\boldsymbol{x}_0, \boldsymbol{x}_1, \ldots]$ of the embedded Markov chain $\tilde{P}$ is given by

$$\boldsymbol{x}_k = \boldsymbol{\phi}(I - R)R^k, \quad \text{for} \quad k \geq 0. \qquad (4.2.14)$$

Proof. The vector $\boldsymbol{\phi}(I - R)$ is clearly a left invariant vector of the matrix $B[R]$, given in (4.2.12). The stochastic matrix $B[R]$ is positive, since R is positive by virtue of the irreducibility of F. The components of $\phi(I - R)$ are therefore nonzero and all of the same sign. The positive vector x_0 is hence of the form $k'\phi(I - R)$, but $x_0(I - R)^{-1}e = 1$ implies that $k' = 1$. The second statement is now evident.

Let now $W_h(x)$, $1 \leq h \leq K$, $x \geq 0$ be the stationary probability that a customer arrives at the beginning of a sojourn in the state h of the Markov renewal arrival process $F(\cdot)$ and has to wait no longer than x before entering service. The vector $W(x)$ has the components $W_h(x)$, $1 \leq h \leq K$.

Theorem 4.2.2. In the stable $SM/M/1$ queue, the vector $W(\cdot)$ is given by

$$W(x) = \phi - \phi R \exp[-\mu x(I - R)], \quad \text{for} \quad x \geq 0. \tag{4.2.15}$$

It may be computed by solving the system of differential equations

$$W'(x) = -\mu W(x)(I - R) + \mu\phi(I - R), \quad \text{for} \quad x \geq 0, \tag{4.2.16}$$

with initial conditions $W(0) = \phi(I - R)$.

Proof. If the arriving customer finds $k \geq 1$ customers already in the system, his waiting time has an Erlang distribution with parameters k and μ. It therefore follows from the law of total probability that

$$\begin{aligned} W(x) &= \phi(I - R) + \phi(I - R) \sum_{k=1}^{\infty} R^k \int_0^x e^{-\mu u} \frac{(\mu u)^{k-1}}{(k-1)!} \mu du \\ &= \phi(I - R) + \mu\phi(I - R)R \int_0^x \exp[-\mu u(I - R)]du \\ &= \phi - \phi R \exp[-\mu x(I - R)], \quad \text{for} \quad x \geq 0. \end{aligned}$$

The differential equations (4.2.16) follow by routine manipulations.

Remarks

a. In the most interesting applications of semi-Markovian arrivals, the lengths of the sojourn times vary considerably with the states h, $1 \leq h \leq K$. This can result in strongly oscillatory behavior of the queue lengths and waiting times. It is then of definite interest to study the conditional waiting time distributions $\phi_h^{-1}W_h(x)$, $1 \leq h \leq K$, of customers of different arrival types h.

b. The average waiting time distribution $W(x)e$, given by

$$W(x)e = 1 - \phi R \exp[-\mu x(I - R)]e, \quad \text{for} \quad x \geq 0, \tag{4.2.17}$$

is very similar to a PH-distribution, since the vector ϕR is nonnegative and satisfies $\phi Re < 1$. The matrix $-\mu(I - R)$ has nonnegative off-diagonal elements and is nonsingular, but the vector $-\mu(I - R)e$ is in general not nonpositive. It is therefore not possible, in general, to construct a finite-state Markov process in which $W(\cdot)e$ is the probability distribution of an absorption time. In [227] we termed such a probability distribution a *generalized PH-distribution.* It may still be computed by solving a simple linear system of differential equations with constant coefficients, but the useful connection with the theory of finite-state Markov processes is lost.

In extending the preceding results to the *SM/M/c queue,* only the more complex boundary behavior raises some difficulties. We shall primarily discuss some useful computational simplifications.

We assume that there are $c \geq 2$ parallel exponential servers, each processing customers at the same rate μ. The arrival process is defined as in the single server case and the same notation is presented throughout.

The Markov chain $\tilde{P}$ embedded at the arrival epochs of the *SM/M/c* queue is given by

$$\tilde{P} = \begin{vmatrix} B_{00} & B_{01} & 0 & \cdots & & 0 & 0 & \cdots \\ B_{10} & B_{11} & B_{12} & \cdots & & 0 & 0 & \cdots \\ \vdots & \vdots & \vdots & & & \vdots & \vdots & \\ B_{c-2,0} & B_{c-2,1} & B_{c-2,2} & \cdots & B_{c-2,c-1} & 0 & 0 & \cdots \\ B_{c-1,0} & B_{c-1,1} & B_{c-1,2} & \cdots & B_{c-1,c-1} & A_0 & 0 & \cdots \\ B_{c,0} & B_{c,1} & B_{c,2} & \cdots & B_{c,c-1} & A_1 & A_0 & \cdots \\ B_{c+1,0} & B_{c+1,1} & B_{c+1,2} & \cdots & B_{c+1,c-1} & A_2 & A_1 & \cdots \\ \vdots & \vdots & \vdots & & \vdots & \vdots & \vdots & \end{vmatrix}. \tag{4.2.18}$$

All elements are $K \times K$ blocks, and the matrices A_k are given by

$$A_k = \int_0^\infty e^{-c\mu u} \frac{(c\mu u)^k}{k!} dF(u), \quad \text{for} \quad k \geq 0. \tag{4.2.19}$$

The matrices B_{kr}, $k \geq 0$, $0 \leq r \leq \min(k + 1, c - 1)$, are more involved. They are given by

$$B_{kr} = \int_0^\infty p_{kr}(u) dF(u), \tag{4.2.20}$$

where

$$p_{kr}(u) = \binom{k+1}{r} e^{-\mu r u}(1 - e^{-\mu u})^{k+1-r}, \qquad \text{for } \begin{array}{l} 0 \le k \le c-1, \\ 0 \le r \le k+1, \end{array}$$

$$= \int_0^u e^{-\mu c\tau} \frac{(\mu c\tau)^{k-c}}{(k-c)!} \binom{c}{r} e^{-\mu r(u-\tau)}[1 - e^{-\mu(u-\tau)}]^{c-r} c\mu d\tau,$$

$$\text{for } k \ge c, \qquad 0 \le r \le c-1. \quad (4.2.21)$$

The expressions for the matrices B_{kr} are obtained by a classical probability argument. Given the length u of the interarrival interval, $p_{kr}(u)$ is the probability that the number of customers in the system decreases from $k + 1$ to r during that interval of length u. If fewer than $c + 1$ customers are present, then all are in service and $p_{kr}(u)$ is given by the binomial probability in the first expression. If $k + 1$ exceeds c, we condition on the first time (τ) at which the number of customers in the system drops to c. This readily leads, by the law of total probability, to the second expression.

The *SM/M/c* queue is *stable* if and only if

$$c\mu(\boldsymbol{\phi\phi}^*) = c\mu\lambda_1' > 1. \quad (4.2.22)$$

The matrix R is then the minimal nonnegative solution to

$$R = \int_0^\infty \exp[-c\mu u(I - R)]dF(u). \quad (4.2.23)$$

By virtue of the discussion in section 1.5, the invariant vector $\mathbf{x} = [\mathbf{x}_0, \mathbf{x}_1, \ldots, \mathbf{x}_{c-1}, \mathbf{x}_c, \ldots]$ of $\tilde{P}$ has a modified matrix-geometric form. It satisfies

$$\mathbf{x}_k = \mathbf{x}_{c-1}R^{k-c+1}, \quad \text{for } k \ge c-1. \quad (4.2.24)$$

The vector $[\mathbf{x}_0, \ldots, \mathbf{x}_{c-1}]$ of dimension cK is the left invariant vector of the stochastic matrix $B[R]$, given by

$$B[R] = \begin{vmatrix} B_{00} & B_{01} & 0 & \cdots & 0 \\ B_{10} & B_{11} & B_{12} & \cdots & 0 \\ \vdots & \vdots & \vdots & & \vdots \\ B_{c-2,0} & B_{c-2,1} & B_{c-2,2} & \cdots & B_{c-2,c-1} \\ B_0[R] & B_1[R] & B_2[R] & \cdots & B_{c-1}[R] \end{vmatrix}, \quad (4.2.25)$$

where

$$B_r[R] = \sum_{k=c-1}^{\infty} R^{k-c+1} B_{kr}, \quad \text{for} \quad 0 \leq r \leq c-1. \tag{4.2.26}$$

The normalizing condition is given by

$$\sum_{k=0}^{c-2} x_k e + x_{c-1}(I-R)^{-1} e = 1. \tag{4.2.27}$$

The following theorem results in major algorithmic simplifications. It eliminates the need for the computation of most of the complicated matrices B_{kr} for $k \geq c$.

Theorem 4.2.3. The matrices $B_r[R]$, $0 \leq r \leq c-1$, satisfy the recurrence relation

$$B_{r+1}[R] = \frac{c}{r+1} B_{c-2,r} + \frac{1}{r+1}[cR - (c-r)I]B_r[R], \tag{4.2.28}$$

for $0 \leq r \leq c-2$. For $r = c-1$, we obtain the accuracy check $B_c[R] = R$.

Proof. We write $B_r[R]$ as $\int_0^\infty V_r(u)dF(u)$, where the matrices $V_r(u)$ are defined for $0 \leq r \leq c$ by

$$V_r(u) = \binom{c}{r} e^{-\mu r u}(1-e^{-\mu u})^{c-r} I + \sum_{k=c}^{\infty} R^{k-c+1} \int_0^u e^{-\mu c \tau} \frac{(\mu c \tau)^{k-c}}{(k-c)!}$$

$$\cdot \binom{c}{r} e^{-\mu r(u-\tau)}(1-e^{-\mu(u-\tau)})^{c-r} c\mu d\tau. \tag{4.2.29}$$

The second term may be rewritten as

$$R \int_0^u \exp[-\mu c \tau(I-R)] \binom{c}{r} e^{-\mu r(u-\tau)}(1-e^{-\mu(u-\tau)})^{c-r} c\mu d\tau$$

$$= -R \int_0^u \binom{c}{r} e^{-\mu r(u-\tau)}(1-e^{-\mu(u-\tau)})^{c-r}$$

$$\cdot d((I-R)^{-1}\exp[-\mu c \tau(I-R)]).$$

Partial integration leads to

$$V_{r+1}(u) = \frac{c}{r+1}\binom{c-1}{r}e^{-\mu r u}(1-e^{-\mu u})^{c-r-1}I$$

$$+ \frac{1}{r+1}[cR - (c-r)I]V_r(u),$$

for $0 \leq r \leq c-1$, and formula (4.2.28) is now immediate. For $r = c$, we obtain directly from (4.2.29) that

$$V_c(u) = e^{-c\mu u}I + e^{-c\mu u}c\mu R\int_0^u \exp(c\mu\tau R)\,d\tau$$

$$= \exp[-c\mu u(I-R)], \quad \text{for} \quad u \geq 0.$$

Equation (4.2.23) readily yields

$$B_c[R] = \int_0^\infty \exp[-c\mu u(I-R)]\,dF(u) = R.$$

4.2b. Examples of Queues with Semi-Markovian Arrivals

Possibly because of the complexity of their discussions by transform methods, semi-Markovian descriptions of arrival or service processes have not yet found extensive practical use. Finite-state Markov renewal processes may be used to represent a great variety of point processes with specific qualitative features in a computationally tractable frame work. We illustrate this by discussing here a number of queueing models in which a semi-Markovian arrival process arises naturally.

A. Markovian Arrivals. An arrival process in which customers arrive at the epochs of transitions of an irreducible K-state Markov process—see, for example, T. Ushijima [299]—is clearly a particular case of a semi-Markovian input process. An $SM/PH/1$ or $SM/M/c$ queue of this type may also be studied as a QBD process. The resulting algorithm is then much simpler. In order to concentrate only on the essential ideas, we indicate the main steps for the case of a single exponential server of rate μ. The irreducible generator of the Markovian arrival process will be denoted by Q. Its stationary probability vector is $\boldsymbol{\pi}$.

Let the diagonal matrix Δ be defined by

$$\Delta = -\operatorname{diag}(Q). \tag{4.2.30}$$

It is then readily verified that the $SM/M/1$ queue with *Markovian* arrivals may be studied as the QBD process with generator $\tilde{Q}$, given by

$$\tilde{Q} = \begin{vmatrix} -\Delta & Q+\Delta & 0 & 0 & \cdots \\ \mu I & -\mu I - \Delta & Q+\Delta & 0 & \cdots \\ 0 & \mu I & -\mu I - \Delta & Q+\Delta & \cdots \\ 0 & 0 & \mu I & -\mu I - \Delta & \cdots \\ \vdots & \vdots & \vdots & \vdots & \end{vmatrix}. \qquad (4.2.31)$$

The classical results on QBD processes are immediately applicable. The queue is *stable* if and only if

$$\mu > \pi \Delta e, \qquad (4.2.32)$$

and the stationary probability vector $x = [x_0, x_1, \ldots]$ of $\tilde{Q}$ is then given by

$$x_k = \pi(I - R)R^k, \quad \text{for} \quad k \geq 0, \qquad (4.2.33)$$

where R is the minimal nonnegative solution to the matrix equation

$$\mu R^2 - R(\mu I + \Delta) + Q + \Delta = 0. \qquad (4.2.34)$$

From the viewpoint of modeling arrival processes, a finite-state Markov process has a serious drawback. A transition from a state to itself is impossible. The arrival types of two successive customers are necessarily different. There is some merit in considering Markov renewal arrival processes *in which the sojourn times have exponential distributions.* Such processes are slightly more general than Markov processes, but allow transitions from a state to itself. Markov renewal processes of this type arise, for example, in the superposition of certain simple PH-renewal processes.

As an interesting example, we now consider the superposition of two independent hyperexponential renewal processes with underlying distributions

$$H_1(x) = \alpha_1(1 - e^{-\lambda_1 x}) + \alpha_2(1 - e^{-\lambda_2 x}),$$

$$H_2(x) = \alpha_3(1 - e^{-\lambda_3 x}) + \alpha_4(1 - e^{-\lambda_4 x}), \quad \text{for} \quad x \geq 0, \qquad (4.2.35)$$

where $0 < \alpha_1 < 1$, $0 < \alpha_3 < 1$, $\alpha_1 + \alpha_2 = \alpha_3 + \alpha_4 = 1$. The same construction may be carried out for several hyperexponential renewal processes and for mixtures of more than two exponential distributions. For a more general discussion, we refer the reader to V. Ramaswami [260].

The two renewal processes are equivalent to two-state Markov renewal processes with transition probability matrices $F_1(\cdot)$ and $F_2(\cdot)$ respectively, given by

$$F_1(x) = \begin{array}{c|cc|} 1 & \alpha_1(1 - e^{-\lambda_1 x}) & \alpha_2(1 - e^{-\lambda_1 x}) \\ 2 & \alpha_1(1 - e^{-\lambda_2 x}) & \alpha_2(1 - e^{-\lambda_2 x}) \end{array}, \tag{4.2.36}$$

and

$$F_2(x) = \begin{array}{c|cc|} 3 & \alpha_3(1 - e^{-\lambda_3 x}) & \alpha_4(1 - e^{-\lambda_3 x}) \\ 4 & \alpha_3(1 - e^{-\lambda_4 x}) & \alpha_4(1 - e^{-\lambda_4 x}) \end{array}.$$

The product process, obtained by considering simultaneous independent realizations of the Markov renewal processes $F_1(\cdot)$ and $F_2(\cdot)$, is—because of the exponential sojourn times—again a Markov renewal process. It has four states (1, 3), (1, 4), (2, 3), and (2, 4), and its transition probability matrix $F(\cdot)$ is given by

$$\begin{aligned}
F_{13,13}(x) &= \left(\alpha_1 \frac{\lambda_1}{\lambda_1 + \lambda_3} + \alpha_3 \frac{\lambda_3}{\lambda_1 + \lambda_3}\right)[1 - e^{-(\lambda_1+\lambda_3)x}], \\
F_{13,14}(x) &= \alpha_4 \frac{\lambda_3}{\lambda_1 + \lambda_3} [1 - e^{-(\lambda_1+\lambda_3)x}], \\
F_{13,23}(x) &= \alpha_2 \frac{\lambda_1}{\lambda_1 + \lambda_3} [1 - e^{-(\lambda_1+\lambda_3)x}], \\
F_{13,24}(x) &= 0, \\
F_{14,13}(x) &= \alpha_3 \frac{\lambda_4}{\lambda_1 + \lambda_4} [1 - e^{-(\lambda_1+\lambda_4)x}], \\
F_{14,14}(x) &= \left(\alpha_1 \frac{\lambda_1}{\lambda_1 + \lambda_4} + \alpha_4 \frac{\lambda_4}{\lambda_1 + \lambda_4}\right)[1 - e^{-(\lambda_1+\lambda_4)x}], \\
F_{14,23}(x) &= 0, \\
F_{14,24}(x) &= \alpha_2 \frac{\lambda_1}{\lambda_1 + \lambda_4} [1 - e^{-(\lambda_1+\lambda_4)x}], \qquad (4.2.37) \\
F_{23,13}(x) &= \alpha_1 \frac{\lambda_2}{\lambda_2 + \lambda_3} [1 - e^{-(\lambda_2+\lambda_3)x}], \\
F_{23,14}(x) &= 0, \\
F_{23,23}(x) &= \left(\alpha_2 \frac{\lambda_2}{\lambda_2 + \lambda_3} + \alpha_3 \frac{\lambda_3}{\lambda_2 + \lambda_3}\right)[1 - e^{-(\lambda_2+\lambda_3)x}], \\
F_{23,24}(x) &= \alpha_4 \frac{\lambda_3}{\lambda_2 + \lambda_3} [1 - e^{-(\lambda_2+\lambda_3)x}],
\end{aligned}$$

$$F_{24,13}(x) = 0,$$

$$F_{24,14}(x) = \alpha_1 \frac{\lambda_2}{\lambda_2 + \lambda_4} [1 - e^{-(\lambda_2+\lambda_4)x}],$$

$$F_{24,23}(x) = \alpha_3 \frac{\lambda_4}{\lambda_2 + \lambda_4} [1 - e^{-(\lambda_2+\lambda_4)x}],$$

$$F_{24,24}(x) = \left(\alpha_2 \frac{\lambda_2}{\lambda_2 + \lambda_4} + \alpha_4 \frac{\lambda_4}{\lambda_2 + \lambda_4}\right)[1 - e^{-(\lambda_2+\lambda_4)x}],$$

for $x \geq 0$. The Markov renewal process $F(\cdot)$ again has exponential sojourn time distributions. In the numerical solution of, for example, an *SM/M/*1 process with this arrival process, this greatly simplifies the computation of the matrices A_k and B_k, for $k \geq 0$. The column vector $\boldsymbol{\phi}^*$ of the row sum means of $F(\cdot)$ has the components $(\lambda_1 + \lambda_3)^{-1}$, $(\lambda_1 + \lambda_4)^{-1}$, $(\lambda_2 + \lambda_3)^{-1}$, $(\lambda_2 + \lambda_4)^{-1}$.

It is not advisable to try to solve the equations $\boldsymbol{\phi} F = \boldsymbol{\phi}$, $\boldsymbol{\phi} \mathbf{e} = 1$, for the invariant probability vector $\boldsymbol{\phi}$ of F by brute force. The vector $\boldsymbol{\phi}$ may, however, be obtained by the following elegant argument. Setting $x_1 = \phi_1(\lambda_1 + \lambda_3)^{-1}$, $x_2 = \phi_2(\lambda_1 + \lambda_4)^{-1}$, $x_3 = \phi_3(\lambda_2 + \lambda_3)^{-1}$, and $x_4 = \phi_4(\lambda_2 + \lambda_4)^{-1}$, we obtain the homogeneous system of linear equations

$$\begin{aligned} -x_1(\alpha_2\lambda_1 + \alpha_4\lambda_3) + x_2\alpha_3\lambda_4 + x_3\alpha_1\lambda_2 &= 0, \\ x_1\alpha_4\lambda_3 - x_2(\alpha_2\lambda_1 + \alpha_3\lambda_4) + x_4\alpha_1\lambda_2 &= 0, \\ x_1\alpha_2\lambda_1 - x_3(\alpha_1\lambda_2 + \alpha_4\lambda_3) + x_4\alpha_3\lambda_4 &= 0, \\ x_2\alpha_2\lambda_1 + x_3\alpha_4\lambda_3 - x_4(\alpha_1\lambda_2 + \alpha_3\lambda_4) &= 0, \end{aligned} \tag{4.2.38}$$

whose coefficient matrix may be written by use of the Kronecker product as

$$\begin{vmatrix} -\alpha_2\lambda_1 & \alpha_2\lambda_1 \\ \alpha_1\lambda_2 & -\alpha_1\lambda_2 \end{vmatrix} \otimes \begin{vmatrix} 1 & 0 \\ 0 & 1 \end{vmatrix} + \begin{vmatrix} 1 & 0 \\ 0 & 1 \end{vmatrix} \otimes \begin{vmatrix} -\alpha_4\lambda_3 & \alpha_4\lambda_3 \\ \alpha_3\lambda_4 & -\alpha_3\lambda_4 \end{vmatrix}. \tag{4.2.39}$$

It is now clear from (4.2.39) that

$$\begin{aligned} x_1 &= c\alpha_1\alpha_3\lambda_2\lambda_4, & x_2 &= c\alpha_1\alpha_4\lambda_2\lambda_3, \\ x_3 &= c\alpha_2\alpha_3\lambda_1\lambda_4, & x_4 &= c\alpha_2\alpha_4\lambda_1\lambda_3. \end{aligned} \tag{4.2.40}$$

Finally, since the components of $\boldsymbol{\phi}$ sum to one, we obtain

$$c = (\lambda_1\lambda_2\lambda_3\lambda_4)^{-1}\left(\sum_{i=1}^{4}\frac{\alpha_i}{\lambda_i}\right)^{-1},$$

$$\phi_1 = \frac{\alpha_1\alpha_3}{\lambda_1\lambda_3}(\lambda_1+\lambda_3)\left(\sum_{i=1}^{4}\frac{\alpha_i}{\lambda_i}\right)^{-1},$$

$$\phi_2 = \frac{\alpha_1\alpha_4}{\lambda_1\lambda_4}(\lambda_1+\lambda_4)\left(\sum_{i=1}^{4}\frac{\alpha_i}{\lambda_i}\right)^{-1}, \tag{4.2.41}$$

$$\phi_3 = \frac{\alpha_2\alpha_3}{\lambda_2\lambda_3}(\lambda_2+\lambda_3)\left(\sum_{i=1}^{4}\frac{\alpha_i}{\lambda_i}\right)^{-1},$$

$$\phi_4 = \frac{\alpha_2\alpha_4}{\lambda_2\lambda_4}(\lambda_2+\lambda_4)\left(\sum_{i=1}^{4}\frac{\alpha_i}{\lambda_i}\right)^{-1},$$

and the equilibrium condition (4.2.11) is seen, after some routine calculations, to be equivalent to

$$\mu > \left(\frac{\alpha_1}{\lambda_1}+\frac{\alpha_2}{\lambda_2}\right)^{-1} + \left(\frac{\alpha_3}{\lambda_3}+\frac{\alpha_4}{\lambda_4}\right)^{-1}. \tag{4.2.42}$$

The right-hand side is clearly the sum of the stationary arrival rates of the two hyperexponential renewal processes, whose superposition provides the input to the exponential server. The general results on the $SM/M/1$ queue may now be routinely applied to provide the various steady-state distributions for this model.

B. A Queue Fed by the Departures from a Finite $M/G/1$ Queue. We consider a tandem queueing system consisting of two units in series. The first unit is described by a finite $M/G/1$ queue of arrival rate λ and service time distribution $H(\cdot)$. Only $K+1$ customers at most may be present at any time in Unit I, and customers who arrive when Unit I is filled to capacity are lost. Let J_n, $n \geq 0$, $0 \leq J_n \leq K$, be the queue length in Unit I immediately after the n-th service completion. We denote by X_n, $n \geq 1$, the length of time between the $(n-1)$st and the n-th service completion in Unit I. For definiteness, time $t = 0$ corresponds to the service completion labeled by $n = 0$, and we also set $X_0 = 0$.

It is now well known that the bivariate sequence $\{(J_n, X_n), n \geq 0\}$ is a Markov renewal sequence. Its transition probability matrix $F(\cdot)$ is of the form

$$F(x) = \begin{vmatrix} F_{00}(x) & F_{01}(x) & F_{02}(x) & \cdots & F_{0,K-1}(x) & F_{0,K}(x) \\ F_{10}(x) & F_{11}(x) & F_{12}(x) & \cdots & F_{1,K-1}(x) & F_{1,K}(x) \\ 0 & F_{21}(x) & F_{22}(x) & \cdots & F_{2,K-1}(x) & F_{2,K}(x) \\ 0 & 0 & F_{31}(x) & \cdots & F_{3,K-1}(x) & F_{3,K}(x) \\ \vdots & \vdots & \vdots & & \vdots & \vdots \\ 0 & 0 & 0 & \cdots & F_{K,K-1}(x) & F_{K,K}(x) \end{vmatrix}, \tag{4.2.43}$$

where

$$F_{0j}(x) = \int_0^x [1 - e^{-\lambda(x-y)}]\,dF_{1j}(y), \qquad \text{for} \quad 0 \le j \le K,$$

$$F_{ij}(x) = \int_0^x e^{-\lambda y} \frac{(\lambda y)^{j-i+1}}{(j-i+1)!}\,dH(y), \qquad \text{for} \quad 1 \le i \le K, \quad 0 \le j \le i-1,$$

$$F_{iK}(x) = \sum_{\nu=K-i+1}^{\infty} \int_0^x e^{-\lambda y} \frac{(\lambda y)^{\nu}}{\nu!}\,dH(y), \qquad \text{for} \quad 1 \le i \le K.$$

The output process of Unit I is now described by the Markov renewal process $F(\cdot)$. It may in turn be considered as the arrival process to Unit II, in front of which an unbounded queue is allowed. If Unit II is staffed by a single server with a PH-service time distribution, we obtain an instance of an *SM/PH/*1 queue. Similarly, when Unit II has c parallel exponential servers of rate μ, an example of an *SM/M/c* queue arises.

The case where Unit I is a *bounded M/M/r queue* deserves special attention. Considered at epochs of departures, the bounded *M/M/r* queue has an embedded $(K + 1)$-state Markov renewal process for which the matrix $F(\cdot)$ may readily be written down. Inferences similar to those above may be drawn for tandem systems involving a first unit of this type.

The output of the bounded *M/M/r* queue, however, may be viewed differently. Recalling the standard description of this queue as a Markov process, we define the input to Unit II by saying that an arrival to II occurs if and only if the queue length in Unit I makes a downward transition. The arrival process to Unit II is now described by a Markov process in which some but not all transitions correspond to arrivals. As such, it becomes a particular case of the point process introduced in M. F. Neuts [232], and the tandem queue with a single server and general service times in Unit II becomes a particular case of the model treated by V. Ramaswami [261].

Three general comments are in order. It is clear that the $SM/PH/1$ or $SM/M/c$ models which are obtained here describe both queues and not merely Unit II. The second state variable fully describes the queue length in Unit I. It is therefore easy to extract from the (modified) matrix-geometric solution much useful information about the *dependence* between the two queue lengths.

The second comment is related to the value of extensive numerical investigations of such tandem queues. One may argue, largely on intuitive grounds, that for a wide range of distributions $H(\cdot)$, the output of Unit I will be so "regular" and its internal dependence so slight that for most practical purposes it may be approximated by, for example, a renewal process. The qualitative range of validity of such an approximation is extremely difficult to delineate by analytic methods, but may be investigated by a well-planned numerical experiment. Representative examples of cases where the approximation satisfies or fails to do so are worth reporting, particularly in the literature addressed to practice. A likely instance of failure is the case where λ is fairly small but $H(\cdot)$ has a large variance. At present, very little information of such a numerical-experimental type is available on queues. The few existing studies give only scant details and tend to report only on expected queue lengths or waiting times for models with particular service time distributions. Similar comments may be made, for example, on the *overflow models,* treated in section 6.4.

Finally, the dearth of analytically tractable results is such that not even the critical arrival rate to the tandem systems under discussion is explicitly expressible in terms of the parameters of the model. Its numerical computation, however, may be routinely carried out for a wide range of parameter values.

C. Platooned Arrivals. A common qualitative feature of arrival streams is the formation of *platoons* or *clusters* of arrivals. Within a platoon, the interarrival times are of comparable size and are typically much shorter than interplatoon arrival times. The following is a semi-Markovian model for platooned arrivals which elegantly represents the essential features of such arrival streams.

Let N_ν, $\nu \geq 1$, be the number of customers in the ν-th platoon. For ease of exposition, we assume that time $t = 0$ corresponds to the last arrival of a platoon. The random variables N_ν will be independent, identically distributed, positive integer-valued. Their common probability density is denoted by $\{p_k\}$ with $p_k = 0$, for $k \leq 0$. The successive partial sums of the random variables N_ν are denoted by S_k for $k \geq 0$, with $S_0 = 0$.

Next, we consider a sequence $\{\tau_h, h \geq 1\}$ of nonnegative random variables, such that given $N_1, N_2, \ldots, N_r, r \geq 1$, the random variables τ_h, $1 \leq h \leq S_r + 1$, are conditionally independent. The conditional distributions of $\tau_{S_0+1}, \tau_{S_1+1}, \ldots, \tau_{S_r+1}$, are all given by the common distribution $F_2(\cdot)$. The conditional distributions of the other random variables τ_k are given by the common distribution $F_1(\cdot)$. Neither of the distributions $F_1(\cdot)$ and $F_2(\cdot)$ is degenerate. $F_1(\cdot)$ will be called the *intraplatoon* (arrival) *distribution*; $F_2(\cdot)$ the *interplatoon* (arrival) *distribution.* We now consider an arrival stream in which individual customers arrive at times generated by the partial sums of the sequence $\{\tau_h, h \geq 1\}$.

If $\{p_k\}$ is a general density *of unbounded support,* the arrival stream is described by a Markov renewal process with infinitely many states, labeled 1, 2, ... In order to display its transition probability matrix $F(\cdot)$, we define the probabilities

$$q_1 = p_1, \qquad q_1' = 1 - q_1,$$

$$q_k = p_k\left(1 - \sum_{\nu=0}^{k-1} p_\nu\right)^{-1}, \qquad q_k' = 1 - q_k, \quad \text{for} \quad k \geq 2. \tag{4.2.44}$$

The matrix $F(x)$, $x \geq 0$, is then given by

$$F(x) = \begin{array}{c|ccccc|} 1 & q_1F_2(x) & q_1'F_2(x) & 0 & 0 & \cdots \\ 2 & q_2F_1(x) & 0 & q_2'F_1(x) & 0 & \cdots \\ 3 & q_3F_1(x) & 0 & 0 & q_3'F_1(x) & \cdots \\ 4 & q_4F_1(x) & 0 & 0 & 0 & \cdots \\ 5 & q_5F_1(x) & 0 & 0 & 0 & \cdots \\ \vdots & \vdots & \vdots & \vdots & \vdots & \end{array}. \tag{4.2.45}$$

Sojourns in the state 1 correspond to interplatoon intervals, while a sojourn in the state k, $k \geq 2$, corresponds to the interarrival time between the $(k - 1)$st and the k-th customer of a platoon.

From an algorithmic viewpoint, the infinite state space is a serious drawback. The approximation by a truncation of the state space raises some quite difficult technical questions, which, however, are unimportant to most situations arising in practical modeling.

If we are willing to impose the mild restriction that the density $\{p_k\}$ is *of phase type,* it becomes possible to represent the platooned arrival process in an alternate manner as a *finite-state* Markov renewal process. As we shall see, the case where $\{p_k\}$ has a bounded support then arises as a particular instance.

We now consider in detail the case where $\{p_k\}$ is of phase type. For notational convenience, we let the density $\{p_k'\}$ with $p_k' = p_{k+1}$, for $k \geq 0$, have the (irreducible) representation $(\boldsymbol{\delta}, D)$, where $\boldsymbol{\delta}$ is a nonnegative K-vector such that $\delta_0 = 1 - \boldsymbol{\delta e} \geq 0$. D is a substochastic matrix of order K such that $I - D$ is nonsingular. We shall also, for convenience, label the absorbing state that arises in the definition of the PH-density $\{p_k'\}$ by 0, rather than by $K + 1$. The sequence $\{p_k'\}$ is then the density of the time until absorption in the Markov chain with states $\{0, 1, \ldots, K\}$ and transition probability matrix

$$\begin{vmatrix} 1 & \mathbf{0} \\ D^o & D \end{vmatrix},$$

with initial probability vector $(\delta_0, \boldsymbol{\delta})$ and where $D^o = \boldsymbol{e} - D\boldsymbol{e}$.

We construct the Markov renewal process $F(\cdot)$ with states $\{0, 1, \ldots, K\}$ and with the transition probability matrix $F(\cdot)$ given by

$$F(x) = \begin{vmatrix} \delta_0 F_2(x) & \boldsymbol{\delta} F_2(x) \\ D^o F_1(x) & D F_1(x) \end{vmatrix}, \quad \text{for} \quad x \geq 0. \tag{4.2.46}$$

Sojourns in the state 0 are now interpreted as interplatoon intervals, while the successive sojourns in the states $1, \ldots, K$, correspond to intraplatoon intervals. Elementary calculations for Markov chains readily show that the probability density of the number of transitions between returns to the state 0 is given by $\{p_k\}$. The mean $EN_r = \nu_1'$ is given by

$$\nu_1' = 1 + \boldsymbol{\delta}(I - D)^{-1}\boldsymbol{e}. \tag{4.2.47}$$

If ϕ_1^* and ϕ_2^* are the means of $F_1(\cdot)$ and $F_2(\cdot)$, then the vector $\boldsymbol{\phi}^*$ of the row sum means of the semi-Markov matrix $F(\cdot)$ is given by $\boldsymbol{\phi}^* = [\phi_2^*, \phi_1^* \boldsymbol{e}]$. The invariant probability vector $\hat{\boldsymbol{\phi}} = [\phi_0, \boldsymbol{\phi}]$ of $F = F(\infty)$ is given by

$$\phi_0 = \frac{1}{\nu_1'},$$

$$\boldsymbol{\phi} = \frac{1}{\nu_1'}\, \boldsymbol{\delta}(I - D)^{-1}, \tag{4.2.48}$$

as is readily seen upon solving the equations $\hat{\boldsymbol{\phi}}F = \hat{\boldsymbol{\phi}}$, $\hat{\boldsymbol{\phi}}\boldsymbol{e} = 1$. For the fundamental mean λ_1' of $F(\cdot)$, we obtain

$$\lambda_1' = \left(1 - \frac{1}{\nu_1'}\right)\phi_1^* + \frac{1}{\nu_1'}\, \phi_2^*. \tag{4.2.49}$$

The key results for the *SM/PH/1* and the *SM/M/c* queues may now be routinely translated to obtain results for the cases where a platooned arrival process serves as the input to a single PH-service unit or to c parallel exponential servers. Rather than doing so, we shall point out a number of parameters and computable stationary probability distributions, which acquire a significance important to the correct physical interpretation of numerical results for this model. In order to keep the analytic expressions simple, we limit our discussion to the case of a single exponential server of rate μ. The model is then a particular $SM/M/1$ queue. Similar expressions may also be obtained for the more complicated versions. The interested reader may find a discussion of algorithmic simplifications and numerical examples in M. F. Neuts and S. Chakravarthy [237], where the case of a single server with PH-service time distribution is treated in detail.

a. The single server queue with platooned arrivals and exponential service times is *stable* if and only if

$$\lambda_1'\mu = \left[\left(1 - \frac{1}{\nu_1'}\right)\phi_1^* + \frac{1}{\nu_1'}\phi_2^*\right]\mu > 1. \tag{4.2.50}$$

b. The stationary probability that an arrival epoch corresponds to the *last* arrival in a platoon is given by

$$\delta_0\phi_0 + \boldsymbol{\phi}\boldsymbol{D}^o = \phi_0 = \nu_1'^{-1}. \tag{4.2.51}$$

To see this we note that the first term on the left-hand side is the probability that the arriving customer is a singleton and therefore ipso facto the last in his platoon. The second term is the probability that the arrival corresponds to a return to the state 0 from one of the states $1, \ldots, K$ in the Markov chain F. Such an arrival is clearly the last in a platoon of size at least two. The equality in (4.2.51) is intuitive. There are indeed as many ends as there are beginnings of platoons.

c. With the matrix R defined as in (4.2.23), the conditional stationary density of the queue length immediately prior to the *first* arrival of a platoon is given by

$$q_0(i) = \phi_0^{-1}[\hat{\boldsymbol{\phi}}(I - R)R^i]_0, \quad \text{for} \quad k \geq 0. \tag{4.2.52}$$

d. With the mass-functions $W_j(\cdot)$, $0 \leq j \leq K$, defined as in theorem 4.2.2, the conditional stationary waiting time distribution of the *first* arrival of a platoon is given by $\phi_0^{-1}W_0(x)$, for $x \geq 0$.

e. The conditional stationary density of the queue length prior to the *last* arrival of a platoon is given by

$$\begin{aligned} q_1(i) &= \nu_1'\left(\delta_0 x_{i0} + \sum_{j=1}^{K} x_{ij}D_j^o\right) \\ &= \nu_1'\left\{\delta_0[\hat{\boldsymbol{\phi}}(I - R)R^i]_0 + \sum_{j=1}^{K} [\hat{\boldsymbol{\phi}}(I - R)R^i]_j D_j^o\right\}, \end{aligned} \tag{4.2.53}$$

for $i \geq 0$. The stationary waiting time distribution of that customer is given by

$$\nu_1' \left[\delta_0 W_0(x) + \sum_{j=1}^{K} W_j(x) D_j^o \right], \quad \text{for} \quad x \geq 0.$$

f. It may be of interest to distinguish between singletons and "genuine" platoons involving at least two arrivals. We may similarly obtain the stationary density of the queue length prior to the *last arrival of a genuine platoon.* It is given by

$$q_2(i) = (1 - \delta_0)^{-1} \nu_1' \sum_{j=1}^{K} [\hat{\boldsymbol{\phi}}(I - R) R^i]_j D_j^o, \quad \text{for} \quad i \geq 0. \quad (4.2.54)$$

The corresponding expression for the waiting time distribution of that customer reads

$$(1 - \delta_0)^{-1} \nu_1' \sum_{j=1}^{K} W_j(x) D_j^o, \quad \text{for} \quad x \geq 0.$$

g. A direct calculation shows that the (conditional) queue length densities and waiting time distributions at the arrivals of singletons or of the first customers of genuine platoons are the same and are given by the expressions in (*c*) above.

h. We may recursively for $k \geq 1$ compute the conditional queue length densities at the arrival epoch of the k-th customer of a platoon. The resulting expressions are somewhat involved, but the underlying probability arguments are straightforward. We shall omit the details.

Two final remarks will conclude our discussion of this model. First, when the density $\{p_k\}$ has a *bounded support,* we clearly obtain a special case of a phase type density. Let p_{K+1} be positive and $p_k = 0$, for $k > K + 1$. The matrix $F(\cdot)$ is then given by

$$F(x) = \begin{vmatrix} q_1 F_2(x) & q_1' F_2(x) & 0 & \cdots & 0 \\ q_2 F_1(x) & 0 & q_2' F_1(x) & \cdots & 0 \\ q_3 F_1(x) & 0 & 0 & \cdots & 0 \\ \vdots & \vdots & \vdots & & \vdots \\ q_K F_1(x) & 0 & 0 & \cdots & q_K' F_1(x) \\ F_1(x) & 0 & 0 & \cdots & 0 \end{vmatrix}, \quad (4.2.55)$$

in terms of the notation defined in formula (4.2.44). Setting $\delta_0 = q_1 = p_1$, $\boldsymbol{\delta} = (q_1', 0, \ldots, 0)$, and

$$D = \begin{vmatrix} 0 & q_2' & 0 & 0 & \cdots & 0 \\ 0 & 0 & q_3' & 0 & \cdots & 0 \\ \cdot & \cdot & \cdot & \cdot & & \cdot \\ \cdot & \cdot & \cdot & \cdot & & \cdot \\ \cdot & \cdot & \cdot & \cdot & & \cdot \\ 0 & 0 & 0 & 0 & \cdots & q_K' \\ 0 & 0 & 0 & 0 & \cdots & 0 \end{vmatrix}, \qquad D^o = \begin{vmatrix} q_2 \\ q_3 \\ \cdot \\ \cdot \\ \cdot \\ q_K \\ 1 \end{vmatrix},$$

we see that $F(\cdot)$ is of the form given in (4.2.46). For this case it is easy to find the conditional distributions of queue length and waiting time, given that the arriving customer is the k-th member, $1 \leq k \leq K + 1$, of the platoon.

Second, it is worth stressing that queues with platooned arrivals may exhibit large random fluctuations, which we shall discuss at greater length in chapter 5. If there are occasional long service times or if there is much random variability in the sizes of platoons, we may see several platoons join the queue before an earlier platoon has cleared. Substantial build-up then occurs and it may require a long time for the queue to dissipate. As we shall see, it requires the consideration of several embedded processes and much more detailed information than the usual mean queue lengths and waiting times in order to gain an adequate quantitative description of such queues.

D. Switches. Semi-Markovian arrival processes arise naturally in models where a switching or channel selection procedure directs arriving customers to queues in front of two or more parallel service units. Under conditions of useful generality, one may then frequently compute marginal stationary distributions by *decomposing* a multi-queue problem into single queues. This *does not* mean that the separate queues are described by independent stochastic processes, but only that for the purpose of evaluating the marginal stationary distributions of each queue, the other units may be ignored. The procedure to be discussed is an example of the wider topic of *network decomposition,* a subject in which stochastic independence results need to be considered with great care. This interesting topic was investigated in E. Çinlar [55], R. L. Disney [79, 82], and W. K. Hall and R. L. Disney [126].

We shall limit our discussion to some fairly simple cases in order to avoid the introduction of the involved notation needed in the more complex examples. Consider an ordinary renewal process with interarrival time distribution $H(\cdot)$ of positive mean γ_1'. Given are two positive integers K_1 and K_2 with $K_1 + K_2 = K$. There are two exponential, single-server Units I and II of service rates μ_1 and μ_2 respectively. A switch directs the n-th arriving customer to Unit I if and only if $n \equiv \nu$ mod(K), with $1 \leq \nu \leq K_1$, and to Unit II otherwise. Once assigned to a unit, a customer waits for service in that unit. The queue length processes in Units I and II are not independent, but the stationary marginal queue length (and waiting time) distributions for each unit may be studied by decomposition.

Clearly Unit I by itself may be described as an $SM/M/1$ queue with service rate μ_1 and a Markov renewal arrival process with the transition probability matrix $F_1(\cdot)$ of order K_1, given by

$$F_1(x) = \begin{vmatrix} 0 & H(x) & 0 & \cdots & 0 \\ 0 & 0 & H(x) & \cdots & 0 \\ \vdots & \vdots & \vdots & & \\ 0 & 0 & 0 & \cdots & H(x) \\ H^{(K_2+1)}(x) & 0 & 0 & \cdots & 0 \end{vmatrix}, \quad \text{for} \quad x \geq 0. \tag{4.2.56}$$

Similarly, Unit II separately is an $SM/M/1$ queue with service rate μ_2 and the matrix $F_2(\cdot)$ of order K_2, given by

$$F_2(x) = \begin{vmatrix} 0 & H(x) & 0 & \cdots & 0 \\ 0 & 0 & H(x) & \cdots & 0 \\ \vdots & \vdots & \vdots & & \\ 0 & 0 & 0 & \cdots & H(x) \\ H^{(K_1+1)}(x) & 0 & 0 & \cdots & 0 \end{vmatrix}, \quad \text{for} \quad x \geq 0. \tag{4.2.57}$$

The choice of K_1, $1 \leq K_1 \leq K - 1$, clearly affects the stability of the queues. The necessary and sufficient conditions for both units to be stable is that

$$K\mu_1\gamma_1' > K_1 > K(1 - \mu_2\gamma_2'). \tag{4.2.58}$$

Provided such values of K_1 exist, we may *numerically* determine, for example, the value of K_1 which balances the queues as well as possible, say by minimizing the absolute difference between mean queue lengths or waiting times.

This example of a *deterministic binary switch* is one of the simplest instances of the use of network decomposition. Its extension to more than two units is straightforward.

We shall now consider the underlying idea from a more general viewpoint. This will enable us to make some algorithmic observations, which are also useful in other contexts. Consider a K-state irreducible Markov renewal arrival process with the transition probability matrix $Q(\cdot)$, partitioned as follows

$$Q(x) = \begin{vmatrix} Q_{11}(x) & Q_{12}(x) \\ Q_{21}(x) & Q_{22}(x) \end{vmatrix}, \tag{4.2.59}$$

where $Q_{11}(\cdot)$ and $Q_{22}(\cdot)$ are square matrices of mass functions, respectively of orders K_1 and K_2. We let each transition into one of the states $1, \ldots, K_1$, correspond to an arrival to Unit I and each transition into one of the remaining states to an arrival to Unit II. The filtered process obtained by considering only the states of successive visits to the set $\{1, \ldots, K_1\}$ and the time intervals between such visits is again a Markov renewal process, whose transition probability matrix will be denoted by $F_1(\cdot)$. A similar construction is applied to the process $Q(\cdot)$, considered only at visits to the set $\{K_1 + 1, \ldots, K\}$.

The matrices $F_1(\cdot)$ and $F_2(\cdot)$ are given by the matrix-convolution equations

$$\begin{aligned} F_1(x) &= Q_{11}(x) + Q_{12}(\cdot)*M_{22}(\cdot)*Q_{21}(x), \\ F_2(x) &= Q_{22}(x) + Q_{21}(\cdot)*M_{11}(\cdot)*Q_{12}(x), \quad \text{for} \quad x \geq 0, \end{aligned} \tag{4.2.60}$$

where the matrices $M_{11}(\cdot)$ and $M_{22}(\cdot)$ themselves are the solutions to the matrix Volterra equations

$$\begin{aligned} M_{11}(x) &= U_1(x) + Q_{11}(\cdot)*M_{11}(x), \\ M_{22}(x) &= U_2(x) + Q_{22}(\cdot)*M_{22}(x), \quad \text{for} \quad x \geq 0. \end{aligned} \tag{4.2.61}$$

The matrices $U_1(\cdot)$ and $U_2(\cdot)$ are diagonal matrices of order K_1 and K_2 with degenerate distributions on the diagonal. Once the matrices $F_1(\cdot)$ and $F_2(\cdot)$ have been determined, the remainder of the discussion proceeds as in the preceding case.

In general, any special structure the matrix $Q(\cdot)$ may have will be destroyed in forming the matrices $F_1(\cdot)$ and $F_2(\cdot)$. The case of the deterministic binary switch is a rare exception. The numerical computation of the matrices $F_1(\cdot)$ and $F_2(\cdot)$ is also, in general, a major task.

There is an alternative approach, which seemingly requires the solution of a problem of larger dimension, but which avoids the computation of $F_1(\cdot)$ and $F_2(\cdot)$ and preserves all simplifying structural properties of the matrix $Q(\cdot)$. We shall give the discussion only for Unit I. The details for Unit II are entirely analogous, but we note the useful fact that all overhead computations for one unit also yield the necessary ingredients for the solution of the other unit.

Consider Unit I with the arrival process Q, except that only transitions into one of the states $1, \ldots, K_1$, correspond to actual arrivals to that unit. The Unit I is now considered immediately prior *to each successive transition of the Markov renewal process* $Q(\cdot)$. The queue length in I and the state of the process $Q(\cdot)$ again define an embedded Markov chain on the state space $\{(i, j), i \geq 0, 1 \leq j \leq K\}$. Its transition probability matrix $\tilde{P}$ is of the canonical form (1.1.5), and the matrices A_k, $k \geq 0$, are given by

$$A_k = \begin{vmatrix} \int_0^\infty e^{-\mu_1 u} \frac{(\mu_1 u)^k}{k!} dQ_{11}(u) & \int_0^\infty e^{-\mu_1 u} \frac{(\mu_1 u)^k}{k!} dQ_{12}(u) \\ \int_0^\infty e^{-\mu_1 u} \frac{(\mu_1 u)^{k-1}}{(k-1)!} dQ_{21}(u) & \int_0^\infty e^{-\mu_1 u} \frac{(\mu_1 u)^{k-1}}{(k-1)!} dQ_{22}(u) \end{vmatrix}, \tag{4.2.62}$$

for $k \geq 0$. We note that the last K_2 rows of A_0 are zero, so that although the matrix R is of order K, only its first K_1 rows need to be computed; its other rows are zero.

The equilibrium condition for Unit I is obtained as follows. Let γ be the invariant probability vector of $Q(\infty)$ and γ^* the vector of row sum means of $Q(\cdot)$. We partition both vectors according to the same indices as in (4.2.59), that is,

$$\gamma = [\gamma_1, \gamma_2], \qquad \gamma^* = \begin{bmatrix} \gamma_1^* \\ \gamma_2^* \end{bmatrix}.$$

The vectors π and β^* of section 1.3 are then given by $\pi = \gamma$, and

$$\beta^* = \begin{bmatrix} \mu_1 \gamma_1^* \\ \mu_1 \gamma_2^* + e \end{bmatrix},$$

so that Unit I is *stable* if and only if

$$\mu_1 \gamma \gamma^* + \gamma_2 e > 1, \tag{4.2.63}$$

or, equivalently,

$$\mu_1 \gamma \gamma^* > \gamma_1 e. \tag{4.2.64}$$

When Unit I is stable, we compute the matrix R, which is of the form

$$R = \begin{vmatrix} R_1 & R_2 \\ 0 & 0 \end{vmatrix},$$

where R_1 is of order K_1 and R_2 is $K_1 \times K_2$. The positive vector x_0 is similarly partitioned as $[x_0', x_0'']$, and the invariant probability vector $x = [x_0, x_1, x_2, \ldots]$ of $\tilde{P}$ is then given by

$$x_k' = x_0' R_1^k, \qquad x_k'' = x_0' R_1^{k-1} R_2, \quad \text{for} \quad k \geq 1. \tag{4.2.65}$$

The normalizing condition

$$x_0'(I - R_1)^{-1} e + x_0'(I - R_1)^{-1} R_2 e + x_0'' e = 1$$

clearly involves the vector x_0'' explicitly.

The stationary probability $z = [z_0, z_1, \ldots]$ of the embedded Markov chain, obtained by considering the chain $\tilde{P}$ only at transitions corresponding to true arrivals to Unit I, is now given by

$$z_k = [x_0'(I - R_1)^{-1} e]^{-1} x_0' R_1^k, \quad \text{for} \quad k \geq 0. \tag{4.2.66}$$

The waiting time distributions may be computed in exactly the same manner as for the $SM/M/1$ queue.

For computational use, this second decomposition is clearly preferable. It has additional advantages that we have not yet stated. By careful programming, the matrices R for both units can be efficiently computed at the same time. This results in some economy of processing time. While not overly significant in a single computation, this becomes very worthwhile when we consider the combinatorial optimization problem of deciding which states of the switch to designate to Unit I and which to Unit II. We are then dealing with a major computational problem for which the preceding algorithm needs to be implemented many times. Even the fact of having to compute the elements of the matrices A_k only once is then sufficient reason to prefer the second approach.

We conclude this discussion by pointing to another appealing use of discrete PH-distributions. Suppose that in our original example of the binary switch, changes from Unit I to Unit II and back occur at the transition epochs of a discrete alternating renewal process. Specifically, the numbers of customers sent to Units I and II are given by the odd- and even-indexed terms in the sequence of independent, positive integer-valued random variables $N_1, N_2, \ldots$. Let the odd-numbered terms have a common PH-density with irreducible representation $(\boldsymbol{\alpha}_1, T_1)$ of order K_1. Similarly, the even-numbered terms have a PH-density with the representation $(\boldsymbol{\alpha}_2, T_2)$ of order K_2.

It is now easily seen that such a switch is of the type specified in (4.2.59) with the matrix $Q(\cdot)$, given by

$$Q(x) = H(x) \begin{vmatrix} T_1 & T_1^o A_2^o \\ T_2^o A_1^o & T_2 \end{vmatrix}. \tag{4.2.67}$$

For this particular case, extensive computational simplifications are clearly at hand.

The discussion in this section evidently supports the common observation that mathematically equivalent solutions may differ widely in their suitability for numerical computation. A broader concern for such issues would clearly enhance the utility of the literature on stochastic models.

4.2c. Group Arrivals and Group Services

Many variants of the queueing models discussed in this book which involve the consideration of *group services* may easily be handled. They typically lead to an analytically more involved definition of the matrices A_k and B_k and therefore require more set-up computation. The dimension m of the matrices A_k is usually the same as for the case of single services, so that the most severely limiting factor in numerical implementation is unaffected by the consideration of group service.

In queues with non-Poisson renewal arrivals in groups, the input process may be viewed as a degenerate form of a semi-Markovian arrival process. *Group arrivals* therefore severely affect the dimension of the matrices A_k when treated by the methods of this book. An alternative is to settle for interarrival time distributions of phase type and to implement the algorithms based on Markov chains of the $M/G/1$ type, proposed in V. Ramaswami [261]. An overview of the effect of group arrivals and/or services on the structure of various classes of queues is given in M. F. Neuts [233].

We shall first describe how a number of models with *group service* may be approached. The details involved in modifying the case of single services are straightforward and will be left to the initiative of the reader.

In the simplest form of $GI/PH/1$ queue with group service, independent identically distributed positive integer-valued random variables N_n, $n \geq 1$, describe the maximum allowable sizes of the successive groups. If the queue length at the beginning of the n-th service exceeds N_n, then a group of N_n customers is served. If the queue length is not zero, but does not exceed N_n, then all customers enter service. If the queue is empty, the first arriving customer is served upon arrival. The successive service times are independent and identically distributed with the PH-distribution $K(\cdot)$ of representation $(\boldsymbol{\beta}, S)$. The service time distribution is the same for all group sizes. It is readily seen that the embedded Markov chain $\tilde{P}$ is exactly as in the single service case of section 4.1, except that the matrices $P(n, t)$, $n \geq 0$, $t \geq 0$, now correspond to the counting process of a PH-renewal process with group renewals. The recursive system of differential equations, analogous to (4.1.1), is easily specified. It may also be obtained by particularizing the differential equations in [232]. The remainder of the analysis of this model is entirely routine.

This model is also highly unrealistic. Of greater interest are various *threshold models* in which a minimum number of customers need to be present for service to start and in which there is an upper bound on the allowable group sizes. It is also desirable to allow the service time distribution to depend on the size of the group served. For the simpler case of Poisson arrivals, this model has been treated, mostly by transform methods, in a large number of papers.

Let the upper and lower threshold values be denoted by M and N respectively. If at the completion of a service fewer than N customers are present, the server waits until the queue length reaches N and then initiates service for a group of size N. If M or more customers are present, a group of size M is served. If at least N but fewer than M customers are present, all enter service as a group. The successive service times are conditionally independent, given the group sizes, but depend on the numbers of customers in the groups. All service time distributions are assumed to be of phase type.

In most cases, the main difficulties of this model consist in defining the states of the embedded Markov chain $\tilde{P}$ and in computing the boundary matrices. In view of the need to consider $M - N + 1$ possibly different phase type distributions, the columns of $\tilde{P}$ corresponding to the boundary states (all those for which fewer than $M - 1$ are present prior to the arrival) require a tedious but straightforward consideration of cases. Once this is accomplished, the analysis proceeds along classical lines.

For definiteness, we shall show the details of the definition of $\tilde{P}$ for the case $M = N > 1$, and for the sake of clarity, the matrix will be displayed for the representative case $M = 3$. The representation $(\boldsymbol{\beta}, S)$ will serve for the service time distribution of a group of M customers.

The state space of $\tilde{P}$ consists of the integers $0, 1, \ldots, M - 1$, and the pairs (i, j) with $i \geq M$, $1 \leq j \leq \nu$. The first index represents the queue length prior to the arrival epoch and the second index, when present, indicates the phase of the service process immediately after the arrival epoch.

An appealing feature of this case is that the matrices A_k, defined in section 4.1, reappear here. No new overhead computations need to be discussed. The matrices A_k, defined by letting $x \to \infty$ in (4.1.3), play a somewhat different role here. We shall keep the distinction clear by denoting these $\nu \times \nu$ matrices by A_k', for $k \geq 0$. It is also convenient to introduce the column vectors $\boldsymbol{b}_k$, by setting

$$b_k = \sum_{r=k+1}^{\infty} A_r{}'e = e - \sum_{r=0}^{k} A_r{}'e, \quad \text{for} \quad k \geq 0. \tag{4.2.68}$$

The matrix $\tilde{P}$, displayed for $M = 3$, is then of the form

$$\tilde{P} = \begin{array}{c|ccccccccccccc}
 & 0 & 1 & 2 & 3 & 4 & 5 & 6 & 7 & 8 & 9 & 10 & 11 & \\
\hline
0 & \cdot & 1 & \cdot & \cdot & \cdot & \cdot & \cdot & \cdot & \cdot & \cdot & \cdot & \cdot & \cdots \\
1 & \cdot & \cdot & 1 & \cdot & \cdot & \cdot & \cdot & \cdot & \cdot & \cdot & \cdot & \cdot & \cdots \\
2 & \cdot & \cdot & 1-{}_{\beta}A_0{}'e & {}_{\beta}A_0{}'e & \cdot & \cdot & \cdot & \cdot & \cdot & \cdot & \cdot & \cdot & \cdots \\
3 & \cdot & b_0 & \cdot & \cdot & A_0{}' & \cdot & \cdot & \cdot & \cdot & \cdot & \cdot & \cdot & \cdots \\
4 & \cdot & \cdot & b_0 & \cdot & \cdot & A_0{}' & \cdot & \cdot & \cdot & \cdot & \cdot & \cdot & \cdots \\
5 & b_1 & \cdot & \cdot & A_1{}' & \cdot & \cdot & A_0{}' & \cdot & \cdot & \cdot & \cdot & \cdot & \cdots \\
6 & \cdot & b_1 & \cdot & \cdot & A_1{}' & \cdot & \cdot & A_0{}' & \cdot & \cdot & \cdot & \cdot & \cdots \\
7 & \cdot & \cdot & b_1 & \cdot & \cdot & A_1{}' & \cdot & \cdot & A_0{}' & \cdot & \cdot & \cdot & \cdots \\
8 & b_2 & \cdot & \cdot & A_2{}' & \cdot & \cdot & A_1{}' & \cdot & \cdot & A_0{}' & \cdot & \cdot & \cdots \\
9 & \cdot & b_2 & \cdot & \cdot & A_2{}' & \cdot & \cdot & A_1{}' & \cdot & \cdot & A_0{}' & \cdot & \cdots \\
10 & \cdot & \cdot & b_2 & \cdot & \cdot & A_2{}' & \cdot & \cdot & A_1{}' & \cdot & \cdot & A_0{}' & \cdots \\
 & \vdots & \vdots & \vdots & \vdots & \vdots & \vdots & \vdots & \vdots & \vdots & \vdots & \vdots & \vdots &
\end{array} . \tag{4.2.69}$$

We see that, in general, for the present model, the matrices A_k are defined by

$$\begin{aligned}
A_{kM} &= A_k{}', && \text{for} \quad k \geq 0, \\
A_{kM+r} &= 0, && \text{for} \quad 1 \leq r \leq M - 1, \quad k \geq 0.
\end{aligned} \tag{4.2.70}$$

It is trivial to obtain the equilibrium condition

$$M\lambda_1{}' > \mu_1{}'. \tag{4.2.71}$$

The matrix R is now the minimal nonnegative solution to the equation

$$R = \sum_{k=0}^{\infty} R^{kM} A_k{}'. \tag{4.2.72}$$

There are a number of obvious simplifications, which the particular structure of $\tilde{P}$ induces into the computation of the vector $(x_0, x_1, \ldots, x_{M-1}, x_M)$, which is given by the system of linear equations

$$x_0 = x_M \sum_{r=1}^{\infty} R^{rM-1} b_r,$$

$$x_i = x_{i-1} + x_M R^{i-1} \sum_{r=0}^{\infty} R^{rM} b_r, \qquad \text{for} \quad 1 \le i \le M-1, \tag{4.2.73}$$

$$x_M = x_{M-1}\beta + x_M \sum_{r=1}^{\infty} R^{rM-1} A_r',$$

$$\sum_{i=0}^{M-1} x_i + x_M (I-R)^{-1} e = 1.$$

By routine calculations using (4.2.68) and (4.2.72), and setting

$$\sum_{r=1}^{\infty} R^{rM-1} A_r' = C,$$

these equations may be rewritten as

$$x_0 = x_M (I - R^M)^{-1} [I - R^{M-1} A_0' - C] e,$$

$$x_i = x_{i-1} + x_M (I - R^M)^{-1} (I - R) R^{i-1} e,$$

$$x_M = x_{M-1} \beta (I - C)^{-1}, \tag{4.2.74}$$

$$\sum_{i=0}^{M-1} x_i + x_M (I-R)^{-1} e = 1.$$

Their numerical solution is now easy. The invariant probability vector x of $\tilde{P}$ is then given by $[x_0, \ldots, x_{M-1}, x_M, x_M R, \ldots]$.

Remarks

a. Many derived quantities, such as the queue length density at an arbitrary time and the various waiting time distributions, may now be computed by methods similar to those in section 4.1. In the interest of brevity we shall omit the details.

b. A model, which is akin to but simpler than the preceding one, is the *GI/PH/*1 queue in which an idle server does not start up unless there are M customers present. Once service is started, it may continue until the queue becomes empty and all customers are served *singly.* This model, under the assumption of Poisson arrivals, has been considered by, among others, D. P. Heyman [138].

The embedded Markov chain, considered again at the successive arrival epochs, has the following structure, displayed here for the case $M = 3$:

$$
\tilde{P} = \begin{array}{c|cccccc|ccc}
 & 0' & 1' & 2' & 1 & 2 & 3 & 4 & 5 & \\
\hline
0' & \cdot & 1 & \cdot & \cdot & \cdot & \cdot & \cdot & \cdot & \cdots \\
1' & \cdot & \cdot & 1 & \cdot & \cdot & \cdot & \cdot & \cdot & \cdots \\
2' & \beta b_2 & \cdot & \cdot & \beta A_2' & \beta A_1' & \beta A_0' & \cdot & \cdot & \cdots \\
1 & b_1 & \cdot & \cdot & A_1' & A_0' & \cdot & \cdot & \cdot & \cdots \\
2 & b_2 & \cdot & \cdot & A_2' & A_1' & A_0' & \cdot & \cdot & \cdots \\
3 & b_3 & \cdot & \cdot & A_3' & A_2' & A_1' & A_0' & \cdot & \cdots \\
4 & b_4 & \cdot & \cdot & A_4' & A_3' & A_2' & A_1' & A_0' & \cdots \\
5 & b_5 & \cdot & \cdot & A_5' & A_4' & A_3' & A_2' & A_1' & \cdots \\
 & \vdots & \vdots & \vdots & \vdots & \vdots & \vdots & \vdots & \vdots &
\end{array} . \qquad (4.2.75)
$$

The matrices $A_k{}'$ are as defined before, with the understanding that (β, S) is now the representation of the service time distribution of individual customers. The vectors b_k, $k \geq M$, are as defined in (4.2.68).

We see that the threshold variable M enters the picture only through the boundary states. This makes it possible to obtain the optimal value of M under a variety of cost criteria with only a modicum of computational effort.

We next consider the more complex algorithmic problems associated with the $GI/PH/1$ and the $GI/M/c$ queues *with group arrivals*. By the device of letting all the sojourn times but one in a semi-Markovian arrival process be degenerate, group arrivals at renewal epochs may be viewed as a degenerate case of a semi-Markovian input process. It is not necessary to do so, but it explains the increase in the dimension of the blocks arising in the partitioning of the matrices $\tilde{P}$.

We shall primarily discuss the case where the group sizes at the successive arrivals are independent, identically distributed, *bounded* random variables. By a device similar to that employed in our discussion of platooned arrivals, it is also possible to construct an algorithmic solution for the case where the group sizes have a *discrete PH-distribution.*

Let the successive groups of arriving customers have the common probability density $\{p_1, p_2, \ldots, p_K\}$, with $p_K > 0$, on the integers $\{1, \ldots, K\}$. The mean group size is denoted by $\eta_1{}'$. The $GI^X/PH/1$ *queue* has an embedded Markov chain $\tilde{P}$ at arrival epochs which is easily constructed from that of the case with single arrivals, treated in section 4.1. The states (i, j), $i \geq 0$, $1 \leq j \leq \nu$, have the same interpretation as in section 4.1, and the same notation as was used there will be preserved.

For the sake of distinction, we adorn the matrices A_k obtained by setting x to infinity in (4.1.3) with primes. The matrix $\tilde{P}$ of section 4.1 will here be denoted by $\tilde{P}'$. Let now $\tilde{P}_r'$, $1 \leq r \leq K$, be the stochastic matrix obtained from $\tilde{P}'$ by the deletion of the first $r - 1$ rows of $\nu \times \nu$ blocks. It is then easily verified that the embedded Markov chain of the $GI^X/PH/1$ queue has the transition probability matrix $\tilde{P}$, given by

$$\tilde{P} = \sum_{r=1}^{K} p_r \tilde{P}_r'. \tag{4.2.76}$$

This matrix has the structure displayed in formula (1.5.6). In order to write $\tilde{P}$ in the canonical form, it is necessary to repartition that matrix into blocks of order $K\nu$, as was discussed in section 1.5. These blocks have the form given in (1.5.7), or, specifically,

$$A_0 = \begin{vmatrix} A_0'' & 0 & 0 & \cdots & 0 \\ A_1'' & A_0'' & 0 & \cdots & 0 \\ \cdot & \cdot & \cdot & & \cdot \\ \cdot & \cdot & \cdot & & \cdot \\ \cdot & \cdot & \cdot & & \cdot \\ A_{K-1}'' & A_{K-2}'' & A_{K-3}'' & \cdots & A_0'' \end{vmatrix},$$

$$A_k = \begin{vmatrix} A_{Kk}'' & A_{Kk-1}'' & A_{Kk-2}'' & \cdots & A_{Kk-K+1}'' \\ A_{Kk+1}'' & A_{Kk}'' & A_{Kk-1}'' & \cdots & A_{Kk-K+2}'' \\ \cdot & \cdot & \cdot & & \cdot \\ \cdot & \cdot & \cdot & & \cdot \\ \cdot & \cdot & \cdot & & \cdot \\ A_{Kk+K-1}'' & A_{Kk+K-2}'' & A_{Kk+K-3}'' & \cdots & A_{Kk}'' \end{vmatrix}, \quad \text{for} \quad k \geq 1, \tag{4.2.77}$$

where the $\nu \times \nu$ matrices A_r'', $r \geq 0$, are given by

$$A_r'' = \sum_{h=\max(1,K-r)}^{K} p_h A_{h+r-K}', \quad \text{for} \quad r \geq 0. \tag{4.2.78}$$

The matrix $A = \sum_{k=0}^{\infty} A_k$, is a *block-circulant* matrix of the form

$$A = \begin{vmatrix} A_0''' & A_{K-1}''' & A_{K-2}''' & \cdots & A_1''' \\ A_1''' & A_0''' & A_{K-1}''' & \cdots & A_2''' \\ \cdot & \cdot & \cdot & & \\ \cdot & \cdot & \cdot & & \\ \cdot & \cdot & \cdot & & \\ A_{K-1}''' & A_{K-2}''' & A_{K-3}''' & \cdots & A_0''' \end{vmatrix}, \tag{4.2.79}$$

where $A_r''' = \sum_{k=0}^{\infty} A_{kK+r}''$, for $0 \leq r \leq K - 1$.

Theorem 4.2.4. The $GI^X/PH/1$ queue is *stable* if and only if $\lambda_1' > \eta_1'\mu_1'$.

Proof. Let θ be the invariant probability vector of $A' = \sum_{k=0}^{\infty} A_k'$, then we have

$$\sum_{r=0}^{K-1} \theta A_r''' = \theta \sum_{r=0}^{K-1} A_r''' = \theta A' = \theta,$$

which readily shows that $K^{-1}[\theta, \theta, \ldots, \theta]$ is the invariant probability vector of the matrix A of formula (4.2.79).

Let $\beta^* = [\beta_0^*, \beta_1^*, \ldots, \beta_{K-1}^*]$ be the vector $\sum_{k=1}^{\infty} kA_k e$, partitioned into ν-vectors, then for $0 \le i \le K - 1$, we have

$$\begin{aligned}\beta_i^* &= \sum_{r=0}^{i} \sum_{h=0}^{\infty} hA_{Kh+r}'' e + \sum_{r=1}^{K-i-1} \sum_{h=0}^{\infty} hA_{Kh-r}'' e \\ &= \sum_{r=0}^{i} \sum_{h=0}^{\infty} hA_{Kh+r}'' e + \sum_{r=i+1}^{K-1} \sum_{h=0}^{\infty} hA_{Kh+r}'' e + \sum_{r=i+1}^{K-1} A_r''' e \\ &= \sum_{r=0}^{K-1} \sum_{h=0}^{\infty} hA_{Kh+r}'' e + \sum_{r=i+1}^{K-1} A_r''' e.\end{aligned}$$

The inner product of the vectors $K^{-1}[\theta, \theta, \ldots, \theta]$ and β^* is therefore given by

$$\begin{aligned}&\theta \sum_{r=0}^{K-1} \sum_{h=0}^{\infty} hA_{Kh+r}'' e + K^{-1} \sum_{i=0}^{K-1} \sum_{r=i+1}^{K-1} \theta A_r''' e \\ &\qquad = K^{-1}\left[\theta \sum_{r=0}^{K-1} \sum_{h=0}^{\infty} KhA_{Kh+r}'' e + \theta \sum_{r=0}^{K-1} r \sum_{h=0}^{\infty} A_{Kh+r}'' e\right] \\ &\qquad = K^{-1}\theta \sum_{h=0}^{\infty} hA_h'' e = K^{-1}\theta \sum_{h=0}^{\infty} (K - h - \eta_1')A_h' e \\ &\qquad = K^{-1}(K - \theta\beta^{*\prime} - \eta_1') = 1 - K^{-1}\left(\frac{\lambda_1'}{\mu_1'} - \eta_1'\right),\end{aligned}$$

since in the $GI/PH/1$ queue with single arrivals $\theta\beta^{*\prime} = \lambda_1'\mu_1'^{-1}$. The final expression exceeds one if and only if $\lambda_1' > \eta_1'\mu_1'$.

Remarks

a. The matrix R for the stable $GI^X/PH/1$ queue is a *positive* matrix of order $K\nu$. This readily follows from the positivity of the matrices A_k', $k \ge 1$. The particular structure of the matrices A_k, $k \ge 0$, shown in (4.2.69), may be exploited to reduce the memory storage requirements, but no simplifying structural properties of the matrix R are generally available. The curse of dimensionality is inherent in the physical description of this model. No substantive algorithmic simplifications are to be expected from alternate general methods of solution.

The computation of all other steady-state features of this queue proceeds along classical lines. For the sake of brevity, we leave the detailed development to the initiative of the reader.

b. It has been observed that in transform solutions of related models, the case of *geometrically* distributed group sizes leads to somewhat more tractable expressions. It is now evident that the underlying reason for this is that the geometric distribution is a particular PH-distribution. A complete matrix-geometric analysis may be given of the $GI^X/PH/1$ queue in which the group size distribution is of *phase type*. We shall not go into the details, but the key to this solution lies in the definition of the state space. The embedded Markov chain has the states (i, j, h), $i \geq 0$, $1 \leq j \leq \nu$, $1 \leq h \leq K$. The index i now represents the number of *groups* of which at least one customer is still in the queue. The index j represents the service phase immediately after an arrival. The index h serves to describe the phase within the group of which a customer is currently in service. K is the order of the representation of the discrete PH-density of the group sizes. The matrices A_k and B_k, $k \geq 0$, are again of order $K\nu$, and the matrix $\tilde{P}$ is of the canonical form. It is easily seen how from this description of the state space and from the invariant vector of $\tilde{P}$, the density of the queue length and the waiting time distributions may be constructed. The same dimensionality problem is also present here.

The $GI^X/M/c$ queue with service rate μ and interarrival time distribution $F(\cdot)$, in which the group sizes are bounded above by K, has an embedded Markov chain at the arrival epochs which is similar in structure to that of the preceding model. There are two points that merit a more detailed discussion. The first deals with the boundary behavior, which needs to be treated somewhat differently when $1 \leq K < c$, than in the case when $K \geq c$. The second deals with the numerical computation of the entries of $\tilde{P}$ for a general probability distribution $F(\cdot)$. The greatly simplified computation for the case where $F(\cdot)$ is *of phase type* was discussed in section 2.5. The $GI^X/M/c$ queue was discussed by B. W. Conolly [68], M. F. Neuts [231], R. A. Restrepo [266], and with algorithmic implementations in D. E. Baily and M. F. Neuts [10].

The transition probability matrix $\tilde{P}$ of the Markov chain embedded at arrivals is given by

$$\tilde{P} = \begin{vmatrix} P_{00} & \cdots & P_{0,c-1} & a_{K-c} & \cdots & a_1 & a_0 & 0 & 0 & \cdots \\ P_{10} & \cdots & P_{1,c-1} & a_{K-c+1} & \cdots & a_2 & a_1 & a_0 & 0 & \cdots \\ P_{20} & \cdots & P_{2,c-1} & a_{K-c+2} & \cdots & a_3 & a_2 & a_1 & a_0 & \cdots \\ P_{30} & \cdots & P_{3,c-1} & a_{K-c+3} & \cdots & a_4 & a_3 & a_2 & a_1 & \cdots \\ P_{40} & \cdots & P_{4,c-1} & a_{K-c+4} & \cdots & a_5 & a_4 & a_3 & a_2 & \cdots \\ P_{50} & \cdots & P_{5,c-1} & a_{K-c+5} & \cdots & a_6 & a_5 & a_4 & a_3 & \cdots \\ \vdots & & \vdots & \vdots & & \vdots & \vdots & \vdots & \vdots & \end{vmatrix}, \tag{4.2.80}$$

provided that $K \geq c$. For $1 \leq K < c$, we obtain instead

$$\tilde{P} = \begin{vmatrix} P_{00} & \cdots & P_{0,c-1} & 0 & 0 & \cdots & 0 & 0 & \cdots \\ \vdots & & \vdots & \vdots & \vdots & & \vdots & \vdots & \\ P_{c-K-1,0} & \cdots & P_{c-K-1,c-1} & 0 & 0 & \cdots & 0 & 0 & \cdots \\ P_{c-K,0} & \cdots & P_{c-K,c-1} & a_0 & 0 & \cdots & 0 & 0 & \cdots \\ \vdots & & \vdots & \vdots & \vdots & & \vdots & \vdots & \\ P_{c-1,0} & \cdots & P_{c-1,c-1} & a_{K-1} & a_{K-2} & \cdots & a_0 & 0 & \cdots \\ P_{c,0} & \cdots & P_{c,c-1} & a_K & a_{K-1} & \cdots & a_1 & a_0 & \cdots \\ P_{c+1,0} & \cdots & P_{c+1,c-1} & a_{K+1} & a_K & \cdots & a_2 & a_1 & \cdots \\ \vdots & & \vdots & \vdots & \vdots & & \vdots & \vdots & \end{vmatrix} \tag{4.2.81}$$

The elements of both matrices can be explicitly defined in terms of the parameters of the queue, but for algorithmic purposes it is more advantageous to show how, as in the case of the $GI^X/PH/1$ model, these matrices are simply related to the transition probability matrix for the embedded Markov chain of the simpler $GI/M/c$ queue *with single arrivals*. It will therefore suffice to show how the latter may be computed.

The transition probability matrix $\tilde{P}'$ for the $GI/M/c$ queue with single arrivals is given by

$$\tilde{P}' = \begin{vmatrix} P_{00}' & P_{01}' & 0 & \cdots & 0 & 0 & 0 & 0 & \cdots \\ P_{10}' & P_{11}' & P_{12}' & \cdots & 0 & 0 & 0 & 0 & \cdots \\ \vdots & \vdots & \vdots & & \vdots & \vdots & \vdots & \vdots & \\ P'_{c-1,0} & P'_{c-1,1} & P'_{c-1,2} & \cdots & P'_{c-1,c-1} & b_0 & 0 & 0 & \cdots \\ P'_{c,0} & P'_{c,1} & P'_{c,2} & \cdots & P'_{c,c-1} & b_1 & b_0 & 0 & \cdots \\ P'_{c+1,0} & P'_{c+1,1} & P'_{c+1,2} & \cdots & P'_{c+1,c-1} & b_2 & b_1 & b_0 & \cdots \\ P'_{c+2,0} & P'_{c+2,1} & P'_{c+2,2} & \cdots & P'_{c+2,c-1} & b_3 & b_2 & b_1 & \cdots \\ \vdots & \vdots & \vdots & & \vdots & \vdots & \vdots & \vdots & \end{vmatrix}, \tag{4.2.82}$$

where

$$b_j = \int_0^\infty e^{-c\mu u} \frac{(c\mu u)^j}{j!} \, dF(u), \quad \text{for} \quad j \geq 0,$$

$$\begin{aligned} P_{ij}' &= 0, & \text{for} \quad i < j \leq c, \\ &= \int_0^\infty \binom{i+1}{j} e^{-j\mu u}(1 - e^{-\mu u})^{i-j+1} dF(u), & \text{for} \quad j \leq i \leq c, \\ &= \int_0^\infty dF(u) \int_0^u e^{-c\mu v} \frac{(c\mu v)^{i-c}}{(i-c)!} \binom{c}{j} e^{-j\mu(u-v)}[1 - e^{-\mu(u-v)}]^{c-j} c\mu dv, \\ & \qquad \text{for} \quad i > c, \quad 0 \leq j \leq c-1. \end{aligned} \tag{4.2.83}$$

The matrices $\tilde{P}$ in (4.2.80) and (4.2.81) are now obtained by forming

$$\tilde{P} = \sum_{r=1}^{K} p_r \tilde{P}_r', \tag{4.2.84}$$

where $\tilde{P}_r'$ is the stochastic matrix obtained by deleting the first $r - 1$ rows of the matrix $\tilde{P}'$.

For $K \geq c$, we partition the matrix $\tilde{P}$ in (4.2.80) into $K \times K$ blocks in the natural manner and we obtain a matrix of the canonical form. For the case where $1 \leq K < c$, a slightly more involved partitioning is required. We now write $\tilde{P}$, as given in formula (4.2.81), into the form

$$\tilde{P} = \begin{vmatrix} B_{00} & B_{01} & 0 & 0 & 0 & \cdots \\ B_{10} & B_{11} & A_0 & 0 & 0 & \cdots \\ B_{20} & B_{21} & A_1 & A_0 & 0 & \cdots \\ B_{30} & B_{31} & A_2 & A_1 & A_0 & \cdots \\ B_{40} & B_{41} & A_3 & A_2 & A_1 & \cdots \\ \vdots & \vdots & \vdots & \vdots & \vdots & \end{vmatrix}, \tag{4.2.85}$$

where the matrices B_{n0}, $n \geq 1$, have dimensions $K \times (c - K)$; B_{00} is a square matrix of order $c - K$, B_{01} is of dimensions $(c - K) \times K$, and the B_{n1}, $n \geq 1$, are square matrices of order K.

The matrices A_k, $k \geq 0$, are given by

$$A_0 = \begin{vmatrix} a_0 & 0 & \cdots & 0 \\ a_1 & a_0 & \cdots & 0 \\ \vdots & \vdots & & \vdots \\ a_{K-1} & a_{K-2} & \cdots & a_0 \end{vmatrix},$$

$$A_k = \begin{vmatrix} a_{Kk} & a_{Kk-1} & \cdots & a_{Kk-K+1} \\ a_{Kk+1} & a_{Kk} & \cdots & a_{Kk-K+2} \\ \vdots & \vdots & & \vdots \\ a_{Kk+K-1} & a_{Kk+K-2} & \cdots & a_{Kk} \end{vmatrix}, \quad \text{for} \quad k \geq 1.$$

A calculation similar to that in theorem 4.2.4 leads to the anticipated equilibrium condition

$$\eta_1' < c\mu\lambda_1', \tag{4.2.86}$$

where, as before, λ_1' is the mean interarrival time of groups and η_1' denotes the mean group size.

For $K \geq c$, the stationary density $\mathbf{x} = [\mathbf{x}_0, \mathbf{x}_1, \ldots]$ of the queue length prior to arrivals is given by the standard equations

$$R = \sum_{k=0}^{\infty} R^k A_k,$$

$$\mathbf{x}_0 = \mathbf{x}_0 B[R],$$

$$x_k = x_0 R^k, \quad \text{for} \quad k \geq 0,$$
$$x_0(I - R)^{-1}e = 1. \tag{4.2.87}$$

For the case $1 \leq K \leq c - 1$, we partition x into a $(c - K)$-vector x_0 and K-vectors x_k, $k \geq 1$. The modified matrix-geometric solution is now given by the formulas

$$R = \sum_{k=0}^{\infty} R^k A_k,$$
$$(x_0, x_1) \begin{vmatrix} B_{00} & B_{01} \\ B_0[R] & B_1[R] \end{vmatrix} = (x_0, x_1), \tag{4.2.88}$$
$$x_k = x_1 R^{k-1}, \quad \text{for} \quad k \geq 1,$$
$$x_0 e + x_1(I - R)^{-1}e = 1,$$

where $B_0[R] = \Sigma_{n=1}^{\infty} R^{n-1} B_{n0}$, $B_1[R] = \Sigma_{n=1}^{\infty} R^{n-1} B_{n1}$.

Once the vector x has been evaluated, it is a matter of (sometimes tedious) routine to derive algorithmically useful expressions for the distributions of the queue length at an arbitrary time and for the means and variances of the queue lengths and waiting times. This is of some interest for the present model, as both indices i and j of the state (i, j) in the partitioned Markov chain $\tilde{P}$ are involved with the queue length. The state (i, j) does indeed correspond to a queue of length $iK + j$, $i \geq 0$, $0 \leq j \leq K - 1$, at least for the case $K \geq c$. For the alternate case, the states $(0, j)$, $0 \leq j \leq c - K - 1$, correspond to a queue length of j, and the states (i, j), $i \geq 1$, $0 \leq j \leq K - 1$, to the presence of $(c - K) + (i - 1)K + j$ customers in the system. As we shall see, this leads to involved but computationally useful moment formulas.

The following results are stated without proofs. For the latter, we refer the interested reader to [10] and [231].

Let $\tilde{\tilde{P}}$ be the stochastic matrix obtained by replacing $dF(u)$ by $\lambda_1'^{-1}[1 - F(u)]du$, throughout in the definition of the elements of the matrix $\tilde{P}$. The new matrix is partitioned in the same manner as $\tilde{P}$. The blocks in $\tilde{\tilde{P}}$ will be distinguished by the addition of a *caret* (ˆ). The vector y, similarly partitioned as $[y_0, y_1, y_2, \ldots]$, represents the stationary probability density of the queue length at an arbitrary time.

Theorem 4.2.5. Provided the queue is stable and the probability distribution $F(\cdot)$ is nonlattice, the vector y is given by

$$y = x\tilde{\tilde{P}},$$

or explicitly by

$$y_0 = x_0 \sum_{n=0}^{\infty} R^n \hat{B}_n = x_0 \hat{B}[R],$$

$$y_i = x_0 R^{i-1} \hat{A}[R], \quad \text{for} \quad i \geq 1, \tag{4.2.89}$$

when $K \geq c$. Similarly, when $1 \leq K < c$,

$$y_0 = x_0 \hat{B}_{00} + x_1 \sum_{n=1}^{\infty} R^{n-1} \hat{B}_{n0} = x_0 \hat{B}_{00} + x_1 \hat{B}_0[R],$$

$$y_1 = x_0 \hat{B}_{01} + x_1 \sum_{n=1}^{\infty} R^{n-1} \hat{B}_{n1} = x_0 \hat{B}_{01} + x_1 \hat{B}_1[R], \tag{4.2.90}$$

$$y_i = x_1 R^{i-2} \hat{A}[R], \quad \text{for} \quad i \geq 2.$$

The matrix $\hat{A}[R]$ is defined by

$$\hat{A}[R] = \sum_{n=0}^{\infty} R^n \hat{A}_n. \tag{4.2.91}$$

By L, $\hat{L}$ and V, $\hat{V}$ we shall denote the means and the variances of the probability densities $\boldsymbol{x}$ and $\boldsymbol{y}$ respectively. In order to express these moments concisely, we introduce the column vectors

$$\boldsymbol{u} = (0, 1, 2, \ldots, K-1),$$
$$\boldsymbol{v} = (0, 1, 2, \ldots, c-K-1),$$
$$\boldsymbol{w} = (c-K, c-K+1, \ldots, c-1),$$
$$\boldsymbol{u}^o = (0, 1, 2^2, \ldots, (K-1)^2),$$
$$\boldsymbol{v}^o = (0, 1, 2^2, \ldots, (c-K-1)^2),$$
$$\boldsymbol{w}^o = ((c-K)^2, (c-K+1)^2, \ldots, (c-1)^2).$$

The following moment formulas then hold:

For $K \geq c$:

$$L = x_0(I-R)^{-1}\boldsymbol{u} + Kx_0(I-R)^{-2}R\boldsymbol{e},$$

$$\hat{L} = x_0\hat{B}[R]\boldsymbol{u} + x_0(I-R)^{-1}\hat{A}[R]\boldsymbol{u} + Kx_0(I-R)^{-2}\hat{A}[R]\boldsymbol{e},$$

$$V = x_0(I-R)^{-1}\boldsymbol{u}^o + x_0(I-R)^{-2}R(2K\boldsymbol{u} - K^2\boldsymbol{e}) + 2K^2x_0(I-R)^{-3}R\boldsymbol{e} - L^2, \tag{4.2.92}$$

$$\hat{V} = x_0\hat{B}[R]\boldsymbol{u}^o + x_0(I-R)^{-1}\hat{A}[R]\boldsymbol{u}^o + 2K^2x_0(I-R)^{-3}\hat{A}[R]\boldsymbol{e} + x_0(I-R)^{-2}\hat{A}[R](2K\boldsymbol{u} - K^2\boldsymbol{e}) - \hat{L}^2.$$

For $1 \leq K < c$:

$$L = x_0 v + x_1(I-R)^{-1}w + Kx_1(I-R)^{-2}Re,$$

$$\hat{L} = x_0\hat{B}_{00}v + x_1\hat{B}_0[R]v + x_0\hat{B}_{01}w + x_1\hat{B}_1[R]w$$
$$+ x_1(I-R)^{-1}\hat{A}[R]w + Kx_1(I-R)^{-2}\hat{A}[R]e,$$

$$V = x_0 v^o + x_1(I-R)^{-1}w^o + x_1(I-R)^{-2}R(2Kw - K^2e)$$
$$+ 2K^2x_1(I-R)^{-3}Re - L^2,$$

$$\hat{V} = x_0\hat{B}_{00}v^o + x_1\hat{B}_0[R]v^o + x_0\hat{B}_{01}w^o + x_1\hat{B}_1[R]w^o$$
$$+ x_1(I-R)^{-1}\hat{A}[R]w^o + x_1(I-R)^{-2}\hat{A}[R](2Kw - K^2e)$$
$$+ 2K^2x_1(I-R)^{-3}\hat{A}[R]e - \hat{L}^2.$$

When $F(\cdot)$ is not of phase type, the numerical evaluation of the quantities in formula (4.2.83) may proceed as follows.

We define $\phi_{i+1,j}(u)$ to be

$$\int_0^u e^{-c\mu v}\frac{(c\mu v)^{i-c-1}}{(i-c-1)!}\binom{c}{j}e^{-j\mu(u-v)}[1-e^{-\mu(u-v)}]^{c-j}c\mu\,dv, \qquad (4.2.93)$$

for $i > c$, $0 \leq j \leq c$. These functions satisfy the system of differential equations

$$\phi'_{i+1,j}(u) = -\mu j\phi_{i+1,j}(u) + \mu(j+1)\phi_{i+1,j+1}(u), \qquad (4.2.94)$$

for $i \geq c$, $0 \leq j \leq c-1$, with the initial conditions $\phi_{i+1,j}(0) = 0$, for $i \geq c$, $0 \leq j \leq c-1$. The system (4.2.94) is inhomogeneous since clearly

$$\phi_{i+1,c}(u) = e^{-c\mu u}\frac{(c\mu u)^{i-c+1}}{(i-c+1)!}, \quad \text{for} \quad i \geq c.$$

It is now most convenient to integrate the differential equations (4.2.94) by standard methods and to integrate the functions $\phi_{i+1,j}(u)$ with respect to $F(\cdot)$ by means of a progressive numerical integration method such as Simpson's rule. This yields the quantities P_{ij}' for $i > c$, $0 \leq j \leq c-1$. The integrals for P_{ij}', $j \leq i \leq c-1$, and for the b_j, $j \geq 0$, may be evaluated by direct numerical integration, but it is just as feasible to set up difference-differential equations for their integrands, as was done in (4.2.94), and to proceed in the same manner as before.

The type of integrals in formula (4.2.83) are but simple examples of those which typically arise in the study of queues with renewal input to a finite state Markovian service unit. The same method, that is, numerical integration of a system of difference-differential equations with a concurrent evaluation of the required Stieltjes integrals, may be widely applied. For a more involved example, we refer to D. M. Lucantoni [195].

The quantities $\hat{b}_j$ and $\hat{P}_{ij}'$, obtained from (4.2.83) by replacing $dF(u)$ by $\lambda_1'^{-1}[1 - F(u)]$, are easily computed by use of the formulas

$$\hat{b}_j = \frac{1}{c\mu\lambda_1'}\left(1 - \sum_{\nu=0}^{j} b_\nu\right), \quad \text{for} \quad j \geq 0,$$

$$\hat{P}_{ij}' = \frac{1}{j\mu\lambda_1'} \sum_{\nu=0}^{j-1} P_{i\nu}', \qquad \text{for} \quad i \geq c, \quad 1 \leq j \leq c-1, \tag{4.2.95}$$

$$\hat{P}_{i0}' = 1 - \sum_{j=1}^{c-1} \hat{P}_{ij}' - \sum_{\nu=0}^{i-c} \hat{b}_\nu, \quad \text{for} \quad i \geq c,$$

but the $\hat{P}_{ij}'$, for $j \leq i \leq c$, require direct computation.

4.3. A DESIGN PROBLEM FOR THE *D/PH/1* QUEUE

One of the most useful byproducts of the algorithmic analysis of stochastic models is the ease with which a number of analytically arduous design and control problems may be studied. This is particularly evident when the algorithms are programmed in a conversational mode. Progressively more belabored, but also more informative, subroutines may then be brought into play as the search for optimal parameter values approaches its target.

This methodological approach is widely used in the practice of statistics and engineering design. Its use in stochastic modeling has largely been limited to the costly process of varying parameters in simulation studies. Such an approach shows little interaction with the extensive theoretical literature on control of stochastic models, such as those for queues, population sizes, or epidemics.

Nearly every model discussed in this book could be used to illustrate the purposeful use of interactive computation to study issues of optimization. The present model was chosen because its discussion may be kept fairly brief while covering all essential points, but also and primarily because the consideration of this model has a concrete and practical origin.

The problem is that of finding the optimal interarrival time a in a single server queue with a PH-service time distribution $(\boldsymbol{\beta}, S)$ of mean μ_1' and constant interarrival times. The particular $D/M/1$ case is analytically somewhat tractable and was treated in detail in B. Jansson [152]. Even in that case the determination of the "optimal" value of a ultimately rests on numerical comparisons. A study that is somewhat in this spirit is reported in A. Soriano [287–88]. There are no realistically useful cost criteria for which the optimal a is available in explicit analytic form.

One practical motivation for this problem may be given as that of "choosing a good appointment system." Ours is specifically related to the problem of timing the arrival of work pieces for processing by a machine tool. If the (cutting) speed of the tool is constant, the work piece spends a constant time in process and there is no problem in finding the appropriate interarrival time. By installing a sensor of, for example, the hardness of the material, it becomes feasible adaptively to increase or decrease the cutting speed during the processing of each item. The former reduces the processing time; the latter may be desirable to reduce tool wear. The service time now becomes random, and its underlying distribution may be represented by a discrete or continuous PH-distribution.

In order to preserve the notation of our discussion of the $GI/PH/1$ queue, we let the service time distribution be *continuous,* although computations for the discrete case are slightly more expeditious. The cost criteria will not be explicitly stated. In fact, practical decisions usually involve *several* quantifiable criteria, together with vaguer ones such as the desirability of allocating a given space for in-process inventory. Our purpose is to have the algorithmic solution provide as much useful numerical information as possible, which may either be summarized into specific cost functions or presented as it is for the consideration of the design engineer.

Useful items, in order of the computational effort required for their evaluation, are:

1. the fraction of time the machine will be idle
2. the (stationary) probability that an arriving work piece does not have to wait
3. the means and standard deviations of the queue length and the waiting time distributions at arrivals and at arbitrary times
4. the distributions of the queue length at arrivals and at arbitrary times
5. the distributions of the waiting time of arriving customers and of the server backlog (or virtual waiting time).

The last four distributions may most conveniently be reported by printing sets of selected percentage points, but these nevertheless require the evaluation of the distributions and involve more computation time than the other items.

Only the first item, the fraction of idle time, is explicitly available and given by $1 - \mu_1'/a$. In order for the queue to be stable, it is of course necessary that $a > \mu_1'$. It is clearly undesirable to have the machine idle most of the time. A minimum value ϵ for the quantity $1 - \mu_1'/a$ leads to

$$a \leq (1 - \epsilon)^{-1}\mu_1' = a_{\max}.$$

This enables us to contain the values of a to be considered within the interval $(\mu_1', a_{\max}]$. A variety of search techniques may now be used to approach optimal or desirable values of a. The classical methods, such as binary or Fibonacci search, may be implemented when one has little prior intuition about where good values of a are to be found.

For the application at hand, one may wish to try systematically the values $a_k = \mu_1'100/(100 - k)$, for $k = 1, 2, \ldots$. For $a = a_k$, the machine is then idle k percent of the time. A larger initial step may also be desirable for reasons that are obvious.

In the $D/PH/1$ queue with interarrival time a, the matrices A_k are given by $A_k = P(k, a)$, for $k \geq 0$. An increasing set of trial values for a has the advantage that one can simply continue the numerical integration of the system of differential equations (4.1.1) to go from one trial value of a to the next. The recursive computation of the matrix $R(a)$ from the equation

$$R(a) = \sum_{k=0}^{\infty} R^k(a)P(k, a) \tag{4.3.1}$$

may also be initialized with the preceding R-matrix.

The matrix $R(a)$, on which all other stationary features of the queue depend, is clearly a highly involved and implicit function of the parameters of the queue. It is not at all astonishing that an explicit determination of a is out of the question.

Values of a close to μ_1' will usually be discarded on the basis of the first two moments of the queue length only. As a increases, we will see the mean queue lengths decrease at the expense of more idle time in the system. It may now occur that the mean queue lengths have satisfactorily decreased, but that the corresponding standard deviation is still large. This is indicative of a service time distribution for which occasionally long service times may arise, even though the mean processing time is small. The rare long services will cause substantial build-up in the queue.

Whether this is to be avoided greatly depends on the application at hand. A long queue length may cause hazardous storage problems or a correspondingly long wait may, for example, cause the work piece to cool to the point that it can no longer be processed. In such a situation, we shall have to determine the smallest value of a for which, for example, the 0.999 quantile of the queue length or waiting time distributions *at arrivals* do not exceed given values. To deal with such design problems, the asymptotic results of theorem 4.1.5 are useful.

If an occasional build-up of the queue is not a serious hazard but merely an inconvenience, the distribution of the queue length at an arbitrary time provides the appropriate information. It enables us to determine the smallest value of a for which the fraction of time that the queue length is in excess of a given value $L_{\max}$ is smaller than a given $\epsilon_1 > 0$.

For cost criteria, which are based on *rare events* related to queue length or waiting time (as opposed to average values), it is usually necessary to compute the stationary distributions themselves. For the queue length distribution, this is readily done by making use of its matrix-geometric form. For the waiting time distribution, we may use the approximations given by theorem 4.1.5 or implement the efficient algorithm discussed in section 4.1.

In conclusion, we offer a comment on the value of certain qualitative results. It may be argued that, in general, we will not know that a given cost or design criterion is, for example, a monotone or a convex function of a and that the "optimal" value of a which we compute rests largely on heuristics and physical intuition. Such indeed is the nature of many real, nonlinear optimization problems. Where they are demonstrable, qualitative results, particularly on the dependence of the quantiles of the queue length or waiting time distributions, are very useful. Few such results are available to date.

4.4. SOME MODELS FOR DAMS

There is great similarity of structure between the Markov chains, which have arisen out of the models for dams, and those of queueing theory. The Markov chains appearing in the study of the dam with Markovian input and unit release, discussed by M. S. Ali Khan and J. Gani [2] and S. Odoom and E. H. Lloyd [244], are of the $M/G/1$ type and have special features that allow for particularly efficacious numerical solution. In that model, the content of the dam is measured from emptiness upward and is conceived as being unbounded above.

In recent papers, R. M. Phatarfod [250, 251] and A. G. Pakes and R. M. Phatarfod [248] and A. G. Pakes [249] have cogently argued for measuring the dam content *downward* from a given level at which it overflows and for conceiving (for mathematical tractability) of the dam as being infinitely deep. Such a model describes a *bottomless dam* with the preceding aptly said to be topless.

We shall discuss the more complex of the models for the bottomless dam in detail. The dam is considered at the epochs $n \geq 0$, immediately *after* the release of $M \geq 1$ units of water. The choice of the states requires special attention. We shall obtain a Markov chain with the state space $E = \{(i, j): i \geq 0, 0 \leq j \leq r\}$. The index i measures the *depletion* (Y_n) in units of water from the maximum possible content immediately after a release. The value $i = 0$ corresponds to M units below the overflow level of the dam. The index j describes the input X_n during the time interval $(n, n + 1)$.

The inputs X_n during the time periods $(n, n + 1)$ have a seasonal behavior, which is modeled by assuming that the sequence $\{X_n, n \geq 0\}$ is an $(r + 1)$-state Markov chain with the states $0, 1, \ldots, r$, and an *irreducible* transition probability matrix A'. Since

$$Y_{n+1} = (Y_n + M - X_n)^+,$$

it readily follows that the bivariate sequence $\{(Y_n, X_n), n \geq 0\}$ is a Markov chain on the state space E with transition probability matrix $\tilde{P}$, given by

$$\tilde{P} = \begin{vmatrix} B'_{M-1} & A'_{M-1} & \cdots & A'_1 & A'_0 & 0 & \cdots & 0 & \cdots \\ B'_M & A'_M & \cdots & A'_2 & A'_1 & A'_0 & \cdots & 0 & \cdots \\ \vdots & \vdots & & \vdots & \vdots & \vdots & & \vdots & \\ B'_{2M-2} & A'_{2M-2} & \cdots & A'_M & A'_{M-1} & A'_{M-2} & \cdots & A'_0 & \cdots \\ B'_{2M-1} & A'_{2M-1} & \cdots & A'_{M+1} & A'_M & A'_{M-1} & \cdots & A'_1 & \cdots \\ \vdots & \vdots & & \vdots & \vdots & \vdots & & \vdots & \\ B'_{3M-2} & A'_{3M-2} & \cdots & A'_{2M} & A'_{2M-1} & A'_{2M-2} & \cdots & A'_M & \cdots \\ B'_{3M-1} & A'_{3M-1} & \cdots & A'_{2M+1} & A'_{2M} & A'_{2M-1} & \cdots & A'_{M+1} & \cdots \\ \vdots & \vdots & & \vdots & \vdots & \vdots & & \vdots & \end{vmatrix}. \quad (4.4.1)$$

The matrices A_ν' of order $r + 1$ are zero for $\nu > r$. For $0 \leq \nu \leq r$, they are zero *except for the row of index* ν, which is given by the corresponding row of the matrix A'. The matrices B_ν' of order $r + 1$ are given by

$$\begin{aligned} B'_\nu &= \sum_{k=\nu+1}^{r} A'_k, \quad \text{for} \quad M - 1 \leq \nu < r, \\ &= 0, \qquad\qquad \text{for} \quad \nu \geq r. \end{aligned} \quad (4.4.2)$$

Since the maximum possible input during any interval $(n, n + 1)$ is at most r, it is necessary for the ergodicity of $\tilde{P}$ that $M < r$. When partitioned in the natural manner into square blocks of order $(r + 1)M$, the matrix $\tilde{P}$ is clearly of canonical form. The matrices A_k, $k \geq 0$, are given by

$$A_0 = \begin{vmatrix} A_0' & 0 & \cdots & 0 \\ A_1' & A_0' & \cdots & 0 \\ \cdot & \cdot & & \cdot \\ \cdot & \cdot & & \cdot \\ \cdot & \cdot & & \cdot \\ A_{M-1}' & A_{M-2}' & \cdots & A_0' \end{vmatrix},$$

$$A_k = \begin{vmatrix} A_{kM}' & A_{kM-1}' & \cdots & A_{kM-M+1}' \\ A_{kM+1}' & A_{kM}' & \cdots & A_{kM-M+2}' \\ \cdot & \cdot & & \cdot \\ \cdot & \cdot & & \cdot \\ \cdot & \cdot & & \cdot \\ A_{kM+M-1}' & A_{kM+M-2}' & \cdots & A_{kM}' \end{vmatrix}, \tag{4.4.3}$$

for $k \geq 1$. Defining the matrices A_k'' by

$$A_k'' = \sum_{\nu=0}^{\infty} A_{\nu M+k}', \quad \text{for} \quad 0 \leq k \leq M-1, \tag{4.4.4}$$

we see that

$$A = \sum_{\nu=0}^{\infty} A_\nu = \begin{vmatrix} A_0'' & A_{M-1}'' & A_{M-2}'' & \cdots & A_1'' \\ A_1'' & A_0'' & A_{M-1}'' & \cdots & A_2'' \\ \cdot & \cdot & \cdot & & \cdot \\ \cdot & \cdot & \cdot & & \cdot \\ \cdot & \cdot & \cdot & & \cdot \\ A_{M-1}'' & A_{M-2}'' & A_{M-3}'' & \cdots & A_0'' \end{vmatrix} \tag{4.4.5}$$

is a block-circulant matrix. Elementary calculations show that its invariant probability vector is given by $M^{-1}[\boldsymbol{\pi}, \boldsymbol{\pi}, \ldots, \boldsymbol{\pi}]$, where $\boldsymbol{\pi}$ satisfies $\boldsymbol{\pi} A' = \boldsymbol{\pi}$, $\boldsymbol{\pi e} = 1$. The equilibrium condition simplifies to

$$\sum_{j=1}^{r} j\pi_j > M. \tag{4.4.6}$$

We exclude from consideration the trivial case where A' is a *permutation matrix.* Except in that case, the Markov chain $\tilde{P}$ is irreducible.

In contrast to the similar case of the $GI^X/PH/1$ queue, there are major algorithmic simplifications in this model. Let the rows of the matrix R be indexed by (j, k), where $0 \leq j \leq M-1$, $0 \leq k \leq r$. Any zero rows of the matrix A_0 result in a zero row with the same indices in R. The matrix R of order $(r+1)M$ therefore has only $\frac{1}{2}M(M+1)$ nonzero rows, corresponding to the indices satisfying $0 \leq j \leq M-1$, $0 \leq k \leq j$.

It is of interest to see what this means in terms of the probabilistic significance of the matrix R, discussed in chapter 1. The state (h, j, k)

in the canonical partition of $\tilde{P}$ signifies that the depletion of the dam is $hM + j$ and that the input during the next period is k. The depletion at the next time point is then given by $hM + j + M - k$. If $k > j$, the chain starting in the level h, $h \geq 1$, in the state (h, j, k) immediately returns to the level h in one transition. All entries $R_{j,k;j',k'}$ of R for which $k > j$ are therefore necessarily zero.

The computation of the $\frac{1}{2}M(M + 1)$ nonzero rows of R can be carried out for fairly large values of $(r + 1)M$, but requires carefully planned computer programming. The method, proposed in [251], is valid as stated only for the case where the nonzero eigenvalues of R are distinct. As was pointed out in section 1.6, the complex analysis approach may even then be numerically hazardous.

Because of the special definition of the matrices B_ν', given in (4.4.2), there are also major simplifications in the computation of the matrix $B[R]$. We notice that the matrices B_ν, $\nu \geq 0$, of order $(r + 1)M$, obtained from the partitioning of the first $(r + 1)M$ columns of $\tilde{P}$, may be written as $B_\nu = A_{\nu+1} + C_\nu$, $\nu \geq 0$, where the matrices C_ν are zero, except possibly for elements in their first $r + 1$ columns. The first $r + 1$ columns of the matrices C_ν are obtained from the matrices

$$\begin{gathered} A'_{M+1} + \cdots + A'_r, \\ A'_{M+2} + \cdots + A'_r, \\ \vdots \\ A'_r. \end{gathered}$$

These matrices require no computation. They are simply obtained from A' by setting an appropriate number of its rows equal to zero. Furthermore, since $A_\nu' = 0$, for $\nu > r$, it follows that only the first $\tau_1 = [r/M]$ matrices C_ν have any nonvanishing elements at all. We may now write the matrix $B[R]$ as

$$B[R] = C_0 + \sum_{\nu=1}^{\tau_1-1} R^\nu C_\nu + \sum_{\nu=0}^{\tau_2} R^\nu A_{\nu+1}, \tag{4.4.7}$$

where $\tau_2 = [(r + 1)/M]$. The matrices $\sum_{\nu=1}^{\tau_1-1} R^\nu C_\nu$ and $\sum_{\nu=0}^{\tau_2} R^\nu A_{\nu+1}$ have zero rows in the same locations as the matrix R, and in addition, only the first $r + 1$ columns of the first matrix are nonvanishing. For those indices which correspond to zero rows of R, the corresponding rows of $B[R]$ vanish, except for their first $r + 1$ elements, which are obtained from the matrix C_0.

By a rather lengthy argument, one may show that, since A' is irreducible and not a permutation matrix, the stochastic matrix $B[R]$ is irreducible. This guarantees the uniqueness and positivity of the $(r + 1)M$-vector $\dot{x}_0$.

The structure of the matrix $B[R]$ is illustrated for $r = 5$, and $M = 3$, in formula (4.4.8) below. Elements that may be positive are indicated by an asterisk.

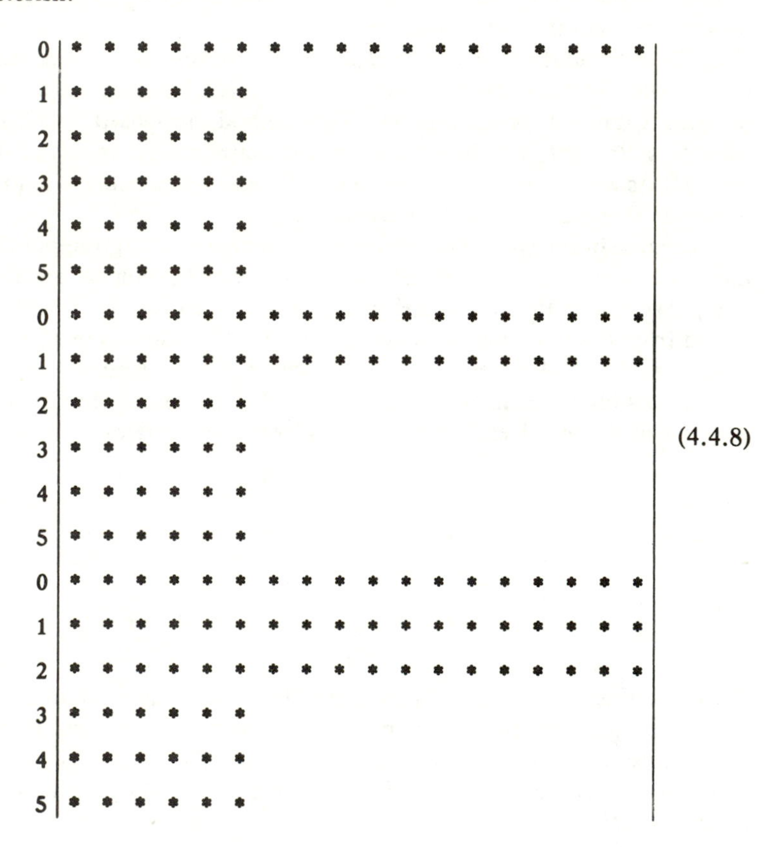

(4.4.8)

Particularly when r is large and M relatively small, this structure lends itself well to an efficient computation of the vector x_0. The components $(x_0)_{ij}$ with $1 \leq i \leq M$, $i < j \leq r$, are given as nonnegative linear combinations of the components $(x_0)_{ij}$ with $0 \leq i \leq M$, $0 \leq j \leq i$, and may readily be eliminated.

The resulting equations with the indices $1 \leq i \leq M$, $0 \leq j \leq i$, may be solved to express the $(x_0)_{ij}$ with indices $1 \leq i \leq M$, $1 \leq j \leq i$, each in the form $(x_0)_{ij} = a_{ij}(x_0)_{00}$. A final substitution leaves a system of $r + 1$ homogeneous linear equations for the components $(x_0)_{0j}$, $0 \leq j \leq r$.

The vector $\nu = (I - R)^{-1}e$ should be computed beforehand by solving the system

$$\nu - R\nu = e.$$

This system may be solved either by reductions similar to those discussed above, or by Gauss-Seidel iteration for which the structure of its coefficient matrix makes it particularly well suited.

We have discussed the algorithmic simplifications of this model at some length because in this case they enable us to reduce a problem of considerable dimensionality down to highly manageable computations. Here again, the conditional densities of the depletion, given that the chain is in one of the states (h, i, j), $h \geq 0$, $0 \leq i \leq M - 1$, $0 \leq j \leq r$, will be particularly illuminating. They will show, for example, how a cyclical pattern built into the inflow process will be reflected by the fluctuations of the depletion of the dam.

All the algorithmic simplifications of this model are particularly dramatic in the case $M = 1$, the unit release rule. The matrix $\tilde{P}$ is then given by

$$\tilde{P} = \left| \begin{array}{ccccc} B_0' & A_0' & 0 & 0 & \cdots \\ B_1' & A_1' & A_0' & 0 & \cdots \\ B_2' & A_2' & A_1' & A_0' & \cdots \\ B_3' & A_3' & A_2' & A_1' & \cdots \\ \vdots & \vdots & \vdots & \vdots & \end{array} \right|. \tag{4.4.9}$$

Only the first row of R, which we shall denote by $[\xi_0, \xi_1, \ldots, \xi_r]$, is nonzero. The equation $R = \sum_{\nu=0}^{r} R^\nu A_\nu'$ is equivalent to the system of nonlinear equations

$$\xi_k = A_{0k}' + \sum_{\nu=1}^{r} \xi_\nu \xi_0^{\nu-1} A_{\nu k}', \quad \text{for} \quad 0 \leq k \leq r, \tag{4.4.10}$$

which may be solved by a variety of numerical methods.

The matrix $B'[R] = \sum_{\nu=0}^{r} R^\nu B_\nu'$ is readily seen to be given by

$$B'[R] = (I - R)^{-1}(A' - R), \tag{4.4.11}$$

so that

$$x_0 = \boldsymbol{\pi}(I - R), \tag{4.4.12}$$

where $\boldsymbol{\pi}$ is the invariant probability vector of A'.

The invariant vector x of $\tilde{P}$ is now explicitly given by

$$x_{0j} = \pi_j - \pi_0\xi_j, \qquad \text{for} \quad 0 \le j \le r,$$
$$x_{ij} = \pi_0\xi_j\xi_0^{i-1}(1 - \xi_0), \quad \text{for} \quad i \ge 1, \qquad 0 \le j \le r. \tag{4.4.13}$$

The conditional densities $q_j(i)$, $i \ge 0$, are given for $0 \le j \le r$ by

$$q_j(0) = 1 - \pi_j^{-1}\pi_0\xi_j,$$
$$q_j(i) = \pi_j^{-1}\pi_0\xi_j\xi_0^{i-1}(1 - \xi_0), \quad \text{for} \quad i \ge 1. \tag{4.4.14}$$

The conditional mean depletion, given that the dam is considered at the beginning of an interval in which the input is j, $0 \le j \le r$, is given by

$$D_j = \pi_j^{-1}\pi_0\xi_j(1 - \xi_0)^{-1}, \quad \text{for} \quad 0 \le j \le r. \tag{4.4.15}$$

This model was first discussed in R. M. Phatarfod [250].

4.5. THE MULTI-SERVER *GI/PH/c* QUEUE

This section is devoted to a discussion of the major dimensionality problems which arise in the numerical solution of multi-server queues in which the service times are no longer exponentially distributed. The general theorems on modified matrix-geometric solutions, proved in chapter 1, give a *complete characterization of the steady-state probability distributions of the GI/PH/c queue with heterogeneous servers.* The strictly positive matrix R arising in these results, however, is of such high dimension that the direct numerical implementation of the resulting algorithms is limited to very small values of c and to particular PH-distributions of low order.

The same can be said of all other exact algorithms for multi-server queues proposed to date. The implementation of other, usually more restrictive, algorithms, as well as the one to be proposed here, may have merit for particular applications involving only two or three servers. It may also be used for the numerical investigation of approximations, such as those proposed, for example, in O. J. Boxma, J. W. Cohen, and N. Huffels [33], G. P. Cosmetatos [69, 70], E. Maaløe [199], or S. A. Nozaki and S. M. Ross [243]. With regards to the solution of models with even as few as five nonexponential servers, the algorithmic feasibility of *all* present methods is limited. The following references deal with direct procedures for the numerical solution of nonexponential, multi-server queues: D. M. Avis [8], A. Brandwajn [35], J. C. Heffer [133], F. S. Hillier and F. D. Lo [142], P. Hokstad [145], A. S. Kapadia [155], J. O. Mayhugh and R. E. McCormick [202], S. Shapiro [283], Y. Takahashi and Y. Takami [295], Y. Takahashi [294, 296], and O. S. Yu [314, 315].

We now consider the *GI/PH/c* queue with interarrival time distribution $F(\cdot)$ of finite mean λ_1'. The service time distribution of the r-th server is of phase type with the irreducible representation $[\beta(r), S(r)]$ of order $\nu(r)$, $1 \le r \le c$. The corresponding mean is denoted by $\mu_1'(r)$. Corresponding to each service time distribution we define the sequence of matrices $P_r(n, t)$, $n \ge 0$, $t \ge 0$, which satisfy the equations (4.1.1) with $\beta(r)$ and $S(r)$, playing for each r, $1 \le r \le c$, the role of β and S.

The states of the Markov chain $\tilde{P}$, embedded at the epochs of arrivals, consist of *ordinary* states and *boundary* states. The *ordinary* states correspond to the situations where all c servers are occupied; they are of the form $(i; j_1, j_2, \ldots, j_c)$, with $i \ge c$, $1 \le j_r \le \nu(r)$, $1 \le r \le c$, and will be listed in the lexicographic order. The chain is in the state $(i; j_1, j_2, \ldots, j_r)$ if prior to an arrival there are i customers in the queue and the r servers are respectively in the phases $j_1, \ldots, j_r$ of service.

The *boundary* states correspond to the situations where, prior to an arrival, one or more servers are idle. This set of states is further partitioned, first according to the *number* of idle servers, and second according to the *choice* of the servers who are occupied immediately after the arrival. In order to avoid introducing much new notation, we shall discuss the boundary states only for the case of two servers. It is clear that the number of boundary states increases very rapidly with c. For $c = 2$, we have the $\nu(1)\nu(2) + \nu(1) + \nu(2)$ boundary states

$$(0, j_1), \quad \text{for} \quad 1 \le j_1 \le \nu(1),$$

$$(0, j_2), \quad \text{for} \quad 1 \le j_2 \le \nu(2),$$

$$(1, j_1, j_2), \quad \text{for} \quad 1 \le j_1 \le \nu(1), \qquad 1 \le j_2 \le \nu(2).$$

For c servers, there will be

$$\sum_{r=1}^{c} \nu(r) + \sum \nu(r_1)\nu(r_2) + \cdots + \nu(1)\nu(2) \cdots \nu(c)$$

boundary states. The successive summations extend over all possible choices of k distinct indices $r_1, \ldots, r_k$ out of the set $\{1, 2, \ldots, c\}$.

The indices j_1 and j_2 in the preceding expressions refer to the service phases the appropriate servers are in immediately after each arrival. The chain is, for example, in the state $(0, j_1)$ when the arriving customer finds the queue empty, chooses the server 1, and starts his service in the phase j_1, $1 \le j_1 \le \nu(1)$.

Let us denote the set of boundary states by E' and let these states be ordered first in groups by increasing number of busy servers and within each group according to some ordering of the busy servers. Finally, within each such ordering, the service phase indices are listed in the lexicograpic order. The last set of boundary states corresponds to the set of indices $(c - 1; j_1, j_2, \ldots, j_c)$ written in the lexicographic order.

The Markov chain $\tilde{P}$ then always has the structure

$$\tilde{P} = \begin{array}{c} \\ E' \\ \\ \\ \\ \\ \\ \end{array} \left|\begin{array}{c|cccc} E' & & & & \\ & & & 0 & \\ & A_0 & & & \\ \hline & A_1 & A_0 & 0 & \cdots \\ & A_2 & A_1 & A_0 & \cdots \\ & A_3 & A_2 & A_1 & \cdots \\ & \vdots & \vdots & \vdots & \end{array}\right|, \qquad (4.5.1)$$

where the matrices A_k, $k \geq 0$, of order $m = \Pi_{r=1}^{c} \nu(r)$, are given by

$$A_k = \Sigma \int_0^\infty P_1(k_1, t) \otimes P_2(k_2, t) \otimes \cdots \otimes P_c(k_c, t) dF(t), \qquad (4.5.2)$$

where the summation extends over the set of all $(k_1, k_2, \ldots, k_c)$ satisfying $k_r \geq 0$, $1 \leq r \leq c$, $k_1 + k_2 + \cdots + k_c = k$.

Whenever $k \geq c$, the matrix A_k is strictly positive. The matrix $A = \Sigma_{k=0}^{\infty} A_k$ is given by

$$\begin{aligned} A = \int_0^\infty & \exp\{[S(1) + S^o(1)B^o(1)]t\} \otimes \cdots \\ & \otimes \exp\{[S(c) + S^o(c)B^o(c)]t\} dF(t), \end{aligned} \qquad (4.5.3)$$

and has the invariant probability vector $\boldsymbol{\theta}(1) \otimes \boldsymbol{\theta}(2) \otimes \cdots \otimes \boldsymbol{\theta}(c)$, where $\boldsymbol{\theta}(r)$ is given by $\boldsymbol{\theta}(r)[S(r) + S^o(r)B^o(r)] = \mathbf{0}$, $\boldsymbol{\theta}(r)\boldsymbol{e} = 1$.

By a direct calculation similar to that for the *GI*/*PH*/1 queue, or by applying a general theorem of S. S. Lavenberg [184], we obtain the equilibrium condition

$$\lambda_1'^{-1} < \sum_{r=1}^{c} \mu_r'^{-1}. \qquad (4.5.4)$$

As in the case of the *GI*/*PH*/1 queue, some of the boundary states may be ephemeral, but these may be conveniently handled by the same notational convention as in section 4.1. It is clear that in the stable case the invariant probability vector $\boldsymbol{x}$ of $\tilde{P}$ is of modified matrix-geometric form with, for $i \geq c - 1$,

$$\boldsymbol{x}_i = \boldsymbol{x}_{c-1} R^{i-c+1}. \tag{4.5.5}$$

In order to find the vector $[\boldsymbol{x}_0, \ldots, \boldsymbol{x}_{c-1}]$, a huge system of linear equations of the same order as there are boundary states needs to be solved. The matrix R is *strictly positive* and of dimension $m = \prod_{r=1}^{c} \nu(r)$.

When the number of servers is small and m of manageable dimension, there are structural properties that may be exploited to facilitate the computation of the initial components of $\boldsymbol{x}$ to some extent. Let us consider these in some detail for the case $c = 2$.

In order to specify the columns of $\tilde{P}$ which correspond to the boundary states in the set E', we need to specify a rule according to which the arriving customer will choose from among the idle servers. A variety of such rules may be prescribed, and some of these may add further complexity to the description of the boundary states. This would happen, for example, if the customer had to join that server which had longest been idle.

For the purpose of the present discussion, we assume that when both servers are idle, the arriving customer joins server 1 with probability ϕ_1 and server 2 with probability $\phi_2 = 1 - \phi_1$, $0 \leq \phi_1 \leq 1$. For systems with more than two heterogeneous servers, the channel selection rule is an important feature of the design of the system. It may have a substantial effect on the distribution of the time spent in the system.

The columns of the matrix $\tilde{P}$ corresponding to the boundary states are now given by

$$\begin{array}{c|cccc|c|} & \mathbf{0} & \mathbf{0'} & \mathbf{1} & & \mathbf{2} \\ \mathbf{0} & B_0 & B_0{}' & A_0{}' & & 0 \\ \mathbf{0'} & B_{0'} & B_{0'}{}' & A_{0'}{}' & & 0 \\ \mathbf{1} & B_1 & B_1{}' & A_1{}' & & A_0 \\ \mathbf{2} & B_2 & B_2{}' & A_2{}' & & A_1 \\ \mathbf{3} & B_3 & B_3{}' & A_3{}' & & A_2 \\ \mathbf{4} & B_4 & B_4{}' & A_4{}' & & A_3 \\ \vdots & \vdots & \vdots & \vdots & & \vdots \end{array} . \tag{4.5.6}$$

The column corresponding to the level **2** is added on the right to show how the present pattern fits in with that displayed in (4.5.1).

The new blocks in (4.5.6) are given by

$$B_0 = \phi_1 \sum_{r=1}^{\infty} \int_0^{\infty} P_1(r, t)dF(t)e \cdot \beta(1),$$

$$B_0' = \phi_2 \sum_{r=1}^{\infty} \int_0^{\infty} P_1(r, t)dF(t)e \cdot \beta(2),$$

$$B_{0'} = \phi_1 \sum_{r=1}^{\infty} \int_0^{\infty} P_2(r, t)dF(t)e \cdot \beta(1),$$

$$B_{0'}' = \phi_2 \sum_{r=1}^{\infty} \int_0^{\infty} P_2(r, t)dF(t)e \cdot \beta(2),$$

$$A_0' = \int_0^{\infty} P_1(0, t)dF(t) \otimes \beta(2),$$

$$A_{0'}' = \beta(1) \otimes \int_0^{\infty} P_2(0, t)dF(t), \tag{4.5.7}$$

$$\begin{aligned} A_i' = \int_0^{\infty} dF(t) \int_0^t \sum_{i_1=0}^{i} [&P_1(i_1, v)S^o(1)B^o(1) \otimes P_2(i-i_1, v)P_2(0, t-v) \\ &+ P_1(i-i_1, v)P_1(0, t-v) \\ &\otimes P_2(i_1, v)S^o(2)B^o(2)]dv, \qquad \text{for} \quad i \geq 1, \end{aligned}$$

$$\begin{aligned} B_i = \phi_1 \int_0^{\infty} dF(t) \int_0^t \int_0^{t-v} \sum_{i_1=0}^{i-1} &[P_1(i_1, v)S^o(1) \cdot \beta(1) \\ &\otimes P_2(i-i_1-1, v)P_2(0, u)S^o(2) \\ &+ P_1(i-i_1-1, v)P_1(0, u)S^o(1) \cdot \beta(1) \\ &\otimes P_2(i_1, v)S^o(2)]dudv, \qquad \text{for} \quad i \geq 1, \end{aligned}$$

$$\begin{aligned} B_i' = \phi_2 \int_0^{\infty} dF(t) \int_0^t \int_0^{t-v} \sum_{i_1=0}^{i-1} &[P_1(i_1, v)S^o(1) \\ &\otimes P_2(i-i_1-1, v)P_2(0, u)S^o(2) \cdot \beta(2) \\ &+ P_1(i-i_1-1, v)P_1(0, u)S^o(1) \\ &\otimes P_2(i_1, v)S^o(2) \cdot \beta(2)]dudv, \qquad \text{for} \quad i \geq 1. \end{aligned}$$

The probability argument that leads, for example, to the formula for B_i, $i \geq 1$, proceeds as follows. Immediately following the arrival who finds i customers present, there are $i + 1$ customers in the queue. All these are served before the next arrival occurs. We need to consider the number of services completed by each of the two servers and also to assure that when one of the servers becomes idle, the other server is processing the last remaining customer. Let the server 1 become idle at time v, $0 \leq v \leq t$, with the completion of his $(i_1 + 1)$st service. During the interval $[0, v]$, server 2 must then have completed $i - i_1 - 1$ services, so that one customer remains to finish service at some time $v + u$, $0 \leq u \leq t - v$. The second term in the integrand is obtained by reversing the roles of the two servers. The last customer is now served by server 1. Upon his arrival to the empty queue, the new customer selects server 1 and chooses his phase of service according to the vector $\boldsymbol{\beta}(1)$. This accounts for the factors ϕ_1 and $\boldsymbol{\beta}(1)$, appearing in the appropriate places in the formula. The summation and the integrals now arise out of the application of the law of total probability. Similar arguments lead to the other expressions.

There are some simplifications, useful in numerical computations. We may write

$$\sum_{r=1}^{\infty} \int_0^{\infty} P_1(r, t)dF(t)\boldsymbol{e} = \boldsymbol{e} - \int_0^{\infty} P_1(0, t)dF(t)\boldsymbol{e},$$

and

$$\int_0^{t-v} P_1(0, u)du = \int_0^{t-v} \exp[S(1)u]du = S^{-1}(1)[\exp S(1)(t - v) - I],$$

and similar formulas for $P_2(\cdot, t)$.

By using the second formula, B_i, $i \geq 1$, may be rewritten as

$$\begin{aligned} B_i = \phi_1 \int_0^{\infty} dF(t) \int_0^t \sum_{i_1=0}^{i-1} & [P_1(i, v)S^o(1) \cdot \boldsymbol{\beta}(1) \\ & \otimes P_2(i - i_1 - 1, v)[I - P_2(0, t - v)]\boldsymbol{e} \\ & + P_1(i - i_1 - 1, v)[I - P_1(0, t - v)]\boldsymbol{e} \cdot \boldsymbol{\beta}(1) \\ & \otimes P_2(i_1, v)S^o(2)]dv, \end{aligned} \quad (4.5.8)$$

with a similar formula for B_i', $i \geq 1$.

The enormous complexity of the transition probabilities corresponding to the boundary states is due to the fact that with heterogeneous servers, it becomes necessary to consider the order in which servers become idle. It is clear that from the viewpoint of computational feasibility, the *GI/PH/c* queue with heterogeneous servers has not yet been solved. Among the classical queues, it presents a major challenge to future research.

We shall conclude this section by brief discussions of several particular cases in which additional useful simplifications may be exploited in the development of algorithms.

The *GI/M/c queue with heterogeneous servers* is a case in which all algorithmic difficulties spring from the boundary states. If we denote the service rates by $\mu(r)$, $1 \le r \le c$, we readily see that the matrices A_k, $k \ge 0$, given in (4.5.2), become *scalars* given by

$$A_k = \int_0^\infty \exp\left[-t \sum_{r=1}^{c} \mu(r)\right] \frac{\left[t \sum_{r=1}^{c} \mu(r)\right]^k}{k!} \, dF(t), \qquad k \ge 0. \quad (4.5.9)$$

The *scalar* R is then the unique solution in (0, 1) of the equation

$$z = f\left[(1 - z) \sum_{r=1}^{c} \mu(r)\right], \quad (4.5.10)$$

where $f(\cdot)$ is the Laplace-Stieltjes transform of $F(\cdot)$.

There are *in general* $2^c - 1$ boundary equations which require an explicit definition of the channel selection rule for all possible configurations of idle and busy servers. In some cases, where the c servers may be grouped in sets having the same rate, it is possible to reduce the number of boundary states considerably by counting only the numbers of occupied servers in each group. There is no systematic way to discuss all such variants at once. One faces the tedious, but basically elementary, task of spelling out all boundary states and of setting up the corresponding transition probabilities. The computation of the invariant probability vector $\mathbf{x}$ is then primarily a matter of solving the boundary equations as efficiently as possible.

The *GI/PH/c queue with identical servers* allows a major simplification in the description of the state space, which was first put to algorithmic use by J. O. Mayhugh and R. E. McCormick [202] for the $M/E_\nu/c$ model. The state of the system is now described by $(i; j_1, j_2, \ldots, j_\nu)$, where i indicates the queue length and j_r the number of *servers* who are in the service phase r, $1 \le r \le \nu$. Clearly $\sum_{r=1}^{\nu} j_r = \min(i, c)$. Numerical implementations based on this approach have been reported only for the particular cases of Erlang and hyperexponential service times by the use of complex variable methods. There is no difficulty in showing that the lumped Markov chain, which we now obtain, still has a modified matrix-geometric solution. The particular PH-distributions treated in the literature do not appreciably simplify the analysis, so that future implementation studies for general PH-distributions would be both feasible and useful. The simplifications in the case of identical servers stem from the fact that only the *number* of service completions between successive arrivals need to be considered. The order in which servers become idle is now immaterial.

The $PH/PH/c$ queue, particularly with identical servers, allows certain simplifications that are not present in the case of a general interarrival time distribution. By appropriately defining the states, one obtains a QBD process for which the matrix R may be efficiently computed when the number of absorption indices of the interarrival time distribution is small. These simplifying features were discussed for the $PH/PH/1$ queue in section 3.7. There is no need to repeat this discussion here.

It may be of interest to show in detail the differences in the matrices A_0, A_1, and A_2 which arise in the QBD processes for the $PH/PH/2$ queue when we distinguish between the two servers and when we do not. The interarrival time distribution has the representation (α, T) of order m_1.

When we distinguish between the servers, as we need to when they have different service time distributions, the matrices A_0, A_1, and A_2 are given by

$$\begin{aligned} A_0 &= T^o A^o \otimes I \otimes I, \\ A_1 &= T \otimes I \otimes I + I \otimes S(1) \otimes I + I \otimes I \otimes S(2), \qquad (4.5.11) \\ A_2 &= I \otimes S^o(1)B^o(1) \otimes I + I \otimes I \otimes S^o(2)B^o(2), \end{aligned}$$

and are of order $m = m_1 \nu(1)\nu(2)$.

When both servers are identical, the matrices A_0, A_1, and A_2 may, by redefining the state space, be represented by

$$\begin{aligned} A_0 &= T^o A^o \otimes I, \\ A_1 &= T \otimes I + I \otimes Z, \qquad (4.5.12) \\ A_2 &= I \otimes Z^o. \end{aligned}$$

The matrices Z and Z^o may be defined in several alternate ways, depending on the description of the state space. Let us represent the ordinary states of the QBD process by (i, h, j_1, j_2), $i \geq 2$, $1 \leq h \leq m_1$, $1 \leq j_1, j_2 \leq \nu$, $j_1 \neq j_2$, and by (i, h, j^*), $i \geq 2$, $1 \leq h \leq m_1$, $1 \leq j^* \leq \nu$. The index i indicates the queue length, and h is the phase of the arrival process. When $j_1 \neq j_2$, one server is in the phase j_1; the other in the phase j_2. States of the form (i, h, j^*) signify that both servers are in the service phase j. The matrices Z and Z^o are then of order $\frac{1}{2}\nu(\nu + 1)$, and their elements are given by

$$\begin{aligned} Z(j^*, j^*) &= 2S_{jj}, \\ Z(j_1, j_2; j_1, j_2) &= S_{j_1 j_1} + S_{j_2 j_2}, \\ Z(j_1^*, j_2^*) &= 0, \end{aligned}$$

$$\begin{aligned}
Z(j_1, j_2; j_3, j_4) &= \delta_{j_1j_3}S_{j_2j_4} + \delta_{j_2j_4}S_{j_1j_3},\\
Z(j_1, j_2; j_3*) &= \delta_{j_1j_3}S_{j_2j_3} + \delta_{j_2j_3}S_{j_1j_3},\\
Z(j_1*; j_2, j_3) &= \delta_{j_1j_2}S_{j_1j_3} + \delta_{j_1j_3}S_{j_1j_2},\\
Z^o(j*, j*) &= 2S_j{}^o\beta_j,\\
Z^o(j_1, j_2; j_1, j_2) &= S_{j_1}{}^o\beta_{j_1} + S_{j_2}{}^o\beta_{j_2},\\
Z^o(j_1*, j_2*) &= 0, \quad \text{for} \quad j_1 \neq j_2,\\
Z^o(j_1, j_2; j_3, j_4) &= \delta_{j_1j_3}S_{j_2}{}^o\beta_{j_4} + \delta_{j_2j_4}S_{j_1}{}^o\beta_{j_3},\\
&\qquad \text{for} \quad j_1 \neq j_3, \quad \text{or} \quad j_2 \neq j_4,\\
Z^o(j_1, j_2; j_3*) &= \delta_{j_1j_3}S_{j_1}{}^o\beta_{j_3} + \delta_{j_2j_3}S_{j_1}{}^o\beta_{j_3},\\
Z^o(j_1*; j_2, j_3) &= \delta_{j_1j_2}S_{j_1}{}^o\beta_{j_3} + \delta_{j_1j_3}S_{j_1}{}^o\beta_{j_2}.
\end{aligned} \tag{4.5.13}$$

These are the matrices corresponding to $\nu = 3$:

$$Z = \begin{array}{r|cccccc|}
1* & 2S_{11} & 0 & 0 & 2S_{12} & 2S_{13} & 0\\
2* & 0 & 2S_{22} & 0 & 2S_{21} & 0 & 2S_{23}\\
3* & 0 & 0 & 2S_{33} & 0 & 2S_{31} & 2S_{32}\\
(1,2) & S_{21} & S_{12} & 0 & S_{11}+S_{22} & S_{23} & S_{13}\\
(1,3) & S_{31} & 0 & S_{13} & S_{32} & S_{11}+S_{33} & S_{12}\\
(2,3) & 0 & S_{32} & S_{23} & S_{31} & S_{21} & S_{22}+S_{33}
\end{array},$$

$$Z^o = \begin{array}{r|cccccc|}
1* & 2S_1{}^o\beta_1 & 0 & 0 & 2S_1{}^o\beta_2 & 2S_1{}^o\beta_3 & 0\\
2* & 0 & 2S_2{}^o\beta_2 & 0 & 2S_2{}^o\beta_1 & 0 & 2S_2{}^o\beta_3\\
3* & 0 & 0 & 2S_3{}^o\beta_3 & 0 & 2S_3{}^o\beta_1 & 2S_3{}^o\beta_2\\
(1,2) & S_2{}^o\beta_1 & S_1{}^o\beta_2 & 0 & S_1{}^o\beta_1+S_2{}^o\beta_2 & S_2{}^o\beta_3 & S_1{}^o\beta_3\\
(1,3) & S_3{}^o\beta_1 & 0 & S_1{}^o\beta_3 & S_3{}^o\beta_2 & S_1{}^o\beta_1+S_3{}^o\beta_3 & S_1{}^o\beta_2\\
(2,3) & 0 & S_3{}^o\beta_2 & S_2{}^o\beta_3 & S_3{}^o\beta_1 & S_2{}^o\beta_1 & S_2{}^o\beta_2+S_3{}^o\beta_3
\end{array}. \tag{4.5.14}$$

We see that the advantage of the reduced state space is somewhat counterbalanced by a more complex definition of the matrices A_0, A_1, and A_2. The same also applies to the matrices describing the transition rates near the boundary states. The QBD process on the reduced state space requires both a more careful analysis and more belabored computer programming. However, it is often the only feasible way to handle problems of larger dimension.

Chapter 5
Buffer Models

5.1. INTRODUCTION

A variety of applications, notably those arising in the study of computer and communication systems, lead to the consideration of an *input queue of unlimited capacity* in series with a system of one or more queues of *finite capacity.* The capacity constraints on this second, subsidiary, system lead to interesting physical phenomena, such as *blocking,* and to substantial conceptual and algorithmic difficulties, particularly when the possible interactions inside this system are complex.

In order to place the material presented in this chapter in a broader framework of applications, we shall briefly discuss the general methodological approaches to networks of queues. In the study of networks of queues, one distinguishes between *closed* and *open* networks; the latter are further subdivided into *saturated* and *unsaturated* systems.

In a *closed* network, a finite, constant number N of customers are imagined to circulate forever within a finite, interconnected system of service units. There are no arrivals to or departures from the system. Under appropriate (mostly exponential) assumptions on the service times and on the routing and queue disciplines, closed networks may be described by finite Markov processes with (in most cases) a very large number of states. The stationary probability vector of the infinitesimal generator of many closed network processes assumes, however, a *product form* of great simplicity. In spite of the huge state space, it is often possible to compute many interesting quantities related to the network by simple algorithms; see, for example, J. P. Buzen [44, 45]. The literature on closed queueing networks and on product form solutions is extensive. We refer the reader to the review papers by A. J. Lemoine [187, 188] for further information. In most models for closed networks it is assumed that the waiting spaces within the network are sufficiently large that any one of them could conceivably accommodate all N customers. In such networks there are no capacity constraints. Upon the introduction of capacity constraints, most network problems present major or even insuperable mathematical difficulties.

In an open network, there are one or more *input queues* and one or more *exit nodes*. From the latter, customers may depart the system. We shall limit our attention to the case where there is only one input queue. The analysis of open networks is often conceptually more difficult than that of closed networks. The unbounded input queue introduces an infinite state space, which is usually seen as a source of great difficulty, particularly in numerical solutions. This somewhat illusory difficulty is avoided by the consideration of *saturated* open networks. In such models, one assumes that the input queue contains at all times an adequate supply of customers, so that customers will instantaneously be added to the subsidiary network whenever such access is permitted by the capacity constraints. It is customary to depict a saturated open network as having an infinite input queue at all times, but this may be misleading. An open network is *saturated* when there is no idle access time. This arises even in as simple a model as the *critical* $M/M/1$ queue. This saturated queue typically *does not* exhibit large queue lengths.

The advantage in considering saturated open networks is that the random fluctuations of the input queue may often be ignored. The subsidiary network may then be viewed as though it were a closed network containing a random number N of customers. For detailed discussions and for applications of this approach, we again refer the reader to the literature on networks of queues. For an accessible introduction, see L. Kleinrock [172].

In this chapter we shall only deal with open unsaturated networks. There is a genuine, randomly varying input queue, and we shall study conditions under which that queue, and hence the entire network, admits steady-state behavior.

There is an interesting relationship between the critical arrival rate of an unsaturated network and the (maximum) departure rate from its saturated counterpart. S. S. Lavenberg [184] considered this for networks in which the arrivals to the input queue form a *renewal process* and in which the subsidiary system has a finite-state Markovian description. For the corresponding saturated system, he defines a natural set of departure rates and shows that the maximum departure rate of that system is *equal* to the critical arrival rate of the unsaturated system.

The class of networks covered by Lavenberg's results is very large. It includes any finite node network with a single input queue with renewal arrivals, leading into a subsidiary network in which all internal queues are bounded and have servers with a PH-service time distribution. Any priority rule that results in a finite-state Markovian description is allowed. When viewed prior to epochs of arrival to the input queue and by considering a complete description of all the internal queues and of the phases of service of all the internal servers, the system generally has an embedded Markov chain that is of the type treated in this book. The stationary probability vector of this Markov chain is then modified matrix-geometric. The overriding limitation on the practical applicability of this result is clearly the dimension of the blocks that appear in the partitioned transition probability matrix $\tilde{P}$. The range of specific models for unsaturated open networks for which a detailed algorithmic solution is feasible is limited by the pervasive curse of dimensionality. Within this range, there are, however, many models of practical interest. The remainder of this chapter is devoted to a selection of these.

5.2. SERVERS IN SERIES WITH A FINITE INTERMEDIATE BUFFER

A considerable number of papers have been devoted to the study of the system consisting of a group of $r \geq 1$ identical parallel servers in series with a second group of $c \geq 1$ identical parallel servers. Both groups are separated by a finite intermediate buffer. The physical phenomenon of primary interest is *blocking*. This occurs when the intermediate buffer fills up to the point of preventing the release of served customers from the first group of servers (Unit I) into the buffer.

The available studies of this system differ widely in the distributional assumptions on the service and interarrival times, in the choices of r and c, in the size of the buffer, in the nature of the *unblocking rule* considered, and in the mathematical methodology employed in the discussions. We shall present a detailed discussion of the model with several exponential servers, treated in G. Latouche and M. F. Neuts [183], as it is directly related to the methods developed earlier in this book. See figure 5.1.

Before doing so, we give a list of the references known to us on buffer models. Following each reference, a few representative key words are listed for the reader's information.

B. Avi-Itzhak and M. Yadin [6]; no buffer, general service time distributions.

O. J. Boxma and A. G. Konheim [34]; approximations, several units, exponential servers.

P. Caseau and G. Pujolle [48]; critical arrival rate, exponential servers.

A. B. Clarke [57–60]; special access rule, models with or without intermediate buffers; see also section 5.3.

J. M. Harrison [128, 130]; diffusion approximation, heavy traffic.

D. K. Hildebrand [139, 140]; general interarrival and service time distributions, stability condition.

F. S. Hillier and R. W. Boling [141]; iterative numerical solution, several units.

G. C. Hunt [146]; Markovian models.

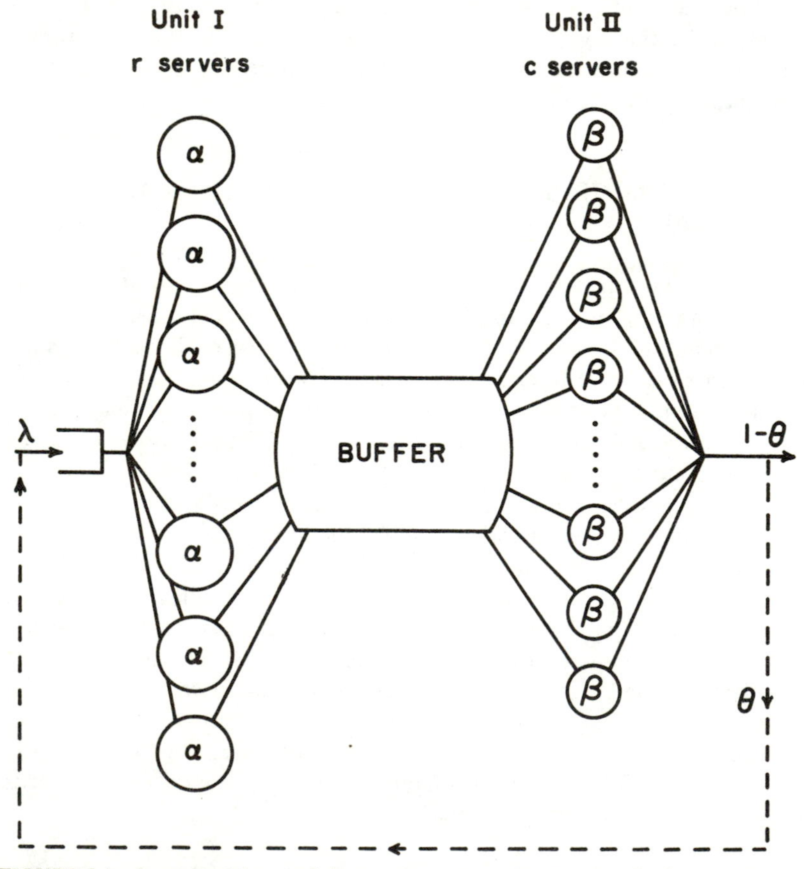

FIGURE 5.1. A system with r parallel servers in tandem with c parallel servers, separated by a finite intermediate buffer.

A. G. Konheim and M. Reiser [174]; exponential servers, $r = c = 1$, numerical solution based on complex analysis, computer applications.

J. Labetoulle and G. Pujolle [178]; deterministic service, computer applications.

G. Latouche [179]; matrix-geometric solution, computer applications, see also Section 5.4.

G. Latouche and M. F. Neuts [183]; discussed in this section.

D. M. Lucantoni and M. F. Neuts [194]; methods applicable to models of the $M/G/1$ type.

M. F. Neuts [216, 217]; $r = 1$, with general service time. Models of the $M/G/1$ type, but treated by transform methods. The second paper requires methodological corrections.

G. F. Newell [239, 240]; extensive treatment of diffusion approximations.

N. U. Prabhu [252]; transient analysis, no intermediate buffer.

G. Pujolle and D. Potier [253]; computer applications.

M. Reiser and A. G. Konheim [265]; $r = c = 1$, exponential servers with feedback loops, computer applications.

T. Suzuki [289]; $r = c = 1$, general service times in Unit I.

B. Wong, W. Giffin, and R. L. Disney [307]; $r = c = 1$, spectral decomposition methods, exponential servers.

The model to be discussed in detail consists of Units I and II, separated by a finite buffer. Unit I has $r \geq 1$, parallel exponential servers each capable of serving customers at a rate α. Unit II consists of c parallel exponential servers of rate β. The arrivals to Unit I occur according to a homogeneous Poisson process of rate λ. The servers in Unit II can be active as long as there are customers who have completed a pass through Unit I. Customers who have completed a service in Unit I and are requesting service from Unit II will be called *II-customers.* The size of the intermediate waiting room is $M - c - 1 \geq 0$, so that there can be at most $M - 1$ customers who are either in process in Unit II or waiting in the buffer.

Blocking. When a service completion occurs in Unit I and the buffer is full, the customer cannot leave his server in Unit I and causes *blocking* of that server. Several servers in Unit I may so become blocked. It may be undesirable to let all servers in I become blocked in this manner. We shall introduce an integer r^*, $1 \leq r^* \leq r$, with the following significance. As long as the number of blocked servers in I does not exceed $r^* - 1$ and the system is not *in the unblocking phase* (to be defined), the unblocked servers in Unit I may continue to process customers. When r^* servers in I become blocked, then *all* servers in Unit I cease service until the system becomes unblocked. This situation is called *full blocking*, in contrast to the earlier situation, which is referred to as *partial blocking*.

In a partially blocked system, a blocked server in I may release his customer into the buffer as soon as a space becomes available. That server may then again initiate a service.

For a fully blocked system, the situation may be more complicated. Repeated occurrences of full blocking close together in time may be highly undesirable. This can be avoided by a substantial clearing of II-customers whenever full blocking occurs. To this effect, we specify a more general *unblocking rule*.

Unblocking. When the system becomes fully blocked, it contains $M + r^* - 1$ II-customers. We specify an integer k^*, with $0 \le k^* \le M + r^* - 2$. No service is performed in Unit I as long as there are more than k^* II-customers present. Services in Unit II continue and cause the departure of II-customers. When the number of these drops to k^*, all interrupted services (in the unblocked servers of Unit I) resume and any free servers may again initiate services. Between the occurrence of full blocking and this resumption of service in I, the system is said to be *in the unblocking phase*. Notice that the total service rate in Unit I depends not merely on the number of blocked servers, but also on whether the system is partially blocked or in the unblocking phase. As discussed in the following, this distinction affects the detailed description of the state space for the queue.

Feedback. As was noted in [174] and [265], it is of interest, for example, to models for computer systems to consider *feedback loops*, which allow a customer to return, upon completion of a service in Unit II, either to the input queue or to the buffer. Feedback loops add only a modicum of additional complexity to the study of the stationary queue length distribution. In contrast, they make the study of waiting time distributions exceedingly difficult.

For the sake of illustration, we shall allow a departing customer from Unit II to return instantaneously to the input queue with probability θ, $0 \le \theta < 1$. With probability $\theta' = 1 - \theta$, such a customer departs the system altogether.

The Markov Process

The queueing model is described by a QBD process $\tilde{Q}$ on the state space $E = \{(i, j): i \ge 0, 0 \le j \le N\}$, where N is a finite, positive integer. The index i specifies the number of customers queued up or in service in Unit I. Such customers are called *I-customers*. Upon completion of a service in Unit I, a customer becomes a II-customer. The role of the index j may be more complicated and will be discussed in what follows.

The matrix $\tilde{Q}$ is, in all cases, of the form

$$\tilde{Q} = \begin{vmatrix} A_{01} & A_{02} & & & & & & & \\ A_{10} & A_{11} & A_{12} & & & & & & \\ & A_{20} & A_{21} & A_{22} & & & & & \\ & & \cdots & & & & & & \\ & & & A_{r-2,0} & A_{r-2,1} & A_{r-2,2} & & & \\ & & & & A_{r-1,0} & A_{r-1,1} & A_0 & & \\ & & & & & A_2 & A_1 & A_0 & \\ & & & & & & A_2 & A_1 & \cdots \\ & & & & & & \vdots & \vdots & \end{vmatrix}, \tag{5.2.1}$$

where all blocks are square and of order $N + 1$. The row with index $i \geq 0$ corresponds to the level $i = \{(i, 0), (i, 1), \ldots, (i, N)\}$. The specific definition of the blocks is most conveniently given by considering models of increasing complexity.

Model A: $r^* = 1$, $k^* = M - 1$. This case arises naturally when $r = 1$, and was considered by A. G. Konheim and M. Reiser in [174] and [265]. In this case, $N = M$, the blocks are of order $M + 1$, and the index j, $0 \leq j \leq M$, denotes the number of II-customers in the system.

The matrices A_0, A_1, and A_2 are given by

$$\begin{aligned} (A_0)_{jj'} &= \lambda, && \text{for } 0 \leq j = j' \leq M, \\ &= \min(c, j)\beta\theta, && \text{for } 1 \leq j \leq M, \quad j' = j - 1, \\ &= 0, && \text{for all other } (j, j'). \\ (A_1)_{jj'} &= -\lambda - r\alpha - \min(c, j)\beta, && \text{for } 0 \leq j = j' \leq M - 1, \\ &= -\lambda - c\beta, && \text{for } j = j' = M, \\ &= \min(c, j)\beta\theta', && \text{for } 1 \leq j \leq M, \quad j' = j - 1, \\ &= 0, && \text{for all other } (j, j'). \end{aligned} \tag{5.2.2}$$

$$
\begin{aligned}
(A_2)_{jj'} &= r\alpha, && \text{for} \quad 0 \le j \le M-1, \quad j' = j+1, \\
&= 0, && \text{for all other } (j, j').
\end{aligned}
$$

The boundary matrices are given by

$$
\begin{aligned}
(A_{i0})_{jj'} &= i\alpha, && \text{for} \quad 0 \le j \le M-1, \quad j' = j+1, \\
&= 0, && \text{for all other } (j, j'). \\
(A_{i1})_{jj'} &= -\lambda - i\alpha - \min(c, j)\beta, && \text{for} \quad 0 \le j = j' \le M-1, \\
&= -\lambda - \min(c, j)\beta, && \text{for} \quad j = j' = M, \qquad (5.2.3) \\
&= \min(c, j)\beta\theta', && \text{for} \quad 1 \le j \le M, \quad j' = j+1, \\
&= 0, && \text{for all other } (j, j'). \\
A_{i2} &= A_0, && \text{for} \quad 0 \le i \le r-1.
\end{aligned}
$$

The matrices A_{i0} and A_{i1} are defined for $1 \le i \le r-1$, and for $0 \le i \le r-1$, respectively.

Theorem 5.2.1. The stationary probability vector $\boldsymbol{\pi}$ of $A = A_0 + A_1 + A_2$ for model A is given by

$$
\begin{aligned}
\pi_0 &= \left[\sum_{j=0}^{c-1} \frac{1}{j!} \left(\frac{r\alpha}{\beta} \right)^j + \frac{c^c}{c!} \sum_{j=c}^{M} \left(\frac{r\alpha}{c\beta} \right)^j \right]^{-1}, \\
\pi_j &= \frac{1}{j!} \left(\frac{r\alpha}{\beta} \right)^j \pi_0, \quad \text{for} \quad 1 \le j \le c, \qquad (5.2.4) \\
\pi_j &= \frac{c^c}{c!} \left(\frac{r\alpha}{c\beta} \right)^j \pi_0, \quad \text{for} \quad c \le j \le M.
\end{aligned}
$$

The equilibrium condition $\boldsymbol{\pi} A_2 \boldsymbol{e} > \boldsymbol{\pi} A_0 \boldsymbol{e}$ is equivalent to

$$
\lambda < (1 - \theta) r\alpha \sum_{j=0}^{M-1} \pi_j = (1 - \theta) r\alpha (1 - \pi_M). \qquad (5.2.5)
$$

Proof. By elementary calculations.

Remarks.

a. The geometric sum in the expression for π_0 is not summed in closed form to avoid writing a separate formula for the case $r\alpha = c\beta$.

b. The right-hand side λ^* of formula (5.2.5) is the critical input rate to the system. By Lavenberg's theorem [184] it is also the output rate of the saturated system. This parameter is also known by the neologism "*throughput*." In spite of its widespread interest in the applied literature on queues, it is only a crude descriptor of the behavior of service systems. We shall illustrate this by numerical examples in chapter 6. A small increase in throughput may be brought about at the (hidden) expense of large queues and long waiting times.

As λ^* is often the only feature of complex models which is easily expressible in terms of the parameters of the queue, it may nevertheless serve as a first, simple basis for the comparison of alternate designs of the system. More extensive computations or simulations should then be performed to study the other implications of the design choices.

For the sake of an example, let us set $r = c$ and assume that it is feasible to interchange the servers in Unit I with those in Unit II. For definiteness, let $\alpha \leq \beta$. In order to maximize λ^*, should the slower servers of rate α be assigned to Unit I or to Unit II?

We denote by λ^{**} the critical arrival rate of the system with α and β interchanged, that is, the faster servers are now in Unit I. For $r = c = 1$, it is readily seen that $\lambda^* = \lambda^{**}$. The order of the servers is immaterial.

For $r = c \geq 2$, we expect on intuitive grounds that $\lambda^* > \lambda^{**}$, for all $\alpha < \beta$. We have the rather drastic blocking rule with $r^* = 1$, which causes all servers in Unit I to cease work as soon as one of them becomes blocked. One anticipates that the configuration in which customers are cleared as quickly as possible out of Unit II and the buffer will have less blocking and correspondingly a smaller loss in service capability.

To verify this analytically is, even in this simplest case, another matter. The inequality $\lambda^* > \lambda^{**}$ is equivalent to

$$\frac{\alpha\phi_{M-1}(\alpha)}{\phi_M(\alpha)} > \frac{\phi_{M-1}(\alpha^{-1})}{\phi_M(\alpha^{-1})}, \quad \text{for} \quad 0 < \alpha < 1, \tag{5.2.6}$$

where for $\nu \geq c$, $\phi_\nu(\alpha)$ is defined by

$$\phi_\nu(\alpha) = \sum_{j=0}^{c-1} \frac{(c\alpha)^j}{j!} + \frac{c^c}{c!} \sum_{j=c}^{\nu} \alpha^j. \tag{5.2.7}$$

The quantities in (5.2.6) correspond to λ^* and λ^{**} for $0 < \alpha < \beta = 1$, and have the advantage of being dimension free. The original values of λ^* and λ^{**} are obtained by multiplying the corresponding expressions by $(1 - \theta)\beta c$.

Both sides of (5.2.6) tend (rapidly) to α as $M \to \infty$. As is to be expected, the advantage becomes negligible for all but small buffer sizes. We were able to prove (5.2.6) only for very small values of c and M, by tedious calculations. In view of the particular nature of model A, we also found the issue to be of less than enduring significance.

We did compute both sides of (5.2.6) for a large number of values for c, α, and M, and found the stated inequality to hold in all cases. Some numerical results are presented in table 5.1.

It may be argued that a fairer comparison should be made by setting $r^* = r = c$. This involves the discussion of model B. The somewhat astonishing result obtained will be stated after we discuss this case.

***Model B:* $2 \leq r^* \leq r$, $k^* = M + r^* - 2$.** Full blocking now occurs as soon as r^* servers in I become blocked, but service resumes at the next departure from Unit II. The index j still denotes the number of II-customers in the system and satisfies $0 \leq j \leq M + r^* - 1$. The blocks in $\tilde{Q}$ are now of order $M + r^*$ and are most conveniently defined by modifying the corresponding blocks for model A.

The first M rows and columns in all blocks are the same as for model A, but they are augmented by zero elements. We only need to define explicitly the additional elements for the rows with indices M through $M + r^* - 1$.

TABLE 5.1

$c = 2;\ \alpha/\beta = 0.95$				$c = 10;\ \alpha/\beta = 0.80$			
M	A	B	$r = r^* = c$	M	A	B	$r = r^* = c$
3	0.6962	0.6952	0.7304	11	0.7290	0.7130	0.7633
4	0.7577	0.7571	0.7790	12	0.7470	0.7360	0.7717
5	0.7968	0.7964	0.8113	13	0.7597	0.7519	0.7780
6	0.8238	0.8235	0.8343	14	0.7690	0.7633	0.7827
7	0.8435	0.8433	0.8514	15	0.7760	0.7717	0.7864
8	0.8586	0.8584	0.8648	16	0.7812	0.7780	0.7893
9	0.8704	0.8703	0.8754	17	0.7853	0.7827	0.7915
10	0.8800	0.8799	0.8840	18	0.7884	0.7864	0.7933
11	0.8878	0.8877	0.8912	19	0.7908	0.7893	0.7947
12	0.8944	0.8943	0.8973	20	0.7927	0.7915	0.7958
13	0.9000	0.8999	0.9024	21	0.7942	0.7933	0.7966
14	0.9047	0.9047	0.9068	22	0.7954	0.7947	0.7973
15	0.9089	0.9088	0.9107	23	0.7963	0.7957	0.7978
16	0.9125	0.9124	0.9141	24	0.7971	0.7966	0.7983
17	0.9156	0.9156	0.9171	25	0.7977	0.7973	0.7986
18	0.9184	0.9184	0.9197	26	0.7981	0.7978	0.7989
19	0.9209	0.9209	0.9221	27	0.7985	0.7983	0.7991
20	0.9232	0.9231	0.9242	28	0.7988	0.7986	0.7993
21	0.9252	0.9251	0.9261	29	0.7990	0.7989	0.7994
22	0.9270	0.9270	0.9278	30	0.7992	0.7991	0.7996

Factors for $(1 - \theta)c\beta$ to obtain λ^* for various numbers of servers, values of α/β, and values of M. The number of places in the buffer is given by $M - c - 1$.

These are given by

$$
\begin{aligned}
(A_0)_{jj'} &= \lambda, && \text{for } M \le j = j' \le M + r^* - 1, \\
&= c\beta\theta, && \text{for } M \le j \le M + r^* - 1, \\
& && \quad j' = j - 1, \\
&= 0, && \text{for all other } (j, j'). \\
(A_1)_{jj'} &= -\lambda - c\beta - (r + M - j - 1)\alpha, && \text{for } M \le j = j' \le M + r^* - 2, \\
&= -\lambda - c\beta, && \text{for } j = j' = M + r^* - 1, \\
&= 0, && \text{for all other } (j, j'). \qquad (5.2.8) \\
(A_2)_{jj'} &= (r + M - j)\alpha, && \text{for } M \le j \le M + r^* - 2, \\
& && \quad j' = j + 1, \\
&= c\beta\theta', && \text{for } M \le j \le M + r^* - 1, \\
& && \quad j' = j - 1, \\
&= 0, && \text{for all other } (j, j').
\end{aligned}
$$

The boundary matrices are similarly defined by

$$
\begin{aligned}
(A_{i0})_{jj'} &= i\alpha, && \text{for } j = M, \quad j' = M + 1, \\
&= \min(i, r + M - j)\alpha, && \text{for } M + 1 \le j \le M + r^* - 2, \\
& && \quad j' = j + 1, \\
&= 0, && \text{for all other } (j, j'). \\
(A_{i1})_{jj'} &= -\lambda - c\beta - i\alpha, && \text{for } j = j' = M, \\
&= -\lambda - c\beta - \min(i, r + M - j)\alpha, && \text{for } M + 1 \le j = j' \le \\
& && \quad M + r^* - 1, \qquad (5.2.9) \\
&= c\beta\theta', && \text{for } M + 1 \le j \le M + r^* - 1, \\
& && \quad j' = j - 1, \\
&= 0, && \text{for all other } (j, j'). \\
(A_{i2})_{jj'} &= \lambda, && \text{for } M \le j = j' \le M + r^* - 1, \\
&= c\beta\theta, && \text{for } M + 1 \le j \le M + r^* - 1, \\
& && \quad j' = j - 1, \\
&= 0, && \text{for all other } (j, j').
\end{aligned}
$$

The matrices A_{i0} are defined for $1 \leq i \leq r - 1$; A_{i1} and A_{i2}, for $0 \leq i \leq r - 1$.

Theorem 5.2.2. For model B, the stationary probability vector π of $A = A_0 + A_1 + A_2$ is given by

$$\pi_0 = \left[\sum_{j=0}^{c-1} \frac{1}{j!}\left(\frac{r\alpha}{\beta}\right)^j + \frac{c^c}{c!}\sum_{j=c}^{M}\left(\frac{r\alpha}{c\beta}\right)^j + \frac{c^c}{c!}\sum_{j=M+1}^{M+r^*-1}\left(\frac{r\alpha}{c\beta}\right)^j \prod_{\nu=1}^{j-M}\left(1 - \frac{\nu}{r}\right)\right]^{-1},$$

$$\begin{aligned}
\pi_j &= \frac{1}{j!}\left(\frac{r\alpha}{\beta}\right)^j \pi_0, && \text{for } 1 \leq j \leq c, \\
\pi_j &= \frac{c^c}{c!}\left(\frac{r\alpha}{c\beta}\right)^j \pi_0, && \text{for } c \leq j \leq M, \\
\pi_j &= \frac{c^c}{c!}\left(\frac{r\alpha}{c\beta}\right)^j \prod_{\nu=1}^{j-M}\left(1 - \frac{\nu}{r}\right)\pi_0, && \text{for } M + 1 \leq j \leq M + r^* - 1.
\end{aligned} \tag{5.2.10}$$

The system is *stable* if and only if

$$\lambda < (1 - \theta)r\alpha\left(1 - \pi_{M+r^*-1} - \sum_{j=1}^{r^*-1} \frac{j}{r}\,\pi_{M+j-1}\right). \tag{5.2.11}$$

Proof. By elementary calculations.

Remarks

a. We compared the two critical arrival rates λ^* and λ^{**}, defined as in model A, but for the choices $r = c = r^* \geq 2$. For the purposes of this comparison, we may again set $\beta = 1$, and $0 < \alpha \leq \beta$. The two expressions to be compared are much more involved than those for model A. A comparison by analytic means appears to be out of the question. We wrote a simple interactive program for a personal computer and implemented it for hundreds of choices of c, α, and M. The interesting result was that in all these cases, the *computed values of* λ^* *and* λ^{**} *agreed to the ten decimal places reported.* We may conjecture that if $r = r^*$, the critical arrival rates are the same for both configurations. A proof, possibly based on a clever symmetry argument, would be of interest, but a direct analytic verification seems unlikely and, even then, would be of limited methodological value.

In table 5.1 the value of $[(1 - \theta)c\beta]^{-1}\lambda^*$ is reported for two choices of α/β and c and for increasing values of M. The rapid approach to the limit value within each column is strikingly apparent. The critical arrival rates are sensitive to M only for smaller values of M.

It should be stressed, however, that when α and β are large in absolute magnitude, as is typically the case in computer or data processing applications, even a small difference in the entries of table 5.1 may signify a substantial difference in the number of jobs handled by the system per unit of real time.

b. The equality $\lambda^* = \lambda^{**}$ does not prevent the system from having very different qualitative behavior under the two configurations. Let β be substantially larger than α. With the fast servers in Unit I, the system will experience frequent blocking and the buffer will typically contain a substantial queue. With the faster servers in Unit II, the buffer will typically contain few II-customers and blocking will be rare. The critical arrival rate is then mostly determined by the slow rate of service in Unit I.

Model C: $\mathbf{1 \leq r^* \leq r, c \leq k^* \leq M + r^* - 3.}$ For model C, we need to distinguish between the cases $k^* \geq c$ and $0 \leq k^* < c$. As the details for both cases are similar, they will be presented only for the case $k^* \geq c$. As the case $k^* = M + r^* - 2$ corresponds to model B, we further assume henceforth that $k^* \leq M + r^* - 3$. We see that upon becoming fully blocked, the system enters an unblocking phase whose conditional distribution is Erlang of order $M + r^* - k^* - 1$ and scale parameter $c\beta$. During each such period only arrivals to the input queue and departures from Unit II, with or without feedback, may occur.

The index j now assumes one of the $2M + 2r^* - k^* - 2$ values

$$0, 1, \ldots, M, \ldots, M + r^* - 1, \overline{M + r^* - 2}, \overline{M + r^* - 3}, \ldots, \overline{k^* + 1},$$

where the indices with a bar represent the number of II-customers in the queue during the *unblocking phase.* In order to describe the system by a Markov process of the type given in (5.2.1), we need to distinguish, for example, between the states $(i, M + r^* - 2)$ and $(i, \overline{M + r^* - 2})$. In the former state, there is ordinary partial blocking and some service in Unit I may be going on; in the latter, the system is recovering from full blocking and no server in I is active. The transition rates for these two states are quite different.

The blocks in the partitioned matrix $\tilde{Q}$ are now no longer Jacobi matrices. They are most conveniently defined by modifying the corresponding matrices for model B.

The matrices A_{i0}, $1 \leq i \leq r - 1$, and A_2 are obtained by augmenting the corresponding matrices for model B with zero rows and columns.

The matrices A_{i1}, $0 \le i \le r-1$, and A_1 are obtained from those in model B as follows. In the row with index $M + r^* - 1$, move the entry $c\beta\theta'$ immediately to the *right* of the diagonal element. Augment the diagonal by adding $M + r^* - k^* - 2$ times the element $-\lambda - c\beta$. Next to and on the right of these entries insert the entry $c\beta\theta'$. In the last row place this entry in the column of index k^*. All other elements of the added rows and columns are zero.

The matrices A_{i2}, $0 \le i \le r-1$, and A_0 are constructed in exactly the same manner as the preceding ones, except that the quantities λ and $c\beta\theta$ are used in place of $-\lambda - c\beta$ and $c\beta\theta'$ respectively.

It is clear that in numerical computations, major economies of storage space and processing times may be realized by exploiting the highly particular and sparse structure of these blocks.

Theorem 5.2.3. For model C with $c \le k^* \le M + r^* - 3$, the vector $\boldsymbol{\pi}$ is obtained by solving the equations

$$\pi_j = \frac{r\alpha}{j\beta}\pi_{j-1}, \qquad \text{for} \quad 1 \le j \le c,$$

$$\pi_j = \frac{r\alpha}{c\beta}\pi_{j-1}, \qquad \text{for} \quad c \le j \le k^*,$$

$$\pi_j = \frac{r\alpha}{c\beta}\pi_{j-1} - \tilde{\pi}_{k^*+1}, \qquad \text{for} \quad k^* + 1 \le j \le M,$$

$$\pi_{M+j} = \left(1 - \frac{j}{r}\right)\left(\frac{r\alpha}{c\beta}\right)\pi_{M+j-1} - \tilde{\pi}_{k^*+1}, \quad \text{for} \quad 1 \le j \le r^* - 2,$$

$$\begin{aligned}\pi_{M+r^*-1} &= \tilde{\pi}_{M+r^*-2} = \cdots = \tilde{\pi}_{k^*+1} \\ &= \frac{c^c}{c!}\left(\frac{r\alpha}{c\beta}\right)^{M+r^*-1}\prod_{\nu=1}^{r^*-1}\left(1 - \frac{\nu}{r}\right)\Phi^{-1}\pi_0. \end{aligned} \qquad (5.2.12)$$

The quantity Φ is defined by

$$\Phi = \sum_{h=1}^{r^*-2}\left(\frac{r\alpha}{c\beta}\right)^h \prod_{\nu=1}^{h}\left(1 - \frac{r^* - \nu}{r}\right) + \prod_{\nu=1}^{r^*-1}\left(1 - \frac{\nu}{r}\right)\sum_{h=0}^{M-k^*-1}\left(\frac{r\alpha}{c\beta}\right)^{h+r^*-1},$$

and π_0 is obtained from the normalizing condition $\boldsymbol{\pi e} = 1$. (The components of $\boldsymbol{\pi}$ which bear a tilde correspond to the indices j with a bar.) The equilibrium condition is given by

$$\lambda < (1-\theta)r\alpha\left[\sum_{j=0}^{M-1}\pi_j + \sum_{j=1}^{r^*-1}\left(1-\frac{j}{r}\right)\pi_{M+j-1}\right]. \tag{5.2.13}$$

Proof. The system of equations $\pi A = \mathbf{0}$ leads, after elementary manipulations, to

$$\begin{aligned}
r\alpha\pi_{j-1} &= \min(j, c)\beta\pi_j, && \text{for} \quad 1 \le j \le k^*,\\
r\alpha\pi_{j-1} &= c\beta\pi_j + c\beta\tilde{\pi}_{k^*+1}, && \text{for} \quad k^*+1 \le j \le M,\\
(r-j)\alpha\pi_{M+j-1} &= c\beta\pi_{M+j} + c\beta\tilde{\pi}_{k^*+1}, && \text{for} \quad 1 \le j \le r^*-2,\\
(r-r^*+1)\alpha\pi_{M+r^*-2} &= c\beta\pi_{M+r^*-1},\\
\pi_{M+r^*-1} &= \tilde{\pi}_{M+r^*-2} = \cdots = \tilde{\pi}_{k^*+1}.
\end{aligned} \tag{5.2.14}$$

We use these equations to evaluate expressions for π_{M+r^*-1} and for $\tilde{\pi}_{k^*+1}$. Upon equating these expressions, the stated formula expressing $\tilde{\pi}_{k^*+1}$ in terms of π_0 is obtained. The remaining derivations are now straightforward.

The inequality $\pi A_0 e < \pi A_2 e$ leads to

$$\begin{aligned}
\lambda < &\sum_{j=1}^{c}(r\alpha\pi_{j-1} - j\beta\theta\pi_j) + \sum_{j=c+1}^{M}(r\alpha\pi_{j-1} - c\beta\theta\pi_j)\\
&+ \sum_{j=1}^{r^*-1}[(r-j)\pi_{M+j-1} - c\beta\theta\pi_{M+j}] - c\beta\theta(M+r^*-k^*-2),
\end{aligned}$$

and the equations (5.2.14) readily yield (5.2.13).

The main theorem now follows immediately from our general results.

Theorem 5.2.4. For the stable queue, the stationary probability vector $\mathbf{x} = [\mathbf{x}_0, \mathbf{x}_1, \ldots, \mathbf{x}_{r-1}, \mathbf{x}_r, \ldots]$, partitioned into $(N+1)$-vectors, is given by

$$\mathbf{x}_i = \mathbf{x}_{r-1}R^{i-r+1}, \quad \text{for} \quad i \ge r-1, \tag{5.2.15}$$

where R is the minimal nonnegative solution to the equation

$R^2A_2 + RA_1 + A_0 = 0$, and where $\mathbf{x}_0, \ldots, \mathbf{x}_{r-1}$, are given by

$$\begin{aligned}
&\mathbf{x}_0A_{01} + \mathbf{x}_1A_{10} = \mathbf{0},\\
&\mathbf{x}_iA_{i2} + \mathbf{x}_{i+1}A_{i+1,1} + \mathbf{x}_{i+2}A_{i+2,0} = \mathbf{0}, \quad \text{for} \quad 0 \le i \le r-3,\\
&\mathbf{x}_{r-2}A_{r-2,2} + \mathbf{x}_{r-1}A_{r-1,1} + \mathbf{x}_{r-1}RA_2 = \mathbf{0},\\
&\sum_{\nu=0}^{r-2}\mathbf{x}_\nu e + \mathbf{x}_{r-1}(I-R)^{-1}e = 1.
\end{aligned} \tag{5.2.16}$$

Remarks

a. The vector $x^* = \Sigma_{\nu=0}^{\infty} x_\nu$, which is given by

$$x^* = \sum_{\nu=0}^{r-2} x_\nu + x_{r-1}(I - R)^{-1}, \tag{5.2.17}$$

is in general not equal to π. The components of x^* have useful interpretations, which depend on the model under consideration. For example, for model C, x^*_{M+j-1}, for $1 \le j \le r^* - 1$, is the fraction of time that the system is in partial blocking with j servers of Unit I blocked. Similarly, the sum $x^*_{M+r^*-1} + \tilde{x}^*_{M+r^*-2} + \cdots + \tilde{x}^*_{k^*+1}$ is the fraction of time that service in Unit I is suspended due to full blocking. The stationary marginal density of the number of II-customers in the system may similarly be expressed in terms of x^*.

b. It is important to compare the stationary probabilities x of the system viewed at an arbitrary time to the invariant probability vector z of the discrete time Markov chain, obtained by considering the system *at service completions* in Unit I.

The easiest way to obtain the vector z is to form the vectors

$$\begin{aligned} z_k &= \tau x_{k+1} A_{k+1,0}, && \text{for} \quad 0 \le k \le r-1, \\ z_k &= \tau x_{k+1} A_2 = \tau x_{r-1} R^{k-r+2} A_2, && \text{for} \quad k \ge r-1, \end{aligned} \tag{5.2.18}$$

where

$$\tau = \left[\sum_{i=1}^{r-1} x_i A_{i0} e + x_{r-1} R (I - R)^{-1} A_2 e \right]^{-1}.$$

The vectors z_k, so obtained, are all of dimension $N + 1$, but their zero components may be ignored. This construction was discussed in general in section 3.5. From the positive components of z, a large number of useful results may be inferred. We may readily evaluate the probability that a customer causes the blocking of the j-th server, $1 \le j \le r^* - 2$, in Unit I under partial blocking or the probability that a service completion causes full blocking. It is a simple matter to enumerate the sets of indices (i, j) which correspond to the occurrence of these events and to sum the computed probabilities z_{ij} accordingly.

c. A variety of formulas for moments of the queue length densities are routinely derived. These involve much notation, mostly to handle the terms arising from the boundary behavior. We shall omit the details.

d. When there are no feedback loops, the distribution $W(\cdot)$ of the waiting time in the input queue *and* that of the time-in-system may be most efficiently computed by implementing the method discussed in section 3.9.

In the presence of feedback loops, the network is no longer acyclic, and the consideration of waiting times requires extreme care. The method of section 3.9 may be adapted to handle the waiting time in the input queue when the customers fed back to the input queue have nonpreemptive priority over those who have not yet received service in Unit I. For all other cases, the complexity of the state description is such that computationally useful insights are presently not available.

Extensions and Variants of the Buffer Model

The buffer model has a large number of variants that may be of interest. For some of these the increase in computational complexity is minimal, but most require different algorithmic approaches. We shall informally discuss a number of such alternate models and indicate feasible numerical approaches. It would be of interest to have reports on implementations of these approaches, particularly where the numerical results indicate unusual qualitative behavior or shed light on the sensitivity of the model to the exponential assumptions.

A. State-dependent Arrival Rates or Feedback Probabilities. As will be illustrated by numerical examples for the models discussed in chapter 6, fluctuations of the length of the input queue may be attenuated by reducing the arrival rate or the feedback probability during intervals of partial or full blocking. The structure of the Markov process $\tilde{Q}$ is not affected by allowing the parameters λ and θ to become dependent on the number of II-customers in the system. There is, however, a major increase in the already large number of parameters of the model. This greater parametric versatility also causes the loss of some minor computational shortcuts in the coding of the algorithm.

B. Renewal Arrivals. We may in principle replace the Poisson arrival process by a renewal process of finite mean interarrival time. We then consider the state of the system at epochs of arrival and obtain an embedded block-partitioned Markov chain $\tilde{P}$. A complete analysis, using the general results on matrix-geometric solutions, is possible but complicated. The overhead computations to evaluate the blocks in $\tilde{P}$ become very involved. It becomes, in essence, necessary to compute time-dependent solutions to a number of finite-state Markov processes and to integrate the resulting transition probabilities with respect to the interarrival time distributions. A discussion of the complex overhead computations for a different model in D. M. Lucantoni [195] serves as an illustration of this technique.

C. General Service Time Distributions. For $r = 1$, we may allow the service time distribution in Unit I to be general. This model, discussed by transform methods in T. Suzuki [289] and M. F. Neuts [216, 217], has an embedded Markov chain of the $M/G/1$ type. It may be solved algorithmically by methods outside the scope of this book.

The case $r = c = 1$, with a general service time distribution in I and a PH-service time distribution in II, is also a model of the $M/G/1$ type. An interesting computational project could take as its subject a study of the effect of variability in the service times on the features of the queue. We anticipate that, for example, the blocking probability is quite sensitive to random variability in the service times of Unit II.

For $c = 1$, the service time distribution in Unit II may be general, provided the $r \geq 1$ servers in I remain exponential. A detailed analysis shows that, considered at the beginnings of services in Unit II, this model has an embedded Markov chain $\tilde{P}$ of the type treated in this book. The overhead computations required to set up the matrix $\tilde{P}$ are again substantial.

D. Several Units in Series. Even for exponential servers, systems with finite intermediate buffers and more than two units in series pose major computational problems that are largely unsolved. Iterative methods, such as those proposed by F. S. Hillier and R. W. Boling [141] and Y. Takahashi [294], are limited by the vast increase in computer time for problems of high dimension. It would be of interest to investigate the R-matrices for fairly simple models with three or four units in series to see if good structural approximations to these matrices may be found.

5.3. A. B. CLARKE'S TANDEM QUEUE

In a series of papers, A. B. Clarke [57–60] has investigated a novel type of tandem queueing model. Servers are placed in series, and each customer will receive service from one and only one server. The novel feature is that a busy service unit prevents the access of new customers to servers further down the line. A departing customer may also be temporarily prevented from leaving by occupied service units down line. We shall specifically deal with the model in [58], which involves two servers and an intermediate buffer. The other of Clarke's models do not directly fall within the scope of this book.

Consider a flow of customers from left to right. There are two servers I and II, with II placed to the right of I. If an arriving customer finds both units free, he proceeds directly to Unit II. If Unit II is busy and I is free, that customer is served in Unit I. Whenever Unit I is busy, arriving customers join a queue in front of Unit I. In that case, a customer *cannot bypass* Unit I to gain access to Unit II. See figure 5.2.

Between both servers there is a buffer of size n. It holds those customers who have completed service in I but cannot leave the system because Unit II is busy. As soon as the customer in II finishes service, he and all those in the buffer depart the system together.

The finiteness of the buffer affects the system as follows. If the buffer is full and a service completion in I occurs, that customer cannot enter the buffer, and *blocks* Unit I. No further service can be performed in Unit I until Unit II completes service. At that time, the blocking customer departs along with those in the buffer.

An interesting transition occurs when Unit I is busy but II is not. When the service in I terminates, that customer passes out of the system, unencumbered by Unit II. If customers are waiting, the first starts service in Unit II and the second, if there is one, enters processing in Unit I. At such a transition, two customers enter service.

Arrivals are according to a Poisson process of rate λ. All service times are mutually independent and exponentially distributed. The service rates in I and II are μ_1 and μ_2 respectively. Since μ_1 and μ_2 may be different, our state space is slightly different from that in [58], where the case $\mu_1 = \mu_2$ is treated by methods of complex analysis.

The queueing model is studied as a Markov process on the state space $E = \{(i, j): i \geq 0, 0 \leq j \leq n + 2\}$. The index i denotes the number of customers waiting or receiving service in Unit I. The index j counts the customers in service in Unit II or waiting in the buffer to leave, having received service in Unit I.

The generator $\tilde{Q}$ of the process is given by

$$\tilde{Q} = \left|\begin{matrix} B_0 & C_0 & 0 & 0 & 0 & 0 & \cdots \\ B_1 & A_1 & A_0 & 0 & 0 & 0 & \cdots \\ A_3 & A_2 & A_1 & A_0 & 0 & 0 & \cdots \\ 0 & A_3 & A_2 & A_1 & A_0 & 0 & \cdots \\ 0 & 0 & A_3 & A_2 & A_1 & A_0 & \cdots \\ 0 & 0 & 0 & A_3 & A_2 & A_1 & \cdots \\ 0 & 0 & 0 & 0 & A_3 & A_2 & \cdots \\ \vdots & \vdots & \vdots & \vdots & \vdots & \vdots & \end{matrix}\right|, \tag{5.3.1}$$

where all blocks are square matrices of order $n + 3$. The specific definitions of these matrices are displayed here for the representative case $n = 4$. The definition for general values of n is immediate by following the same pattern.

We have

$$B_0 = \begin{vmatrix} -\lambda & \lambda & \cdot & \cdot & \cdot & \cdot & \cdot \\ \mu_2 & -a_1 & \cdot & \cdot & \cdot & \cdot & \cdot \\ \mu_2 & \cdot & -a_1 & \cdot & \cdot & \cdot & \cdot \\ \mu_2 & \cdot & \cdot & -a_1 & \cdot & \cdot & \cdot \\ \mu_2 & \cdot & \cdot & \cdot & -a_1 & \cdot & \cdot \\ \mu_2 & \cdot & \cdot & \cdot & \cdot & -a_1 & \cdot \\ \mu_2 & \cdot & \cdot & \cdot & \cdot & \cdot & -a_1 \end{vmatrix},$$

with $a_1 = \lambda + \mu_2$,

$$C_0 = \begin{vmatrix} 0 & \cdot & \cdot & \cdot & \cdot & \cdot & \cdot \\ \cdot & \lambda & \cdot & \cdot & \cdot & \cdot & \cdot \\ \cdot & \cdot & \lambda & \cdot & \cdot & \cdot & \cdot \\ \cdot & \cdot & \cdot & \lambda & \cdot & \cdot & \cdot \\ \cdot & \cdot & \cdot & \cdot & \lambda & \cdot & \cdot \\ \cdot & \cdot & \cdot & \cdot & \cdot & \lambda & \cdot \\ \cdot & \cdot & \cdot & \cdot & \cdot & \cdot & \lambda \end{vmatrix},$$

$$B_1 = \begin{vmatrix} \mu_1 & \cdot & \cdot & \cdot & \cdot & \cdot & \cdot \\ \cdot & \cdot & \mu_1 & \cdot & \cdot & \cdot & \cdot \\ \cdot & \cdot & \cdot & \mu_1 & \cdot & \cdot & \cdot \\ \cdot & \cdot & \cdot & \cdot & \mu_1 & \cdot & \cdot \\ \cdot & \cdot & \cdot & \cdot & \cdot & \mu_1 & \cdot \\ \cdot & \cdot & \cdot & \cdot & \cdot & \cdot & \mu_1 \\ \cdot & \mu_2 & \cdot & \cdot & \cdot & \cdot & \cdot \end{vmatrix},$$

and

$$A_0 = \lambda I,$$

$$A_1 = \begin{vmatrix} -\lambda-\mu_1 & \cdot & \cdot & \cdot & \cdot & \cdot & \cdot \\ \mu_2 & -a_2 & \cdot & \cdot & \cdot & \cdot & \cdot \\ \mu_2 & \cdot & -a_2 & \cdot & \cdot & \cdot & \cdot \\ \mu_2 & \cdot & \cdot & -a_2 & \cdot & \cdot & \cdot \\ \mu_2 & \cdot & \cdot & \cdot & -a_2 & \cdot & \cdot \\ \mu_2 & \cdot & \cdot & \cdot & \cdot & -a_2 & \cdot \\ \cdot & \cdot & \cdot & \cdot & \cdot & \cdot & -\lambda-\mu_2 \end{vmatrix},$$

with $a_2 = \lambda + \mu_1 + \mu_2$,

$$A_2 = \begin{vmatrix} \cdot & \cdot & \cdot & \cdot & \cdot & \cdot & \cdot \\ \cdot & \cdot & \mu_1 & \cdot & \cdot & \cdot & \cdot \\ \cdot & \cdot & \cdot & \mu_1 & \cdot & \cdot & \cdot \\ \cdot & \cdot & \cdot & \cdot & \mu_1 & \cdot & \cdot \\ \cdot & \cdot & \cdot & \cdot & \cdot & \mu_1 & \cdot \\ \cdot & \cdot & \cdot & \cdot & \cdot & \cdot & \mu_1 \\ \cdot & \mu_2 & \cdot & \cdot & \cdot & \cdot & \cdot \end{vmatrix},$$

$$A_3 = \begin{vmatrix} \cdot & \mu_1 & \cdot & \cdot & \cdot & \cdot & \cdot \\ \cdot & \cdot & \cdot & \cdot & \cdot & \cdot & \cdot \\ \cdot & \cdot & \cdot & \cdot & \cdot & \cdot & \cdot \\ \cdot & \cdot & \cdot & \cdot & \cdot & \cdot & \cdot \\ \cdot & \cdot & \cdot & \cdot & \cdot & \cdot & \cdot \\ \cdot & \cdot & \cdot & \cdot & \cdot & \cdot & \cdot \\ \cdot & \cdot & \cdot & \cdot & \cdot & \cdot & \cdot \end{vmatrix}. \tag{5.3.2}$$

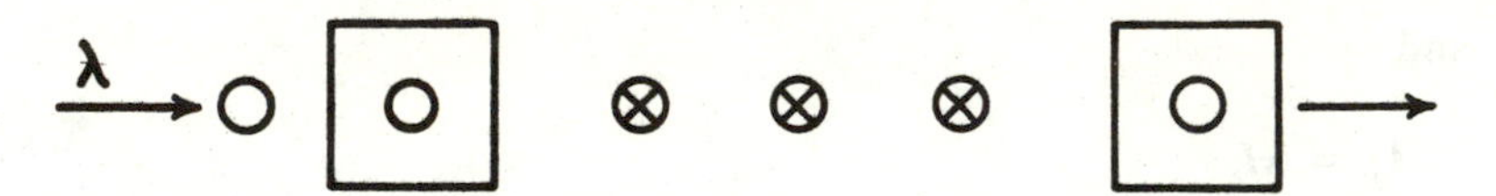

FIGURE 5.2. The configuration of the tandem queue, studied by A. B. Clarke.

The specific form of the matrices given in (5.3.2) is obtained by a systematic consideration of all possible transitions in the system. For example, the entries μ_2 in the first columns of B_0 and A_1 correspond to a departure of the customer who completes service in Unit II and all customers behind him in the buffer. If there is an arrival to an empty system, that customer starts service in Unit II. This accounts for the element λ in the first row and second column of B_0.

The element μ_1 in the first row of A_3 corresponds to the transition where two customers, one in each unit, start service. This is brought about by a service completion in I, when both Unit II and the buffer are empty.

The element μ_2 in the last row of A_2 corresponds to the clearing of a full buffer and the departure of a customer who blocks Unit I. At such a transition, two customers will also start service, but the index i is reduced by one unit only. To see this, we note that the blocking customer is no longer included in the count given by the index i. At the transition, one customer goes to Unit II. The second customer starting service in I remains among those counted by the first index.

The matrix $A = A_0 + A_1 + A_2 + A_3$ is given by

$$A = \begin{vmatrix} -\mu_1 & \mu_1 & \cdot & \cdot & \cdot & \cdot & \cdot \\ \mu_2 & -\mu_1-\mu_2 & \mu_1 & \cdot & \cdot & \cdot & \cdot \\ \mu_2 & \cdot & -\mu_1-\mu_2 & \mu_1 & \cdot & \cdot & \cdot \\ \mu_2 & \cdot & \cdot & -\mu_1-\mu_2 & \mu_1 & \cdot & \cdot \\ \mu_2 & \cdot & \cdot & \cdot & -\mu_1-\mu_2 & \mu_1 & \cdot \\ \mu_2 & \cdot & \cdot & \cdot & \cdot & -\mu_1-\mu_2 & \mu_1 \\ \cdot & \mu_2 & \cdot & \cdot & \cdot & \cdot & -\mu_2 \end{vmatrix}. \tag{5.3.3}$$

By elementary calculations one finds the stationary probability vector $\boldsymbol{\pi}$ of A to be given by

$$\begin{aligned} \pi_0 &= (1-p)(1-p^{n-1})[1-(1-p)p^{n+1}]^{-1}, \\ \pi_j &= (1-p)p^j[1-(1-p)p^{n+1}]^{-1}, \quad \text{for} \quad 1 \le j \le n+1, \\ \pi_{n+2} &= p^{n+2}[1-(1-p)p^{n+1}]^{-1}, \end{aligned} \tag{5.3.4}$$

with $p = \mu_1(\mu_1+\mu_2)^{-1}$.

Theorem 5.3.1. The queue is *stable* if and only if

$$\lambda < (\mu_1 + \mu_2)[p^2 + 2p(1-p) - p^{n+2}][1 - (1-p)p^{n+1}]^{-1}. \tag{5.3.5}$$

In this case, the stationary probability vector $x = [x_0, x_1, x_2, \ldots]$ of $\tilde{Q}$ satisfies

$$x_i = x_1 R^{i-1}, \quad \text{for} \quad i \geq 1, \tag{5.3.6}$$

where the positive matrix R is the minimal nonnegative solution to the equation

$$A_0 + RA_1 + R^2A_2 + R^3A_3 = 0. \tag{5.3.7}$$

The matrix $\lambda^{-1}R(A_2 + A_3 + RA_3)$ is stochastic. The vector $[x_0, x_1]$ is obtained by solving the equations

$$[x_0, x_1] \left| \begin{array}{cc} B_0 & C_0 \\ B_1 + RA_3 & A_1 + RA_2 + R^2A_3 \end{array} \right| = [0, 0],$$

$$x_0 e + x_1(I - R)^{-1} e = 1. \tag{5.3.8}$$

Proof. The general equilibrium condition of theorem 1.7.1 is given here by $\pi A_0 e < \pi(2A_3 + A_2)e$, or $\lambda < 2\mu_1\pi_0 + \mu_1(1 - \pi_0 - \pi_{n+2}) + \mu_2\pi_{n+2}$. Upon substitution for π_0 and π_{n+2}, we obtain (5.3.5).

Formula (1.7.12) yields $\lambda e = (RA_2 + RA_3 + R^2A_3)e$, so that the matrix $\lambda^{-1}R(A_2 + A_3 + RA_3)$ is stochastic. By writing equation (5.3.7) as

$$R = -A_0A_1^{-1} - (R^2A_2 + R^3A_3)A_1^{-1}, \tag{5.3.9}$$

we see first that $-A_0A_1^{-1}$ is a lower triangular matrix with positive diagonal elements. Second, all but the last element in the first column are positive. The other elements vanish. Upon successive substitutions, we see that the successive iterates gain positive elements in a systematic manner. After a small (computable) number of iterations, they and hence also R are positive matrices. It is now readily verified by inspection that the coefficient matrix in the first equation of (5.3.8) is an irreducible generator. The vectors x_0 and x_1 are therefore positive.

Remarks

a. For $\mu_1 = \mu_2 = \mu$, (5.3.5) reduces to

$$\lambda < 2\mu \frac{3 - 2^{-n}}{4 - 2^{-n}}, \quad \text{for} \quad n \geq 0, \tag{5.3.10}$$

which agrees with the result obtained in [58]. Even with an infinite buffer, the critical arrival rate for this queue is only 3/4 that of the corresponding

$M/M/2$ queue with the two servers placed in parallel. This reflects, of course, the fact that access to the second server is impossible whenever Unit I is occupied.

b. As $n \to \infty$, the general equilibrium becomes

$$\lambda < (\mu_1 + \mu_2)p(2 - p), \quad \text{for} \quad 0 < p < 1. \tag{5.3.11}$$

The factor $p(2 - p)$ increases from zero to one on [0, 1]. This suggests that, at least for large buffer sizes, it is very important that the faster of the two servers be assigned to Unit I.

The same conclusion also holds for finite n. To see this, we let $\phi_n(p)$ be the factor of $\mu_1 + \mu_2$ in formula (5.3.5) and show that the inequality

$$\phi_n(p) < \phi_n(1 - p) \tag{5.3.12}$$

holds for $0 < p < 1/2$. It is readily proved by induction that

$$(1 - p)^2 - p^2 > (1 - p)^{n+2} - p^{n+2}, \quad \text{for} \quad 0 < p < \frac{1}{2}, \qquad n \geq 1.$$

Furthermore, $1 - (1 - p)p^{n+1} > 1 - p(1 - p)^{n+1}$, for $0 < p < 1/2$. We now have

$$\begin{aligned}\phi_n(1 - p) &= \frac{(1 - p)^2 + 2p(1 - p) - (1 - p)^{n+2}}{1 - p(1 - p)^{n+1}} \\ &> \frac{(1 - p)^2 + 2p(1 - p) - (1 - p)^{n+2}}{1 - (1 - p)p^{n+1}} \\ &> \frac{p^2 + 2p(1 - p) - p^{n+2}}{1 - (1 - p)p^{n+1}} = \phi_n(p).\end{aligned}$$

(I am indebted to S. Chakravarthy for this proof.)

c. There is another interesting comparison. Let the sum $\mu_1 + \mu_2$ be constant and let us find that value of μ_1 for which the factor $\phi_n(p)$ is maximum. Without loss of generality, we may set $\mu_1 + \mu_2 = 1$, so that $p = \mu_1$.

The maximum values $\phi_n(p^*)$ of $\phi_n(p)$, as well as the corresponding $\phi_n(1 - p^*)$, are listed in table 5.2. We see that, provided a given combined service rate can be freely allocated to the two servers, it is possible to bring the factor $\phi_n(p)$ quite close to one. For example, with $n = 10$, it is possible to achieve a critical arrival rate that is more than 91 percent of that of the configuration with the two servers in parallel. In order to do so, the server in I should be three times as fast as the one in Unit II. The importance of correctly assigning the faster server to Unit I is evident upon consideration of table 5.2 of values of $\phi_n(1 - p^*)$.

Algorithmic Simplifications. It appears unlikely that many practical situations will be encountered for which extensive numerical solu-

TABLE 5.2

n	p^*	$\phi_n(p^*)$	$\phi_n(1-p^*)$
1	0.58	0.7319	0.6567
2	0.62	0.7783	0.6157
3	0.64	0.8121	0.5907
4	0.68	0.8377	0.5378
5	0.70	0.8579	0.5100
6	0.70	0.8739	0.5100
7	0.72	0.8875	0.4816
8	0.74	0.8987	0.4524
9	0.74	0.9076	0.4524
10	0.76	0.9160	0.4224
15	0.80	0.9428	0.3600
20	0.82	0.9576	0.3276
25	0.84	0.9670	0.2944
30	0.86	0.9737	0.2604

The values p^* of p for which $\phi_n(p)$, $0 < p < 1$, is maximum for various n. Also the corresponding values of $\phi_n(p^*)$ and $\phi_n(1-p^*)$.

tions of the present model would be needed. The model does, however, have a large number of features that may be exploited to simplify computation. We propose to discuss these for the sake of example. Similar ideas may often be used in other, more complex problems.

a. The matrix R is easily computed by successive substitutions in the equation (5.3.9), written as

$$R = [A_0 + R^2(A_2 + RA_3)](-A_1^{-1}).$$

In evaluating the factor inside the square brackets, one takes the sparsity of the matrices A_0, A_2, and A_3 into account. One notes, for example, that

$$A_2 + RA_3 = \begin{vmatrix} \cdot & \mu_1 R_{00} & \cdot & \cdot & \cdot & \cdot & \cdot \\ \cdot & \mu_1 R_{10} & \mu_1 & \cdot & \cdot & \cdot & \cdot \\ \cdot & \mu_1 R_{20} & \cdot & \mu_1 & \cdot & \cdot & \cdot \\ \cdot & \mu_1 R_{30} & \cdot & \cdot & \mu_1 & \cdot & \cdot \\ \cdot & \mu_1 R_{40} & \cdot & \cdot & \cdot & \mu_1 & \cdot \\ \cdot & \mu_1 R_{50} & \cdot & \cdot & \cdot & \cdot & \mu_1 \\ \cdot & \mu_2 + \mu_1 R_{60} & \cdot & \cdot & \cdot & \cdot & \cdot \end{vmatrix}.$$

There is no need to store the matrix $-A_1^{-1}$, which is given by

$$-A_1^{-1} = \begin{vmatrix} (\lambda+\mu_1)^{-1} & \cdot & \cdot & \cdot & \cdot & \cdot & \cdot \\ a_3 & a_2^{-1} & \cdot & \cdot & \cdot & \cdot & \cdot \\ a_3 & \cdot & a_2^{-1} & \cdot & \cdot & \cdot & \cdot \\ a_3 & \cdot & \cdot & a_2^{-1} & \cdot & \cdot & \cdot \\ a_3 & \cdot & \cdot & \cdot & a_2^{-1} & \cdot & \cdot \\ a_3 & \cdot & \cdot & \cdot & \cdot & a_2^{-1} & \cdot \\ \cdot & \cdot & \cdot & \cdot & \cdot & \cdot & (\lambda+\mu_2)^{-1} \end{vmatrix},$$

where $a_3 = \mu_2 a_2^{-1}(\lambda + \mu_1)^{-1}$. The displayed matrices for $n = 4$ are representative of the general case.

The successive substitutions in equation (5.3.9) can be efficiently carried out by using a simple special purpose subroutine. The modicum of extra care needed in the coding is largely rewarded by the reduced execution times.

b. The system (5.3.8) for $\boldsymbol{x}_0$ and $\boldsymbol{x}_1$ is efficiently solved as follows. The vector $\boldsymbol{\nu} = (I - R)^{-1}\boldsymbol{e}$ is obtained as the solution of $\boldsymbol{\nu} - R\boldsymbol{\nu} = \boldsymbol{e}$. The vector $\boldsymbol{x}_0$ is eliminated by writing the system as

$$\boldsymbol{x}_1[A_1+RA_2+R^2A_3+(B_1+RA_3)(-B_0^{-1})C_0]=\boldsymbol{0},$$

$$\boldsymbol{x}_1\boldsymbol{\nu}+\boldsymbol{x}_1(B_1+RA_3)(-B_0^{-1})\boldsymbol{e}=1,$$

$$\boldsymbol{x}_0=\boldsymbol{x}_1(B_1+RA_3)(-B_0^{-1}).$$

The vector $\boldsymbol{x}_1$ is now found by solving the $n + 3$ homogeneous equations, subject to the normalizing condition. The vector $\boldsymbol{x}_0$ is then immediately obtained.

c. In addition to easily derived and computed expressions for the moments of various marginal and conditional queue length distributions, we may obtain the following formulas, which require little additional computation.

The stationary elementary probability that a service completion occurs in Unit I during $(t, t + dt)$ is given by $c_1 dt$, where

$$c_1 = \mu_1\boldsymbol{x}_1(I - R)^{-1}\boldsymbol{e} = \mu_1(1 - \boldsymbol{x}_0\boldsymbol{e}). \tag{5.3.13}$$

The corresponding probability of a departure from Unit II is given by $c_2 dt$, where

$$c_2 = \mu_2 \sum_{\nu=1}^{n+2} x_{0\nu} + \mu_2 \sum_{\nu=1}^{n+2} [\boldsymbol{x}_1(I - R)^{-1}]_\nu$$

$$= \mu_2\{1 - x_{00} - [\boldsymbol{x}_1(I - R)^{-1}]_0\}. \tag{5.3.14}$$

The probability that a customer receives service from Unit I is then equal to $c_1(c_1 + c_2)^{-1}$.

The stationary density d_j, $1 \leq j \leq n + 2$, of the size of the groups departing together from Unit II and the buffer is given by

$$d_j = c_2^{-1}\mu_2\{x_{0j} + [x_1(I - R)^{-1}]_j\}, \quad \text{for} \quad 1 \leq j \leq n + 2. \tag{5.3.15}$$

For the sake of comparison, the marginal density of the number of customers in the buffer, possibly including a customer blocking Unit I, is given by

$$\begin{aligned} b_0 &= x_{00} + x_{01} + [x_1(I - R)^{-1}]_0 + [x_1(I - R)^{-1}]_1, \\ b_j &= x_{0j} + [x_1(I - R)^{-1}]_j, \quad \text{for} \quad 1 \leq j \leq n + 2. \end{aligned} \tag{5.3.16}$$

The elementary probability that during $(t, t + dt)$ Unit I becomes blocked is given by $\beta_1 dt$, where

$$\beta_1 = \mu_1[x_1(I - R)^{-1}]_{n-1}. \tag{5.3.17}$$

Of the customers served in Unit I, a fraction $c_1^{-1}\beta_1$ will cause blocking. This is to be compared to the fraction of time Unit I is blocked. This quantity is given by

$$\beta_1' = x_{0,n+2} + [x_1(I - R)^{-1}]_{n+2}. \tag{5.3.18}$$

Stationary Waiting Time Distributions. It is often possible to adapt the procedures for the computation of waiting time distributions in QBD processes, discussed in section 3.9, to the case where the generator $\tilde{Q}$ is a *band matrix.* As the present model provides us with an example, we shall discuss the elementary ideas underlying this adaptation here. Only waiting times under the first come, first served, discipline will be considered. Many waiting time distributions, such as those at arrival epochs, at an arbitrary time, at epochs of blocking, and so on, may then be related to those of the time until absorption in the process $\tilde{Q}^o$, with generator

$$\tilde{Q}^o = \left|\begin{matrix} 0 & 0 & 0 & 0 & 0 & 0 & 0 & \cdots \\ 0 & 0 & 0 & 0 & 0 & 0 & 0 & \cdots \\ A_3 & A_2 & D & 0 & 0 & 0 & 0 & \cdots \\ 0 & A_3 & A_2 & D & 0 & 0 & 0 & \cdots \\ 0 & 0 & A_3 & A_2 & D & 0 & 0 & \cdots \\ 0 & 0 & 0 & A_3 & A_2 & D & 0 & \cdots \\ 0 & 0 & 0 & 0 & A_3 & A_2 & D & \cdots \\ 0 & 0 & 0 & 0 & 0 & A_3 & A_2 & \cdots \\ \vdots & \vdots & \vdots & \vdots & \vdots & \vdots & \vdots & \end{matrix}\right|, \tag{5.3.19}$$

with $D = A_0 + A_1$, and with an appropriately defined initial probability vector $y = [y_0, y_1, y_2, \ldots]$.

We see that upon repartitioning $\tilde{Q}^o$ in the obvious manner into blocks of order $2(n + 3)$ and the vector y accordingly into vectors of dimension $2(n + 3)$, the discussion of the time until absorption into one of the states $(0, j)$ or $(1, j)$, $0 \le j \le n + 2$, proceeds in exactly the same manner as in section 3.9. The special structure of the larger blocks is, of course, to be exploited in the numerical computations.

Let $W_{0j}(x)$ and $W_{1j}(x)$ denote the probability that absorption occurs no later than time x into the state $(0, j)$ or $(1, j)$ respectively. The vectors $W_0(x) = \{W_{0j}(x)\}$, and $W_1(x) = \{W_{1j}(x)\}$, are then computable by the procedure of section 3.9.

Depending on the distribution that is required, it may be possible that the solution of an additional finite system of differential equations involving the vectors $W_0(\cdot)$ and $W_1(\cdot)$ is necessary.

Let us illustrate this by discussing in some detail the computation of the distribution of the time-in-system of a customer arriving to the stationary queue for the model at hand. The steady-state probabilities of the states of the system found by that customer are given by the components of the vector $x = [x_0, x_1, x_1R, x_1R^2, \ldots]$.

The Laplace-Stieltjes transform $[w_0(s), w_1(s)]$ of the vector $[W_0(\cdot), W_1(\cdot)]$ is given by

$$[w_0(s), w_1(s)] = \sum_{i=0}^{\infty} [x_0, x_1] \begin{vmatrix} 0 & 0 \\ R & R^2 \end{vmatrix}^i \cdot \left\{ \begin{vmatrix} sI - D & 0 \\ -A_2 & sI - D \end{vmatrix}^{-1} \begin{vmatrix} A_3 & A_2 \\ 0 & A_3 \end{vmatrix} \right\}^i. \qquad (5.3.20)$$

The expression on the right in (5.3.20) is of the form (3.9.12). The efficient procedure, appropriate to this form, may be implemented to compute $[W_0(\cdot), W_1(\cdot)]$ by the numerical solution of a system of $4(n + 3)^2$ linear differential equations with constant coefficients. This system is highly structured.

Using the notation of theorem 3.9.2, the vector $[W_0(\cdot), W_1(\cdot)]$ is computed by means of the following algorithmic steps.

Step 1. Evaluate the partitioned matrix V^o of order $2(n + 3)$ by solving the equation

$$\begin{vmatrix} V_{00}^o & V_{01}^o \\ V_{10}^o & V_{11}^o \end{vmatrix} \begin{vmatrix} D & 0 \\ A_2 & D \end{vmatrix} + \begin{vmatrix} 0 & 0 \\ R & R^2 \end{vmatrix} \begin{vmatrix} V_{00}^o & V_{01}^o \\ V_{10}^o & V_{11}^o \end{vmatrix} \begin{vmatrix} A_3 & A_2 \\ 0 & A_3 \end{vmatrix}$$

$$= \begin{vmatrix} -I & 0 \\ 0 & -I \end{vmatrix}. \quad (5.3.21)$$

Upon examination of (5.3.21), we find that $V_{01}^o = 0$ and $V_{00}^o = -D^{-1}$. The matrices V_{10}^o and V_{11}^o are found by solving the equations

$$V_{11}^oD + R^2V_{10}^oA_2 = RD^{-1}A_2 + R^2D^{-1}A_3 - I,$$

$$V_{10}^oD + R^2V_{10}^oA_3 + V_{11}^oA_2 = RD^{-1}A_3.$$

This may be done by several readily apparent methods and involves only manipulations with matrices of order $n + 3$.

Step 2. Integrate the matrix-differential equation

$$\begin{vmatrix} \Theta_{00}'(x) & \Theta_{01}'(x) \\ \Theta_{10}'(x) & \Theta_{11}'(x) \end{vmatrix} = \begin{vmatrix} \Theta_{00}(x) & \Theta_{01}(x) \\ \Theta_{10}(x) & \Theta_{11}(x) \end{vmatrix} \begin{vmatrix} D & 0 \\ A_2 & D \end{vmatrix} + \begin{vmatrix} 0 & 0 \\ R & R^2 \end{vmatrix} \begin{vmatrix} \Theta_{00}(x) & \Theta_{01}(x) \\ \Theta_{10}(x) & \Theta_{11}(x) \end{vmatrix} \begin{vmatrix} A_3 & A_2 \\ 0 & A_3 \end{vmatrix}, \quad (5.3.22)$$

with $\Theta(0) = V^o$. Upon examination of (5.3.22), we find that $\Theta_{01}(x) = 0$, for $x \geq 0$. The remaining differential equations may be solved by standard numerical methods and require only manipulations with matrices of order $n + 3$.

Step 3. Evaluate the vectors $W_0(x)$ and $W_1(x)$ by

$$W_0(x) = x_0 - x_1(V_{10}^oD + V_{11}^oA_2) - x_1R[\Theta_{00}(x)A_3 + R\Theta_{10}(x)A_3],$$

$$W_1(x) = -x_1V_{11}^oD - x_1R[\Theta_{00}(x)A_2 + R\Theta_{10}(x)A_2 + R\Theta_{11}(x)A_3],$$

$$\text{for} \quad x \geq 0. \quad (5.3.23)$$

We note that the computation up to this point depends only on the structure of the band matrix $\tilde{Q}^o$, given in (5.3.19), and *not* on the other special features of the model.

The remaining computations depend strongly on the nature of the system. The path leading to absorption in $\tilde{Q}^o$ traces out the "history" of the services performed during the wait of the customer under consideration, up to the point that one of the states $(0, j)$ or $(1, j)$, $0 \leq j \leq n + 2$, is reached. The further progress of the customer depends on which of these states is reached. The conditional distributions of the *residual* time-in-system from the time of absorption are given in table 5.3.

To see how, for example, the entry for $(1, j)$, $1 \leq j \leq n + 1$, is obtained, we need to consider the following three alternatives. (*a*) The cus-

TABLE 5.3

State	Conditional Distribution
(0, 0)	$E_1(\mu_2, x)$
(0, j) $1 \le j \le n+1$	$pE_1(\mu_1+\mu_2, \cdot)*E_1(\mu_2, x) + (1-p)E_1(\mu_1+\mu_2, \cdot)*E_1(\mu_1, x)$,
(0, $n+2$)	$E_2(\mu_2, x)$,
(1, 0)	$E_1(\mu_1, \cdot)*E_1(\mu_2, x)$,
(1, j) $1 \le j \le n+1$	$p^2E_2(\mu_1+\mu_2, \cdot)*E_1(\mu_2, x) + p(1-p)E_2(\mu_1+\mu_2, \cdot)*E_1(\mu_1, \cdot) +$ $(1-p)E_1(\mu_1+\mu_2, \cdot)*E_1(\mu_1, \cdot)*E_1(\mu_2, x)$,
(1, $n+2$)	$pE_2(\mu_2, \cdot)*E_1(\mu_1+\mu_2, x) + (1-p)E_1(\mu_2, \cdot)*E_1(\mu_1+\mu_2, \cdot)*E_1(\mu_1, x)$.

Conditional distributions of the residual time-in-system.

tomer in service in Unit I and the customer under consideration both spend time in the buffer. (b) The customer in service, but not the customer under consideration, spend time in the buffer. (c) The buffer clears out before the customer in service leaves. After he does, the customer under consideration is served by Unit II. It is now easily verified that the three terms in the entry in table 5.3 correspond to these alternatives.

If we denote the six PH-distributions in table 5.3 by $\Phi_i(\cdot)$, $1 \le i \le 6$, we see that the distribution of the time in the system is given by

$$W_{00}(\cdot)*\Phi_1(x) + \sum_{j=1}^{n+1} W_{0j}(\cdot)*\Phi_2(x) + W_{0,n+2}(\cdot)*\Phi_3(x)$$

$$+ W_{10}(\cdot)*\Phi_4(x) + \sum_{j=1}^{n+1} W_{1j}(\cdot)*\Phi_5(x) + W_{1,n+2}(\cdot)*\Phi_6(x).$$

In order to evaluate this distribution numerically, we need an algorithm for the convolution of a general mass-function $W(\cdot)$ on $[0, \infty)$ and a PH-distribution $F(\cdot)$ with representation $(\boldsymbol{\alpha}, T)$.

We have

$$W(\cdot)*F(x) = W(0+)\alpha_{m+1} + \int_0^x W(u)\boldsymbol{\alpha}\exp[T(x-u)]\,\boldsymbol{T}^o du, \quad \text{for} \quad x \ge 0. \tag{5.3.24}$$

Setting

$$\boldsymbol{\nu}(x) = \boldsymbol{\alpha}\int_0^x W(u)\exp[T(x-u)]\,du,$$

we see that $\boldsymbol{\nu}(x)$ satisfies the system of differential equations

$$\boldsymbol{\nu}'(x) = \boldsymbol{\nu}(x)T + W(x)\boldsymbol{\alpha}, \tag{5.3.25}$$

with $\nu(0) = \mathbf{0}$. The numerical solution of this system is performed by classical methods, and the formula

$$W(\cdot)*F(x) = \alpha_{m+1}W(0+) + \nu(x)T^o, \quad \text{for} \quad x \geq 0, \qquad (5.3.26)$$

yields the desired convolution.

It is a simple matter to set up representations for the six PH-distributions in table 5.3. All are continuous at zero, so that the corresponding first terms in (5.3.26) all vanish.

5.4. A MULTIPROGRAMMING MODEL

A queueing model, representative of those arising in the study of computer operations, is the multiprogramming-multiprocessor system, discussed in G. Latouche [179]. The system consists of an unlimited input queue to which customers arrive according to a homogeneous Poisson process of rate λ. The subsidiary network consists of two queues, one in front of J parallel exponential servers of rate α, the other in front of C parallel exponential servers of rate β. The capacity constraint states that the total number of customers waiting or receiving service in the subsidiary system should not exceed M, a given positive integer that in most cases of interest satisfies $M \geq J + C$. The former J servers are thought of as *input-output* (I/O) *devices*; the latter C servers as *central processing units* (CPU). The parameter M is called the *maximum degree of multiprogramming*. It represents the maximum number of customers (jobs) directly under the control of the monitoring device at any time.

Customers in the input queue will, when allowed to enter the subsidiary system, join the queue in front of the J I/O devices. Upon completion of service there, a customer leaves the system with probability $1 - \theta$, $0 < \theta < 1$, or joins the queue in front of the C central processors with probability θ.

In the latter case, the customer will return to the I/O queue upon completion of his service by a CPU. A job may in this manner make a geometrically distributed number of passes through the CPU portion of the system before his eventual departure. Upon a departure from the subsidiary system, a customer will enter from the input queue, provided that queue is not empty. A diagram of the system is given in figure 5.3.

The *saturated* version of this and related models for multiprogramming queueing networks have received wide attention, for example, in the papers by I. Adiri, M. Hofri, and M. Yadin [1], D. P. Gaver [106], D. P. Gaver and G. Humfeld [108], D. P. Gaver and G. S. Shedler [107], J. P. Lehoczky and D. P. Gaver [186], P. A. W. Lewis and G. Shedler [189], and I. Mitrani [206]. The discussions by B. Avi-Itzhak and D. P. Heyman

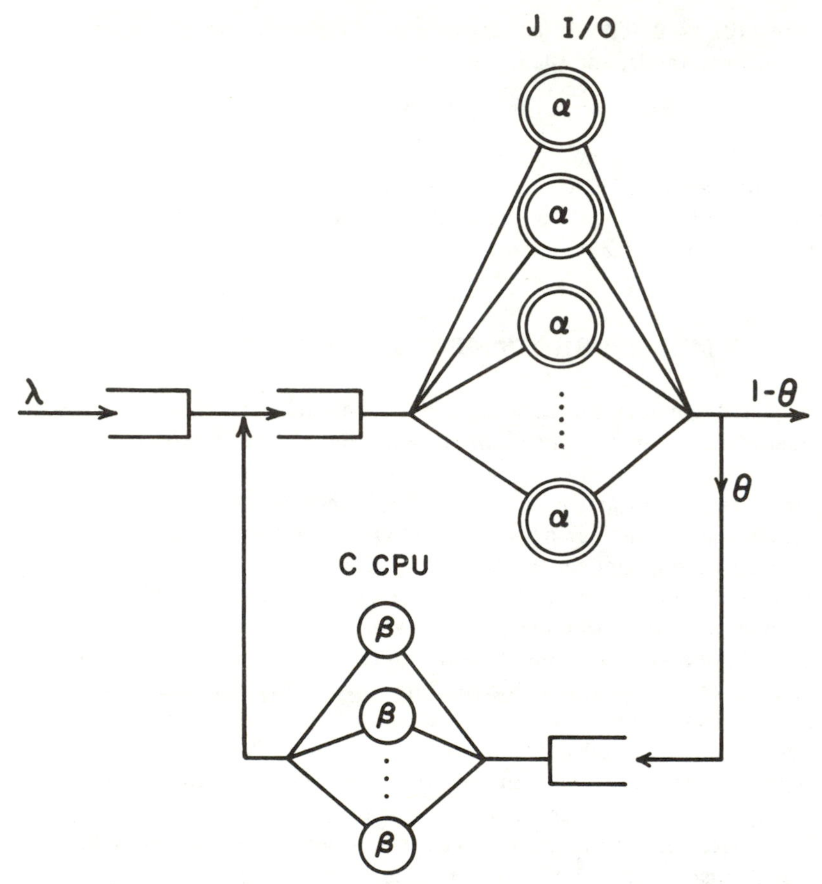

FIGURE 5.3. The multiprogramming model with J I/O devices and C central processors.

[7] and by G. Latouche [179] deal with the *unsaturated* version, which more appropriately describes the operation of a computer system receiving less frequent, but longer, jobs.

Standard assumptions on the independence of the interarrival times, the processing times in both types of service units, and the branching events at departures from the I/O units will be made. The branching probability θ could be made to depend on the number of customers in each of the two queues of the subsidiary system. This involves only greater notational complexity; the same method of solution is applicable.

The model leads to a continuous-parameter Markov process $\tilde{Q}$, whose state space may be defined, as in [179], in a most effective manner.

We first consider states described by the pairs (j, k) with $0 \le j + k \le M - 1$, and ordered as follows

$$\begin{array}{llllll}
(0, 0) & & & & & \\
(0, 1) & (1, 0) & & & & \\
(0, 2) & (1, 1) & (2, 0) & & & \\
(0, 3) & (1, 2) & (2, 1) & (3, 0) & & \\
\vdots & \vdots & \vdots & \vdots & & \\
(0, M-1) & (1, M-2) & (2, M-3) & (3, M-4) & \cdots & (M-1, 0).
\end{array} \tag{5.4.1}$$

These states provide the indices for submatrices defined in the following, which describe the boundary behavior of the process $\tilde{Q}$. For this first set of states, the pair (j, k) signifies that the input queue is *empty* and that the I/O and CPU queues hold respectively j and k customers, where $0 \le j + k \le M - 1$.

When $i = 0$, and there are M customers in the subsidiary system, or when $i \ge 1$, we indicate the state of the system by the pair (i, j). The first index i is now the size of the input queue, and j, with $0 \le j \le M$, denotes the number of customers in the I/O queue. The number of jobs in the CPU queue is then clearly $M - j$.

The set of indices $\{(i, j): i \ge 0, 0 \le j \le M\}$ is ordered lexicographically and, as usual, i is the set $\{(i, j): 0 \le j \le M\}$. Note that some pairs of numbers occur in both sets, but with entirely different physical interpretations. As the boundary states are treated separately, this should not cause confusion.

The matrix $\tilde{Q}$ is now partitioned as follows

$$\tilde{Q} = \begin{vmatrix}
A_{01} & A_{02} & & & & & & \\
A_{10} & A_{11} & A_{12} & & & & & \\
 & A_{20} & A_{21} & A_{22} & & & & \\
 & & \cdots & & & & & \\
 & & & A_{M-1,0} & A_{M-1,1} & A_{M-1,2} & & \\
 & & & & A_{M,0} & A_1 & A_0 & \\
 & & & & & A_2 & A_1 & A_0 \\
 & & & & & & & \cdots
\end{vmatrix}, \tag{5.4.2}$$

where A_0, A_1, and A_2 are square matrices of order $M + 1$, defined by

$$
\begin{aligned}
A_0 &= \lambda I, \\
(A_1)_{jj'} &= 0, && \text{for } |j - j'| > 1, \\
&= C\beta, && \text{for } j' = j + 1, \\
& && \quad 0 \le j \le M - C, \qquad (5.4.3) \\
&= (M - j)\beta, && \text{for } j' = j + 1, \\
& && \quad M - C \le j \le M - 1, \\
&= \min(j, J)\alpha\theta, && \text{for } j' = j - 1, \quad 1 \le j \le M, \\
&= -\lambda - C\beta - \min(j, J)\alpha, && \text{for } j' = j, \quad 0 \le j \le M - C, \\
&= -\lambda - J\alpha - (M - j)\beta, && \text{for } j' = j, \quad M - C \le j \le M, \\
A_2 &= \operatorname{diag}\{\min(j, J)\alpha\theta', \ 0 \le j \le M\}.
\end{aligned}
$$

The definition of the boundary matrices is somewhat more involved. The matrix A_{m0}, $1 \le m \le M$, is of dimensions $(m + 1) \times m$. It corresponds to transitions from states (ν, k) with $\nu + k = m$ to states (ν', k') with $\nu' + k' = m - 1$, and is given by

$$
\begin{aligned}
(A_{m0})_{jj'} &= \min(j, J)\alpha\theta', && \text{for } j' = j + 1, \quad 1 \le j \le m, \\
&= 0, && \text{otherwise.} \qquad (5.4.4)
\end{aligned}
$$

The matrix A_{m1}, $0 \le m \le M - 1$, is a Jacobi matrix of order $m + 1$, whose relevant elements are defined by

$$
\begin{aligned}
(A_{m1})_{jj'} &= \min(m - j, J)\beta, && \text{for } j' = j + 1, \\
& && \quad 0 \le j \le m - 1, \\
&= -\lambda - \min(m - j, J)\beta - \min(j, C)\alpha, && \text{for } j' = j, \\
& && \quad 0 \le j \le m, \\
&= \min(j, C)\alpha\theta, && \text{for } j' = j - 1, \\
& && \quad 1 \le j \le m. \qquad (5.4.5)
\end{aligned}
$$

The matrix A_{m2}, $0 \le m \le M - 1$, is of dimensions $(m + 1) \times (m + 2)$ and corresponds to transitions from states (ν, k) with $\nu + k = m$ to states (ν', k') with $\nu' + k' = m + 1$. We have

$$
\begin{aligned}
(A_{m2})_{jj'} &= \lambda, && \text{for } 0 \le j \le m, \quad j' = j + 1, \\
&= 0, && \text{otherwise.} \qquad (5.4.6)
\end{aligned}
$$

The matrix $\tilde{Q}$ is clearly of the classical QBD form. The vector $\boldsymbol{\pi}$ and the equilibrium condition $\boldsymbol{\pi} A_2 \mathbf{e} > \boldsymbol{\pi} A_0 \mathbf{e}$ may be explicitly expressed in terms of the parameters of the model.

Theorem 5.4.1. With $\gamma = C\beta(J\alpha\theta)^{-1}$, the vector $\boldsymbol{\pi}$ is given by

$$\pi_0 = \left[\sum_{\nu=0}^{J} \frac{J^\nu}{\nu!}\gamma^\nu + \frac{J^J}{J!}\sum_{\nu=J+1}^{M-C}\gamma^\nu + \frac{J^J C!}{J!}\sum_{k=1}^{C}\frac{\gamma^{M-C+k}}{C^k(C-k)!}\right]^{-1},$$

$$\begin{aligned} \pi_\nu &= \frac{J^\nu}{\nu!}\gamma^\nu \pi_0, && \text{for} \quad 0 \le \nu \le J, \\ &= \frac{J^J}{J!}\gamma^\nu \pi_0, && \text{for} \quad J \le \nu \le M - C, \\ &= \frac{J^J}{J!}\frac{C!}{(M-\nu)!}C^{M-C-\nu}\gamma^\nu \pi_0, && \text{for} \quad M - C \le \nu \le M. \end{aligned} \tag{5.4.7}$$

The queue is *stable* if and only if

$$\lambda < \alpha(1-\theta)J - \left[\pi_0 \sum_{\nu=0}^{J-1} \frac{J-\nu}{\nu!} J^\nu \gamma^\nu\right]. \tag{5.4.8}$$

Proof. The system of linear equations $\boldsymbol{\pi} A = \mathbf{0}$ is explicitly written as

$$\begin{aligned} -C\beta\pi_0 + \alpha\theta\pi_1 &= 0, \\ C\beta\pi_{\nu-1} - (C\beta + \nu\alpha\theta)\pi_\nu + (\nu+1)\alpha\theta\pi_{\nu+1} &= 0, \\ &\text{for} \quad 1 \le \nu \le J-1, \\ C\beta\pi_{\nu-1} - (C\beta + J\alpha\theta)\pi_\nu + J\alpha\theta\pi_{\nu+1} &= 0, \\ &\text{for} \quad J \le \nu \le M - C, \\ (M-\nu+1)\beta\pi_{\nu-1} - [(M-\nu)\beta + J\alpha\theta]\pi_\nu + J\alpha\theta\pi_{\nu+1} &= 0, \\ &\text{for} \quad M - C + 1 \le \nu \le M - 1, \\ \beta\pi_{M-1} - J\alpha\theta\pi_M &= 0. \end{aligned} \tag{5.4.9}$$

Adding the first ν equations for $1 \le \nu \le M - 1$ leads to the equivalent system

$$\begin{aligned} C\beta\pi_{\nu-1} &= \nu\alpha\theta\pi_\nu, && \text{for} \quad 1 \le \nu \le J, \\ C\beta\pi_{\nu-1} &= J\alpha\theta\pi_\nu, && \text{for} \quad J+1 \le \nu \le M - C, \\ (M-\nu+1)\beta\pi_{\nu-1} &= J\alpha\theta\pi_\nu, && \text{for} \quad M - C + 1 \le \nu \le M, \end{aligned} \tag{5.4.10}$$

from which the formulas (5.4.7) and (5.4.8) are readily obtained.

Remarks

a. The equations (5.4.10), together with $\boldsymbol{\pi} \mathbf{e} = 1$, are clearly the best suited for the numerical computation of the vector $\boldsymbol{\pi}$.

b. As proved in [179], the critical arrival rate λ^*, given by the right-hand side of (5.4.8), has several interesting, intuitive monotonicity properties. It is easy to show that, with all the other parameters fixed, there exists a *minimum* value M^* of M for which the queue is *stable.* If the inequality (5.4.8) holds for M^*, then it also holds for larger M. We recall that $M \geq J + C$.

The expression for π_0 in (5.4.7) is now considered as a function of M. It is easily verified that

$$\pi_0(J + C) = \left[\sum_{\nu=0}^{J} \frac{J^\nu}{\nu!} \gamma^\nu + \frac{J^J C!}{J!} \sum_{k=1}^{C} \frac{\gamma^{J+k}}{C^k (C - k)!} \right]^{-1},$$

and that

$$[\pi_0(M + 1)]^{-1} = [\pi_0(M)]^{-1} + a_M,$$

where

$$a_{J+C} = \frac{J^J}{J!} \gamma^J \left[1 + C! \sum_{k=0}^{C} \frac{(\gamma - 1)\gamma^k}{C^k (C - k)!} \right],$$

$$a_{M+1} = \gamma a_M, \quad \text{for} \quad M \geq J + C.$$

In order to determine the smallest value $M^*(\lambda)$ for which the queue is stable, we may proceed as follows. We first verify whether the queue can *ever* be stable by considering the case where M is infinite. If M is infinite, the formal expression for $\pi_0(\infty)$ vanishes for $\gamma \geq 1$, and is given by

$$\pi_0(\infty) = \left[\sum_{\nu=0}^{J-1} \frac{J^\nu}{\nu!} \gamma^\nu + \frac{J^J}{J!} \gamma^J (1 - \gamma)^{-1} \right]^{-1}, \quad \text{for} \quad 0 < \gamma < 1.$$

If $\lambda \geq \lambda^*(\infty)$, the queue is never stable. If $\lambda < \lambda^*(\infty)$, we evaluate $\lambda^*(M)$ recursively for $M \geq J + C$ until the first index $M = M^*(\lambda)$, for which $\lambda < \lambda^*(M)$ is obtained. The recurrence formulas for $\pi_0(M)$ make the necessary calculations particularly easy.

c. The dependence of the critical arrival rate λ^* on the parameters C and J may also be investigated. In order to make nontrivial comparisons, the maximum service rates $C\beta$ and $J\alpha$ for the CPU and I/O servers are both kept constant.

By rather lengthy manipulations, G. Latouche [179] has then shown that $\lambda^*(J, C)$ is nonincreasing in each of the variables J and C separately. This result suggests that for the purpose of maximizing the critical input rate, one ought to replace the J I/O servers and the C CPU servers by single devices of service rates $J\alpha$ and $C\beta$ respectively. This is clearly not always feasible, and, moreover, numerical evidence indicates that the resulting increase in the critical arrival rate is usually small.

This again has an intuitive reason. The advantage of a single, fast server over a set of parallel servers of the same combined rate is felt only when in the latter system, some servers are idle while others are not. Such a system occasionally processes *work* at a slower rate than a single server who maintains the same rate whenever the queue is not empty.

The stationary probability vector $\mathbf{x}$ of $\tilde{Q}$ is most conveniently computed by considering the partitioning $[\mathbf{y}_0, \mathbf{y}_1, \ldots, \mathbf{y}_{M-1}, \mathbf{x}_0, \mathbf{x}_1, \mathbf{x}_2, \ldots]$, where $\mathbf{y}_i$, $0 \le i \le M-1$, is a vector of dimension $i+1$, with indices corresponding to the $(i+1)$-st row in the array (5.4.1). The vectors $\mathbf{x}_i$, $i \ge 0$, are of dimension $M+1$ and correspond to the second set of states $\{(i, j), i \ge 0, 0 \le j \le M\}$ described above.

The *positive* matrix R is the minimal nonnegative solution to the matrix equation $R^2A_2 + RA_1 + A_0 = 0$, and the stationary probability vector satisfies $\mathbf{x}_k = \mathbf{x}_0 R^k$, for $k \ge 0$. Rather than repeating the classical results as they apply to the present model, we shall provide additional details on the computation of the vectors $\mathbf{y}_0, \mathbf{y}_1, \ldots, \mathbf{y}_{M-1}$ and $\mathbf{x}_0$, which are determined up to a common multiplicative constant by the system of homogeneous equations

$$
\begin{aligned}
\mathbf{y}_0 A_{01} + \mathbf{y}_1 A_{10} &= 0, \\
\mathbf{y}_i A_{i2} + \mathbf{y}_{i+1} A_{i+1,1} + \mathbf{y}_{i+2} A_{i+2,0} &= \mathbf{0}, \quad \text{for} \quad 0 \le i \le M-3, \\
\mathbf{y}_{M-2} A_{M-2,2} + \mathbf{y}_{M-1} A_{M-1,1} + \mathbf{x}_0 A_{M,0} &= \mathbf{0}, \qquad (5.4.11) \\
\mathbf{y}_{M-1} A_{M-1,2} + \mathbf{x}_0 (A_1 + RA_2) &= \mathbf{0},
\end{aligned}
$$

Since M is usually large, it is indicated to take the particular structure of this system of $\tfrac{1}{2}(M+1)(M+2)$ linear equations into account. This may be conveniently done as follows.

Consider $W_{M-1} = A_1 + RA_2$, which is a square matrix of order $M+1$, and let $\mathbf{v}_{M-1}$ be its first column and V_{M-1} be the matrix obtained from W_{M-1} by deleting that first column. The last equation in (5.4.11) then leads to

$$
\mathbf{x}_0 \mathbf{v}_{M-1} = 0, \qquad \mathbf{y}_{M-1} = -\lambda^{-1} \mathbf{x}_0 V_{M-1}. \qquad (5.4.12)
$$

Upon substitution into the penultimate equation in (5.4.11), we obtain

$$
\mathbf{y}_{M-2} A_{M-2,2} + \mathbf{x}_0 (A_{M,0} - \lambda^{-1} V_{M-1} A_{M-1,1}) = \mathbf{0}.
$$

We partition the coefficient matrix of $\mathbf{x}_0$ again in the same manner and, continuing, we obtain recursively the matrices W_m, defined by

$$W_{M-1} = A_1 + RA_2,$$
$$W_{M-2} = -\lambda^{-1} V_{M-1} A_{M-1,1} + A_{M,0}, \tag{5.4.13}$$
$$W_m = -\lambda^{-1}(V_{m+1}A_{m+1,1} + V_{m+2}A_{m+2,0}), \quad \text{for} \quad 0 \le m \le M-3.$$

The matrices W_m, $0 \le m \le M-3$, are of dimensions $(M+1) \times (m+2)$. The matrices V_m of dimensions $(M+1) \times (m+1)$ are obtained by deleting the first column $\boldsymbol{v}_m$ of W_m. The equations (5.4.11) then yield

$$\boldsymbol{y}_m = -\lambda^{-1}\boldsymbol{x}_0 V_m, \quad \text{for} \quad 0 \le m \le M-1,$$
$$\boldsymbol{x}_0 \boldsymbol{v}_m = 0, \quad \text{for} \quad 0 \le m \le M. \tag{5.4.14}$$

The $M + 1$ equations $\boldsymbol{x}_0 \boldsymbol{v}_m = 0$, $0 \le m \le M$, determine $\boldsymbol{x}_0$ up to a multiplicative constant, which is determined from the normalizing condition

$$\boldsymbol{x}_0 \left[-\lambda^{-1} \sum_{m=0}^{M-1} V_m \boldsymbol{e} + (I-R)^{-1}\boldsymbol{e} \right] = 1. \tag{5.4.15}$$

Once the vector $\boldsymbol{x}_0$ has been computed, the vectors $\boldsymbol{y}_m$, $0 \le m \le M-1$, are routinely obtained. We note that it is not necessary to store all the matrices W_m. Their computation is sufficiently easy that it is worth evaluating them twice. The first time, we evaluate the W_m recursively *using a single storage array* for the purpose of retaining the vector $\boldsymbol{v}_m$ and to accumulate the vector $\sum_{m=0}^{M-1} V_m \boldsymbol{e}$. This vector and the $\boldsymbol{v}_m$ are the only items needed in the computation of $\boldsymbol{x}_0$. The second time, we compute the W_m, starting from the index M and again using a single storage array of size $(M+1)^2$, for the purpose of evaluating the vectors $\boldsymbol{y}_m$. It is also not necessary to evaluate the inverse of $I - R$; it suffices to solve the system of linear equations $\boldsymbol{\zeta} - \boldsymbol{\zeta} R = \boldsymbol{e}$.

Although the construction of an efficient code requires some extra care in programming, it becomes imperative to do so when several computer runs involving high values of M are to be performed. For this model, the storage requirements, more so than the computation time, call for a well-planned algorithm.

By applying arguments that are by now standard, it is possible to study the waiting time distribution for the *incoming* queue, as well as moments of the queue length in the various queues of the system. The marginal distributions of the number of jobs in the I/O and CPU queues are easily derived from the stationary probability vector of $\tilde{Q}$. Simplifying formulas for this purpose may be found in [179]. The computation of the distribution of the *total time* spent in the system is another matter. The model under discussion is a simple example of a queueing network in which a customer may return to a group of servers many times. It is well known that the sojourn time in such a network has a very involved stochastic description, which makes the algorithmic evaluation of this distribution a difficult open problem. This situation is typical of those encountered in studying the time-in-system for cyclic queueing networks.

Chapter 6
Queues in a Random Environment

6.1. THE RANDOM ENVIRONMENT

In this chapter we study a number of queueing models in which certain parameters vary randomly over time. This random variation may come about through external causes, as in the case of queues exhibiting occasional rush-hour behavior, or it may be brought about by factors inherent in the queueing system itself. Among the latter are breakdowns and repairs of servers, certain priority rules, overflows, set-up times for customers calling on an idle server, and customers of two or more types competing for the same group of servers.

The qualitative behavior of such queues is an extremely interesting subject for study. They exhibit during certain time intervals the behavior typical of unstable queues. The path functions of the queue length and waiting time processes therefore show substantial random fluctuations, less commonly seen in the so-called classical queueing models.

Average measures of the queue, such as the mean queue length or waiting time, may become quite meaningless as descriptors of the behavior of the system. The study of the behavioral, design, and control aspects of the queue requires detailed numerical information on the exact distributions of the features of the queue. By the use of a number of fairly complex models, we shall discuss qualitative aspects of queue behavior to a greater extent than was done in the earlier chapters. This will require the detailed interpretation of several numerical examples. Interactive computation will be further shown to be a useful tool to gain insight into the design and control aspects of several service systems whose study by unguided simulations would be very difficult.

This first section is devoted to the definition of algorithmically tractable models for a "random" environment. We need point processes, which define a sequence of consecutive time intervals and assign to each interval a label (preferably) chosen from a finite set of labels. In queues in a random environment, the parameters of the queue, such as the arrival and service rates or the number of servers, will remain constant during such intervals but may depend on the label of each interval.

When the nature of the sequence of labeled intervals is not dictated by the queue discipline itself, we may choose the environment process by the criteria of realism and mathematical tractability. As these criteria are commonly in conflict, the choice of practically useful environment processes is narrow.

The finite-state Markov renewal process—see, for example, R. Pyke [255, 256] or E. Çinlar [56]—is an appealing choice as a descriptor of an environment process. It is constructively defined by the bivariate sequence $\{(J_n, X_n), n \geq 0\}$, where $J_n \in \{1, \ldots, m\}$ and where the *sojourn times* X_n are nonnegative random variables. The independence assumptions arising in the definition of the Markov renewal process are classical and need not be repeated here. The probability structure of the Markov renewal process is fully determined by its initial conditions and by the transition probability matrix $\hat{Q}(\cdot)$, where

$$\hat{Q}_{jh}(x) = P\{J_{n+1} = h, X_{n+1} \leq x \mid J_n = j\}, \quad \text{for} \quad 1 \leq j, h \leq m, x \geq 0.$$

The matrix $\hat{Q}(\infty)$ is stochastic and, in most cases of interest, also irreducible. It is the transition probability matrix of the embedded Markov chain of labels $\{J_n, n \geq 0\}$.

In order to obtain detailed computational results, even for the simple model of section 6.2, the general Markov renewal environment leads to arduous algorithmic effort. In most cases, this effort may be considerably reduced by assuming instead the more restrictive *Markovian environment.* This description is more limited in mathematical generality, but still offers, as we shall see, considerable flexibility in qualitative modeling. It should be emphatically stressed that some results on asymptotic mean queue lengths or waiting times can be obtained under far weaker assumptions on the environment process. Taken by themselves, such results produce qualitative information that is, however, limited and potentially misleading.

A *Markovian environment* is described by an m-state Markov process with generator Q. In most cases, Q is irreducible, although a useful exception occurs, for example, in M. Yadin [310]. In some cases, overly special choices of Markovian environments have been used. It is likely to be of doubtful merit to represent, for example, the "environment" generated by the four seasons of the year by a string of independent exponential random variables with four periodically repeated parameters.

By using the PH-distributions discussed in chapter 2 as building blocks, it is possible to construct a variety of useful Markovian environments. Although these rapidly lead to matrices Q of high order m, they usually have many algorithmically useful structural properties.

An easy example is provided by the *alternating renewal process* with underlying distributions $F_1(\cdot)$ and $F_2(\cdot)$ of phase type. Let $F_i(\cdot)$, $i = 1, 2$, each have an irreducible representation $[\boldsymbol{\alpha}(i), T_i]$ of order m_i. The matrix Q is then defined by

$$Q = \begin{vmatrix} T_1 & T_1^o A_2^o \\ T_2^o A_1^o & T_2 \end{vmatrix}, \tag{6.1.1}$$

where the matrices $T_1^o A_2^o = \{T_{1i}^o \alpha_j(2),\ 1 \leq i \leq m_1,\ 1 \leq j \leq m_2\}$ and $T_2^o A_1^o = \{T_{2i}^o \alpha_j(1),\ 1 \leq i \leq m_2,\ 1 \leq j \leq m_1\}$ are of dimensions $m_1 \times m_2$ and $m_2 \times m_1$ respectively. Note that we assume $\alpha_{m_1+1}(1) = \alpha_{m_2+1}(2) = 0$. The matrix Q is readily seen to be an *irreducible* generator.

For $m_1 = m_2 = 1$, we obtain the simple alternating Poisson process with

$$Q = \begin{vmatrix} -\sigma_1 & \sigma_1 \\ \sigma_2 & -\sigma_2 \end{vmatrix}, \tag{6.1.2}$$

which has been used in a number of applications, for example, in M. Eisen and M. Tainiter [85], H. Heffes [134], and P. Naor and U. Yechiali [214]. As we shall show by examples, a queue in a random environment may be sensitive to the distributional assumptions of the environment process. A more versatile Markovian environment than the alternating Poisson process is usually required.

Considering the matrix Q in (6.1.1), it should be noted that only the transitions between the sets of states $\{1, \ldots, m_1\}$ and $\{m_1 + 1, \ldots, m_1 + m_2\}$ have a physical significance. All other transitions are part of the Markovian chance mechanism, used to generate the two PH-distributions. Neither the representations of $F_1(\cdot)$ and $F_2(\cdot)$ nor their orders are uniquely determined. This is important to the valid interpretation of numerical results, as further discussed in section 6.2.

A Markov renewal process, which passes cyclically through r states $\{1', 2', \ldots, r'\}$ and has sojourn times distributed according to the PH-distributions $F_i(\cdot)$, $1 \leq i \leq r$, may in a similar manner be studied in terms of the Markov process with generator Q, given by

$$Q = \left|\begin{array}{cccccc} T_1 & T_1^o A_2^o & 0 & \cdots & 0 & 0 \\ 0 & T_2 & T_2^o A_3^o & \cdots & 0 & 0 \\ 0 & 0 & T_3 & \cdots & 0 & 9 \\ \cdot & \cdot & \cdot & & \cdot & \cdot \\ \cdot & \cdot & \cdot & & \cdot & \cdot \\ \cdot & \cdot & \cdot & & \cdot & \cdot \\ 0 & 0 & 0 & \cdots & T_{r-1} & T_{r-1}^o A_r^o \\ T_r^o A_1^o & 0 & 0 & \cdots & 0 & T_r \end{array}\right|. \qquad (6.1.3)$$

While the matrix Q is of order $m = \Sigma_{i=1}^r m_i$, its highly special structure is easily exploited in numerical computations.

In using this matrix Q to model queues with a diurnal cycle or inventories of seasonal products, some caution is needed. The total time required for the process Q to cycle through the r stages is clearly a random variable. The beginnings of cycles are by no means periodic. Such a model should therefore be considered to be an easily computable approximation to situations in which there is a strictly periodic cycle. By varying parameters in one or more of the representations $[\alpha(i), T_i]$, $1 \leq i \leq r$, one may numerically assess the sensitivity to the variance of the various sojourn time distributions. An alternative is to perform the more belabored analysis necessary for *periodic queues,* the study of which has only recently been undertaken. See, for example, J. M. Harrison and A. J. Lemoine [129] and M. H. Rothkopf and S. S. Oren [271].

It is clear that the distributions of phase type may be used to construct other forms of Markovian environments. The theoretical analysis of, for example, the model to be discussed in section 6.2 does not depend on the specific form of the matrix Q. The practical feasibility of an algorithmic analysis for large m depends primarily on the use of its special structure to alleviate the burden of computation.

6.2. THE $M/M/1$ QUEUE AND RELATED MODELS IN A MARKOVIAN ENVIRONMENT

We now assume that the environment is described by an m-state, irreducible Markov process Q. During sojourns in the state j, $1 \leq j \leq m$, customers arrive to a single server queue according to a Poisson process of rate λ_j. During sojourn intervals in the state j, the service rate μ_j prevails. When, in the process Q, a transition from j to h occurs, the arrival and service rates instantaneously change to the new values λ_h *and* μ_h. In order to avoid triviality, at least one λ_j and at least one μ_k are assumed to be positive.

By a standard construction, the queue may be studied as a QBD process with generator $\tilde{Q}$, given by

$$\tilde{Q} = \begin{vmatrix} Q - \Delta(\lambda) & \Delta(\lambda) & 0 & 0 & \cdots \\ \Delta(\mu) & Q - \Delta(\lambda + \mu) & \Delta(\lambda) & 0 & \cdots \\ 0 & \Delta(\mu) & Q - \Delta(\lambda + \mu) & \Delta(\lambda) & \cdots \\ 0 & 0 & \Delta(\mu) & Q - \Delta(\lambda + \mu) & \cdots \\ \vdots & \vdots & \vdots & \vdots & \end{vmatrix}, \tag{6.2.1}$$

where $\Delta(\boldsymbol{a})$ denotes, for $\boldsymbol{a} = (a_1, a_2, \ldots, a_m)$, the diagonal matrix $\operatorname{diag}(a_1, \ldots, a_m)$. The pair (i, j), $i \geq 0$, $1 \leq j \leq m$, corresponds to a queue length i and an environmental state j.

Let $\boldsymbol{\pi}$ be the stationary probability vector of Q. Then the queue is *stable* if and only if

$$\boldsymbol{\pi\lambda} < \boldsymbol{\pi\mu}. \tag{6.2.2}$$

The matrix R is the minimal solution of the equation

$$R^2\Delta(\mu) + R[Q - \Delta(\lambda + \mu)] + \Delta(\lambda) = 0, \tag{6.2.3}$$

and by (1.7.12), provided (6.2.2) holds, we have

$$R\boldsymbol{\mu} = \boldsymbol{\lambda}. \tag{6.2.4}$$

Theorem 6.2.1. The stationary probability vector $\boldsymbol{x} = [\boldsymbol{x}_0, \boldsymbol{x}_1, \ldots]$ of the stable queue is given by

$$\boldsymbol{x}_k = \boldsymbol{\pi}(I - R)R^k, \quad \text{for} \quad k \geq 0. \tag{6.2.5}$$

Proof. It only remains to verify that the vector $\boldsymbol{x}_0 = \boldsymbol{\pi}(I - R)$ is strictly positive and satisfies

$$\boldsymbol{\pi}(I - R)[Q - \Delta(\lambda) + R\Delta(\mu)] = \mathbf{0}. \tag{6.2.6}$$

The latter equality readily follows from (6.2.3) and $\boldsymbol{\pi}Q = \mathbf{0}$.

The equation (6.2.3) may be rewritten as

$$R = [R^2\Delta(\mu) + \Delta(\lambda)][\Delta(\lambda + \mu) - Q]^{-1}. \tag{6.2.7}$$

The nonnegative matrix $\Delta(\lambda + \mu)[\Delta(\lambda + \mu) - Q]^{-1}$ has the positive left invariant vector $\boldsymbol{\pi}$. It is now easily verified that the sequence of matrices $\{R(n)\}$, obtained by successive substitutions starting with $R(0) = 0$, is nondecreasing and satisfies $\boldsymbol{\pi}R(n) \leq \boldsymbol{\pi}$, for $n \geq 0$. Since $R(n) \to R$, it follows that $\boldsymbol{\pi}R \leq \boldsymbol{\pi}$.

The inverse in (6.2.7) is strictly positive, since Q is irreducible and $\lambda + \mu \neq \mathbf{0}$. It is easy to see that (6.2.6) is equivalent to

$$\pi(I - R) = \pi(I - R^2)\Delta(\mu)[\Delta(\lambda + \mu) - Q]^{-1}.$$

The vector $\pi(I - R^2)\Delta(\mu)$ is nonnegative. It does not vanish, otherwise R would have spectral radius one. It therefore follows that $\pi(I - R)$ is positive and, by direct verification, that x is the stationary vector of $\tilde{Q}$.

Remarks

a. The matrix R has positive rows corresponding to those and only those indices j for which $\lambda_j > 0$.

b. The vector $\sum_{k=0}^{\infty} x_k$ *equals* π. This is intuitive since π_j is the steady-state probability that the queue is in the environmental state j, $1 \leq j \leq m$, at time t.

c. For the present model, the conditional queue length densities $\{q_k(j), k \geq 0\}$ for $1 \leq j \leq m$, given by

$$q_k(j) = \pi_j^{-1}[\pi(I - R)R^k]_j, \quad \text{for} \quad k \geq 0, \tag{6.2.8}$$

are of particular interest. They show how the queue length, which is typically highly oscillatory, depends on the environmental state j.

d. The traffic intensity $\rho = (\pi\lambda)(\pi\mu)^{-1}$ may clearly be less than one, even when $\lambda_j \geq \mu_j$ for several indices j. Such queues are very interesting in that periods of substantial buildup alternate with periods of recovery. Some numerical examples are discussed in the following.

e. The probability $\pi R e$ that the server is busy is, in general, not equal to ρ. In the particular case where $\mu = \mu e$, it is, since then

$$\pi R e = \mu^{-1}\pi R \mu = \mu^{-1}\pi\lambda = \rho.$$

f. It is not possible, in general, to express R explicitly in terms of λ, μ, and Q. It is of interest to note the particular case where $\lambda = \rho\mu$, with $0 < \rho < 1$. The matrix equation (6.2.3) may then be written as

$$R = R^2 D + \rho D, \tag{6.2.9}$$

where $D = \Delta(\mu)[(1 + \rho)\Delta(\mu) - Q]^{-1}$. The equation (6.2.9) has the same formal solution as the scalar equation $r = r^2 d + \rho d$. The positive solution of the latter equation is given by

$$r = \frac{1}{2d}\left[1 - (1 - 4\rho d^2)^{1/2}\right] = \frac{1}{2}\sum_{\nu=1}^{\infty}\binom{2\nu}{\nu}\rho^{\nu} d^{2\nu-1}.$$

Direct verification now shows that

$$R = \frac{1}{2}\sum_{\nu=1}^{\infty}\binom{2\nu}{\nu}\rho^{\nu} D^{2\nu-1} \tag{6.2.10}$$

is a positive solution of (6.2.9). Moreover, since $\pi D = (1 + \rho)^{-1}\pi$, we have $\pi R = \rho\pi$, so that R has the maximal eigenvalue $\rho < 1$. It also readily follows that $x_k = (1 - \rho)\rho^k\pi$, for $k \geq 0$. All m conditional queue length densities $\{q_k(j)\}$ are *geometric* with parameter $1 - \rho$. This result shows that if the service rate μ_j *can always be chosen to be* $\rho^{-1}\lambda_j$, then the stationary queue length will be indistinguishable from that of a queue with constant rates λ and μ. Such an adaptability of the service rate to the arrival rate is, of course, not always economically and practically feasible.

g. Our treatment of the $M/M/1$ queue in a Markovian environment follows that in M. F. Neuts [229]. Particular cases were discussed in M. M. Eisen [84], M. Eisen and M. Tainiter [85], and P. Naor and U. Yechiali [214]. General analytic results are given in U. Yechiali [313], who also pointed out the special case in (f) above, while P. Purdue [254] devotes particular attention to the busy period of this queueing model. A general discussion of randomly varying rates is given in M. Yadin and R. Syski [312].

The following is an elementary but useful result. We consider the invariant probability vectors z and y of the Markov chains obtained by considering the $M/M/1$ queue in a Markovian environment *immediately before* or *immediately after transitions in the environmental process Q*. Simple relationships between the vectors x, y, and z will be established.

Let ξ_n *be the queue length and* J_n *the phase state immediately after* the n-th transition in the process Q. The sequence $\{(\xi_n, J_n), n \geq 0\}$ is then a discrete parameter Markov chain on the state space $\{(i, h)\colon i \geq 0, 1 \leq h \leq m\}$. Let $p_{ii'}(h, u)$, $u \geq 0$, $1 \leq h \leq m$, $i \geq 0$, $i' \geq 0$, be the conditional probability that in an ordinary $M/M/1$ queue with parameters λ_h, μ_h, the queue length at time u is i', given that there are i customers in the system at time 0. Unwieldy explicit formulas for the quantities $p_{ii'}(h, u)$ are known and given, for example, in J. W. Cohen [61], L. Takács [293], and T. L. Saaty [272]. These formulas will not be needed in our discussion.

The transition probability matrix P of the chain $\{(\xi_n, J_n), n \geq 0\}$ is then given by

$$\begin{aligned} P_{hh'}(i, i') &= \int_0^\infty p_{ii'}(h, u)e^{-\sigma_h u}du \cdot Q_{hh'}, \quad \text{for} \quad i \geq 0, \quad i' \geq 0, \quad h \neq h', \\ &\qquad\qquad\qquad\qquad\qquad\qquad\qquad\qquad 1 \leq h, h' \leq m, \\ &= 0 \quad \text{for} \quad i \geq 0, \quad i' \geq 0, \quad h = h', \end{aligned} \tag{6.2.11}$$

where $\sigma_h = -Q_{hh}$, for $1 \leq h \leq m$.

Partitioning P into $m \times m$ blocks $P(i, i') = \{P_{hh'}(i, i')\}$, we may write

$$P(i, i') = V(i, i')[Q + \Delta(\sigma)], \tag{6.2.12}$$

where $\Delta(\sigma) = \text{diag}(\sigma_1, \ldots, \sigma_m)$. The matrix $V(i, i')$ is defined by

$$V(i, i') = \text{diag}\left\{\int_0^\infty p_{ii'}(h, u)e^{-\sigma_h u}du, \ 1 \le h \le m\right\},$$
$$\text{for} \quad i \ge 0, \qquad i' \ge 0. \tag{6.2.13}$$

Theorem 6.2.2. The vectors $\boldsymbol{x}$, $\boldsymbol{y}$, and $\boldsymbol{z}$ are related by

$$\begin{aligned} \boldsymbol{y}_i &= (\boldsymbol{\pi\sigma})^{-1}\boldsymbol{x}_i[Q + \Delta(\sigma)], \\ \boldsymbol{z}_i &= (\boldsymbol{\pi\sigma})^{-1}\boldsymbol{x}_i\Delta(\sigma), \quad \text{for} \quad i \ge 0. \end{aligned} \tag{6.2.14}$$

Proof. The equations $\boldsymbol{y} = \boldsymbol{y}P$ may be written as

$$\boldsymbol{y}_i = \sum_{i'=0}^{\infty} \boldsymbol{y}_{i'}V(i', i)[Q + \Delta(\sigma)] = \boldsymbol{w}_i[Q + \Delta(\sigma)], \quad \text{for} \quad i \ge 0. \tag{6.2.15}$$

The vectors $\boldsymbol{w}_i$ *are defined by*

$$\boldsymbol{w}_i = \sum_{i'=0}^{\infty} \boldsymbol{y}_{i'}V(i', i), \quad \text{for} \quad i \ge 0. \tag{6.2.16}$$

First setting $\sum_{i=0}^{\infty} \boldsymbol{y}_i = \boldsymbol{u}$, the equations (6.2.15) lead to

$$\boldsymbol{u}\Delta^{-1}(\sigma)[Q + \Delta(\sigma)] = \boldsymbol{u}.$$

Since $\boldsymbol{ue} = 1$, it readily follows that $\boldsymbol{u} = (\boldsymbol{\pi\sigma})^{-1}\boldsymbol{\pi}\Delta(\sigma)$. Setting

$$\int_0^\infty p_{i'i}(h, u)e^{-\sigma_h u}du = p^*_{i'i}(h), \quad \text{for} \quad 1 \le h \le m,$$

the birth-and-death equations for the $M/M/1$ queue immediately yield that

$$\begin{aligned} p^*_{i'0}(h)(\sigma_h + \lambda_h) &= \delta_{i'0} + p^*_{i'1}(h)\mu_h, \\ p^*_{i'i}(h)(\sigma_h + \lambda_h + \mu_h) &= \delta_{i'i} + p^*_{i',i-1}(h)\lambda_h + p^*_{i',i+1}(h)\mu_h, \end{aligned}$$

for $i' \ge 0$, $i \ge 1$, $1 \le h \le m$. This may be rewritten in matrix form as

$$\begin{aligned} V(i', 0)\Delta(\sigma + \lambda) &= \delta_{i'0}I + V(i', 1)\Delta(\mu), \\ V(i', i)\Delta(\sigma + \lambda + \mu) &= \delta_{i'i}I + V(i', i - 1)\Delta(\lambda) \\ &\quad + V(i', i + 1)\Delta(\mu), \end{aligned} \tag{6.2.17}$$

for $i' \geq 0, i \geq 1$. From the equations (6.2.15), (6.2.16), and (6.2.17), we obtain

$$w_0\Delta(\sigma + \lambda) = w_0[Q + \Delta(\sigma)] + w_1\Delta(\mu),$$

$$w_i\Delta(\sigma + \lambda + \mu) = w_i[Q + \Delta(\sigma)] + w_{i-1}\Delta(\lambda) + w_{i+1}\Delta(\mu), \quad \text{for} \quad i \geq 1.$$

Clearly, the vectors w_i, $i \geq 0$, satisfy the same equations

$$w_0[Q - \Delta(\lambda)] + w_1\Delta(\mu) = \mathbf{0},$$

$$w_{i-1}\Delta(\lambda) + w_i[Q - \Delta(\lambda + \mu)] + w_{i+1}\Delta(\mu) = \mathbf{0}, \quad \text{for} \quad i \geq 1,$$

as the vectors x_i, $i \geq 0$, except for the normalizing condition. Since the vectors w_i, $i \geq 0$, are positive and the sum $\Sigma_{i=0}^{\infty} w_i e$ is finite, it follows by the uniqueness (up to a multiplicative constant) of the stationary vector of an irreducible positive recurrent generator that $w_i = kx_i$, for $i \geq 0$, and some positive constant k. It now follows that

$$y_i = k\pi(I - R)R^i[Q + \Delta(\sigma)], \quad \text{for} \quad i \geq 0.$$

Upon normalization, we obtain $k = (\pi\sigma)^{-1}$.

The invariant probability vector z of the Markov chain P', obtained by considering the queue length ξ_n' and the phase state J_n', immediately *prior* to the n-th transition in the process Q, is related to the vector y by

$$z_i\Delta^{-1}(\sigma)[Q + \Delta(\sigma)] = y_i, \quad \text{for} \quad i \geq 0.$$

Since the positive vectors $z_i = (\pi\sigma)^{-1}x_i\Delta(\sigma)$, $i \geq 0$, satisfy the preceding equations, and since $\Sigma_{i=0}^{\infty} z_i e = 1$, it follows that z is the desired probability vector.

Corollary 6.2.1. For $i \geq 0$, $1 \leq j \leq m$,

$$\pi_j^{-1}x_{ij} = (\pi\sigma)(\pi_j\sigma_j)^{-1}z_{ij}. \tag{6.2.18}$$

Remark. The relation (6.2.18) is obvious from (6.2.14). The left-hand side is the conditional probability that there are i customers in the system, given that the steady-state queue is in the phase state j. The right-hand side is the conditional probability that there are i customers present *at the end* of a sojourn in the phase state j.

The equality of these two conditional queue length densities depends crucially on the exponential nature of the sojourn times in the various phase states.

An Application. Let the matrix Q be given by the formula (6.1.1), describing an alternating renewal process of phase type. Let the m_1-vector $\pi(1)$ and the m_2-vector $\pi(2)$ be given by the equations

$$\pi(1)(T_1 + T_1{}^o A_1{}^o) = 0, \qquad \pi(1)e = 1,$$

and

$$\pi(2)(T_2 + T_2{}^o A_2{}^o) = 0, \qquad \pi(2)e = 1.$$

If the means of $F_1(\cdot)$ and $F_2(\cdot)$ are $\mu_1{}'$ and $\mu_2{}'$ respectively, then the vector π of Q is given by

$$\pi = \left[\frac{\mu_1{}'}{\mu_1{}' + \mu_2{}'}\pi(1), \quad \frac{\mu_2{}'}{\mu_1{}' + \mu_2{}'}\pi(2)\right]. \tag{6.2.19}$$

Let us partition the vectors x_i into an m_1-vector $x_i(1)$ and an m_2-vector $x_i(2)$. Of particular interest are the conditional queue length densities $\{q_1(i)\}$ and $\{q_2(i)\}$, given that the Q-process is in one of the states $\{1, \ldots, m_1\}$ or in one of the states $\{m_1 + 1, \ldots, m_1 + m_2\}$ respectively. These densities are given by

$$q_1(i) = \left(\frac{\mu_1{}'}{\mu_1{}' + \mu_2{}'}\right)^{-1} x_i(1)e,$$

$$q_2(i) = \left(\frac{\mu_2{}'}{\mu_1{}' + \mu_2{}'}\right)^{-1} x_i(2)e, \quad \text{for} \quad i \geq 0. \tag{6.2.20}$$

Of equal interest are the conditional queue length densities $\{\tilde{q}_1(i)\}$ and $\{\tilde{q}_2(i)\}$ *at transitions* from the set $\{1, \ldots, m_1\}$ to $\{m_1 + 1, \ldots, m_1 + m_2\}$ and vice versa. These densities are given by

$$\tilde{q}_1(i) = \left[\frac{\mu_1{}'}{\mu_1{}' + \mu_2{}'}\pi(1)T_1{}^o\right]^{-1} x_i(1)T_1{}^o = (\mu_1{}' + \mu_2{}')x_i(1)T_1{}^o,$$

$$\tilde{q}_2(i) = (\mu_1{}' + \mu_2{}')x_i(2)T_2{}^o, \quad \text{for} \quad i \geq 0, \tag{6.2.21}$$

since $\pi(1)T_1{}^o = \mu_1{}'^{-1}$, and $\pi(2)T_2{}^o = \mu_2{}'^{-1}$. The formulas (6.2.21) follow from theorem 6.2.2 by a straightforward conditioning argument. The corresponding conditional queue length densities in (6.2.20) and (6.2.21) are in general not equal to each other.

Remark. There is an interesting use of the conditional densities discussed above in the study of the *time dependent* behavior of queues described by Markovian models. It is often stated that the numerical behavior of the time-dependent state-probabilities of such queues could shed light on behavioral features useful to the practitioner. Indeed, a number of papers have been devoted to the numerical solution of the enormous systems of differential equations satisfied by the state-probabilities of Markovian systems.

We believe that the value of such a computational effort is mainly dependent on the choice of *meaningful* initial conditions. This choice is rarely discussed, but constantly arises when one attempts to formalize the informal but pertinent issues raised by practitioners. The question might be, for example, how long it will take for the queue built up during the absence of a server to dissipate.

We may approach this question by choosing the conditional stationary density of the queue length (and other appropriate state variables) *at the times of returns of the server* as the vector of *initial conditions* to the (infinite) system of differential equations which describes the time-dependent behavior of the queue. This choice of the initial conditions reflects the randomness that the queue *typically* exhibits at such times. The corresponding transient solutions then have useful and elucidating interpretations. Without such a specific choice, much effort can be wasted by arbitrarily varying the initial conditions.

Although the mathematical methods required to solve the time-dependent equations of (moderately large) Markovian systems are classical, we believe it would be of practical interest to have examples of qualitative behavior elucidated by intelligent choices of the initial conditions. As this issue is marginal to the subject of this book, we shall, however, leave such examples to the initiative of the reader.

Numerical Examples

In M. F. Neuts [229], four related numerical examples illustrate the interesting behavior of $M/M/1$ queues in which the arrival rate exceeds the service rate in some of the environmental phase states. Different examples are used here to exhibit other qualitatively interesting features of this model.

Example 1: Random Drift. It is well known that the queue length of a *critical* $M/M/1$ queue with $\lambda = \mu$ exhibits slow long-lived random fluctuations upon which are superimposed small rapid variations about whatever typical queue length happens to prevail during any short interval of time. This behavior is similar to that observed in coin tossing, as discussed, for example, in W. Feller [90, chap. 3].

This same behavior can be seen in *stable* $M/M/1$ queues in a Markovian environment if there are fairly long intervals during which the arrival and service rates are equal to each other. This is illustrated by the following example.

With $m = 5$, we define the matrix Q by

$$Q = \begin{vmatrix} -1 & 1 & 0 & 0 & 0 \\ 0 & -1 & 1 & 0 & 0 \\ 0 & 0 & -1 & 1 & 0 \\ 0 & 0 & 0 & -1 & 1 \\ 1 & 0 & 0 & 0 & -1 \end{vmatrix},$$

and let $\lambda = (4, 10, 6, 6, 6)$, $\mu = (7, 15, 6, 6, 6)$. With this simple choice of the parameters, $\rho = (\pi\lambda)(\pi\mu)^{-1} = 0.8$. This example will illustrate the slow random drift of the queue length during the phases 3, 4, and 5 during which the arrival and service rates are matched. It will also be interesting to compare the numerical results for this case to those of the queue in which Q is replaced by $0.1\,Q$. All other parameters are the same; the vector π and ρ are unchanged. In tables 6.1 and 6.2, the left-hand columns refer to the first Q-matrix; the right-hand columns to the second.

The overall mean queue lengths are given respectively by

$$L_1 = 4.86, \qquad L_1 = 8.81,$$

but these are poor descriptors of the queue behavior. The conditional means and standard deviations of the queue lengths during (or at the end of) each of the five phase intervals are displayed in table 6.1.

The entries in table 6.1 illustrate the steady increase in random drift during the phases 3, 4, and 5. It is interesting to note that for the recovery phases 1 and 2, the standard deviation is considerably larger than the corresponding mean. The "overdispersed" queue lengths during these phases are due to the considerable random variation in the queue lengths at the start of phase 1. We further observe that the entries for the second queue differ roughly by a factor of two from those for the first. This is a mild increase if we consider that in the second queue the average sojourn times in the various phases are increased by a factor of *ten*.

TABLE 6.1

Phase	Mean		Standard Deviation	
1	4.63	5.79	5.49	9.63
2	3.62	2.93	4.79	5.29
3	4.61	8.52	5.08	8.85
4	5.39	12.04	5.41	11.06
5	6.05	14.80	5.75	12.87

The conditional means and standard deviations of the queue length in various phases, corresponding to the matrices Q and $0.1\,Q$.

This example suggests that it is entirely feasible to "control" the random drift of a critical queue by allowing occasional periods of recovery during which the service rate considerably exceeds the arrival rate. This is important in cases where it is desirable to occupy the server fully during certain portions of the overall work cycle. Note also in table 6.2 the columns marked by a dagger. They give the conditional probability that the server is *idle* at an arbitrary time at which the environment is in the given phase state.

The behavior of most highly variable random models is only poorly described by means and second moments. Whenever possible, an appropriately chosen set of quantiles of the distributions of interest is far more descriptive. We illustrate this by displaying the first indices for which the conditional queue length distributions exceed $a/100$, where a is chosen to be 75, 90, 95, and 99 respectively. This information is displayed, again for both queues, in table 6.2.

TABLE 6.2

Phase	Empty†		75%		90%		95%		99%	
1	0.226	0.300	7	7	12	18	16	27	24	44
2	0.266	0.314	5	3	10	7	13	11	22	28
3	0.165	0.093	6	12	11	20	15	26	23	41
4	0.129	0.059	8	17	12	27	16	34	25	50
5	0.110	0.046	9	21	14	32	17	40	26	57

The conditional probability of emptiness and selected percentiles of the conditional queue length distributions in various phases, corresponding to the matrices Q and $0.1\,Q$.
† Indicates the conditional probability that the server is *idle* at an arbitrary time at which the environment is in the given phase state.

The left-hand columns in table 6.2 show how remarkably smooth the first queue is. The differences in the mean queue lengths are largely accounted for by differences in the lower indexed terms of the conditional distributions. Beyond the upper quartile (75 percent), the distributions are essentially the same. For the second queue, both the periods with equal rates and the recovery periods are sufficiently long to cause substantial fluctuations in the queue length. The effect of random drift shows up well in the increase in the percentiles for the phases 3, 4, and 5.

Example 2: Rush-Hour Behavior. Queues in which the arrival rate greatly exceeds the service rate during one or more phases exhibit behavior that is radically different from their counterparts with time-homogeneous parameters. The commonly used descriptors of the queue, such as the traffic intensity and the overall mean and standard deviation of the queue length, become totally uninformative. One surge of arriving customers may not yet have dissipated during the recovery period, when the next surge begins. The queue may therefore exhibit long periods of severe deterioration. At other times, due to the inherent randomness of the arrival and service process, only mild build-up occurs, which is adequately cleared during the recovery period. The numerical results, to be discussed next, raise interesting questions for both the methodologies of computer simulation and statistical inference for queues. For the first, it is clear that long simulation runs will be needed to exhibit the various types of behavior which account for the numerical results displayed in tables 6.3 and 6.4. Routine estimation procedures and the common concern with averages of simulation data will obviously lead to erroneous conclusions.

From a statistical viewpoint, there is as yet no adequate inferential methodology to fit models to highly fluctuating queues and other stochastic models. Although the parameters in our examples are chosen to illustrate a point and are not taken from a specific application, the qualitative results that are obtained are strongly reminiscent of behavior seen in many real queues. Imaginative new approaches leading to the understanding of these types of stochastic behavior are highly desirable.

For the present examples, the Q-matrix is similar to that of the first queue in the preceding example, except that $m = 8$. The mean sojourn time in each phase is one. The service rates $\mu_1, \ldots, \mu_8$ are all equal to 6.25, so that with an adequate supply of customers, an average of fifty customers are served during each cycle through the eight states of the Q-process.

In the four queues A, B, C, and D to be considered, only the vectors λ differ. They are chosen in such a manner that in all cases the traffic intensity ρ is 0.8. The tables 6.3 and 6.4 give the same computed quantities as discussed in the preceding example. The vectors λ are given by

$$\begin{aligned} &\text{Queue } A\text{:} \quad 30, 2, 2, 2, 1, 1, 1, 1. \\ &\text{Queue } B\text{:} \quad 21, 5, 5, 5, 1, 1, 1, 1. \\ &\text{Queue } C\text{:} \quad 15, 7, 7, 7, 1, 1, 1, 1. \\ &\text{Queue } D\text{:} \quad 9, 9, 9, 9, 1, 1, 1, 1. \end{aligned}$$

The overall mean queue lengths are given by

$$A: \; L_1 = 41.43, \qquad C: \; L_1 = 13.94,$$

$$B: \; L_1 = 21.63, \qquad D: \; L_1 = 10.91.$$

These figures, which provide little information, should nevertheless be contrasted with the mean queue length $L_1 = 4$ of a time-homogeneous $M/M/1$ queue with $\rho = 0.8$.

A long list of interpretive comments on the numerical results in tables 6.3 and 6.4 may be given. In the interest of brevity, only the most salient of these will be presented.

The conditional mean queue lengths for Queue A decrease slowly from the high value 53.81 at the end of phase 1. There is a *much slower* decrease in the corresponding standard deviations from 54.15 to 48.36. The queue length in phase 8 is highly overdispersed. A clue to the behavior of the queue which accounts for this is given by the high conditional probability 0.384 of emptiness in phase 8. The high variability of Queue A is due to the high frequency with which successive surges of input will overlay each other, causing occasional build-ups of the queue to well over two hundred customers. The high probability of emptiness in the higher phases, on the other hand, shows that small to medium build-ups during phase 1 mostly dissipate before phase 8 is over. Queue A is highly variable indeed.

In Queue B the arrival rate in phase 1 is decreased and that in phases 2, 3, and 4 is increased, but in such a manner that the service rate in those three phases still exceeds the arrival rate. This queue is clearly much tamer than A, although the slow decrease in the standard deviations indicates that overlaying build-ups are still frequent, but to a lesser extent than in Queue A.

Queue C has a rapid build-up during phase 1, followed by a slower growth during the next three phases. This queue exhibits far less variability than A or B, but the decrease in the conditional standard deviations is still slow. Of the four cases, Queue D, in which the build-up extends evenly over the first four phases, is also the least variable. By the end of phase 8, the server is usually idle. The four arrival patterns are increasingly smoother. The ideal case is, of course, the time-homogeneous case where $\lambda_j = 5$, for $1 \leq j \leq 8$. This is the classical $M/M/1$ queue for

TABLE 6.3

Phase	Mean				Standard Deviation			
	A	B	C	D	A	B	C	D
1	53.81	27.83	15.94	8.64	54.15	28.19	16.39	10.32
2	49.98	26.88	16.95	11.68	53.97	28.12	16.55	11.13
3	46.43	25.98	17.91	14.58	53.55	28.00	16.74	11.96
4	43.14	25.13	18.84	17.42	52.94	27.85	16.96	12.78
5	39.36	21.21	14.72	13.11	52.03	27.21	16.66	12.78
6	35.93	17.95	11.43	9.69	50.93	26.17	15.76	12.03
7	32.81	15.20	8.86	7.08	49.69	24.92	14.57	10.91
8	29.96	12.89	6.85	5.13	48.36	23.56	13.26	9.63

The conditional mean queue lengths and standard deviations in the various phases for the queues A, B, C, and D with differing arrival patterns.

TABLE 6.4

Phase	Idle†				75%				90%				95%				99%			
	A	B	C	D	A	B	C	D	A	B	C	D	A	B	C	D	A	B	C	D
1	0.016	0.030	0.054	0.121	75	39	22	12	124	64	37	22	162	84	48	30	249	128	74	47
2	0.068	0.047	0.041	0.047	71	38	23	16	120	63	38	26	158	83	50	34	245	127	75	51
3	0.113	0.057	0.034	0.024	67	37	25	20	116	62	40	30	154	82	51	38	241	126	77	55
4	0.154	0.064	0.029	0.014	63	36	26	23	112	61	41	34	150	81	52	42	237	125	78	59
5	0.234	0.212	0.181	0.151	58	31	21	19	107	57	36	30	145	76	48	38	232	120	73	55
6	0.291	0.318	0.315	0.294	53	26	17	15	102	52	32	26	140	71	43	34	227	116	69	51
7	0.340	0.401	0.428	0.422	48	21	13	11	97	47	28	22	135	67	39	30	222	111	65	47
8	0.384	0.470	0.519	0.528	43	17	8	6	92	42	23	18	130	62	35	25	217	106	60	43

The conditional probability of emptiness and selected percentiles of the queue length distributions in various phases for the queues A, B, C, and D with differing arrival patterns.

† Indicates the conditional probability that the server is *idle* at an arbitrary time at which the environment is in the given phase state.

which the mean queue length is 4, with a corresponding standard deviation of 4.47.

The *mean* number of arrivals per cycle through the eight phases is the same and equal to 50 for all four queues, as it is for the corresponding $M/M/1$ queue. The standard deviations σ of the number of arrivals per cycle are, however, very different for the various cases. With

$$\sigma = \left[\sum_{j=1}^{8} (\lambda_j + \lambda_j^2)\right]^{1/2},$$

the numerical values of σ for the queues A, B, C, D, and for the $M/M/1$ version are given by

$$A: \quad \sigma = 30.91, \qquad C: \quad \sigma = 20.40,$$

$$B: \quad \sigma = 23.66, \qquad D: \quad \sigma = 19.18,$$

$$M/M/1: \quad \sigma = 15.49.$$

It is to be expected that increased random variability of the arrival (or service) process will be reflected by increased variability in the behavior of the queue. Through algorithmic investigations this variability may be quantified precisely. We also gain insight into the types of behavior which are most likely to manifest themselves.

The model is also quite sensitive to the variability of the durations of the phases 1, 2, 3, 4. This may be seen by replacing the exponential sojourn time distributions in these phases, for example, by Erlang distributions of the same means but smaller standard deviations. An example that illustrates this point is given in [229].

In concluding the discussion of these examples, we note that the total computation time required for their generation was truly small. Even by using an APL program not designed to exploit the particular features of the matrix Q, each example required only a few seconds of computation. This suggests that, with an efficient algorithm, the interactive study of qualitative features of a model such as this is to be highly recommended.

Waiting Time Distributions. The computation of the waiting time distributions for the $M/M/1$ queue in a Markovian environment will be only briefly discussed. It proceeds entirely along the general lines indicated in section 3.9. For the virtual waiting time distribution $W(\cdot)$ with Laplace-Stieltjes transform

$$W^*(s) = \pi(I - R) \sum_{k=0}^{\infty} R^k \{[sI + \Delta(\mu) - Q]^{-1} \Delta(\mu)\}^k e,$$

and for the waiting time distribution at times of transition in the Q-process, the elegant method of section 3.9 may be implemented. These probability distributions are averages over the various phases, however, and as such are not always very informative.

The *conditional* waiting time distributions, given the phase of the Q-process, will reflect the random fluctuations, as did the conditional queue length densities. These distributions may, however, have slowly decaying tails and require much computation, particularly for phases during which the queue grows rapidly. These distributions may be evaluated by the general purpose method of section 3.9.

The Effective Service and Interarrival Times. The actual service time of a customer may extend over several phases of the Q-process and does not have an exponential distribution in general. Consider the conditional probability distribution of a service time starting at time $t = 0$, with the Q-process in the state j. This is the distribution of the *effective service time*, starting in the state j. Analogous definitions and results hold for interarrival times and will be stated without discussion.

Let $\psi_{jh}(\nu, t)$, $\nu \geq 0, t \geq 0, 1 \leq j, h \leq m$, be the probability that a service, starting at time 0 in the phase j, lasts for a time *at least* t, that during $(0, t]$ there are ν new arrivals and that the Q-process is in the phase h at time t. A direct birth-and-death argument then yields that

$$\psi'_{jh}(\nu, t) = \delta_{jh}(-\lambda_j - \mu_j + Q_{jj})\psi_{jj}(\nu, t) + \sum_{r \neq j} Q_{jr}\psi_{rh}(\nu, t)$$
$$+ (1 - \delta_{\nu 0})\lambda_j \psi_{jh}(\nu - 1, t),$$

for $t \geq 0, \nu \geq 0, 1 \leq j, h \leq m$, with the initial conditions $\psi_{jh}(\nu, 0) = \delta_{\nu 0}\delta_{jh}$, for $\nu \geq 0, 1 \leq j, h \leq m$. In matrix notation, we obtain the recursive system of differential equations

$$\psi'(0, t) = [Q - \Delta(\lambda + \mu)]\psi(0, t),$$
$$\psi'(\nu, t) = [Q - \Delta(\lambda + \mu)]\psi(\nu, t) + \Delta(\lambda)\psi(\nu - 1, t), \qquad (6.2.22)$$

for $\nu \geq 1, t > 0$, with $\psi(\nu, 0) = \delta_{\nu 0}I$, for $\nu \geq 0$. These immediately lead to

$$\psi^*(z, t) = \sum_{\nu=0}^{\infty} \psi(\nu, t)z^\nu = \exp\{[Q - (1 - z)\Delta(\lambda) - \Delta(\mu)]t\}, \qquad (6.2.23)$$

for $0 \leq z \leq 1, t \geq 0$.

Most of the statements in the following theorem are now immediate consequences of formula (6.2.23).

Theorem 6.2.3. The elementary conditional probability that a service, starting at time 0 with the Q-process in the state j, ends during

$(t, t + dt]$ with the Q-process in the state h is the (j, h)-element of the matrix

$$\exp\{[Q - \Delta(\mu)]t\}\Delta(\mu)dt.$$

For any initial probability vector γ on $\{1, \ldots, m\}$, the effective service time has a PH-distribution with the representation $[\gamma, Q - \Delta(\mu)]$, and its mean is given by $E_s = \gamma[\Delta(\mu) - Q]^{-1}e$.

The probability generating function $p_j(z)$ of the number of arrivals during the effective service time starting in the state j is the j-th component of the m-vector

$$p(z) = \int_0^\infty \psi^*(z, t)\mu dt = [(1 - z)\Delta(\lambda) + \Delta(\mu) - Q]^{-1}\mu. \qquad (6.2.24)$$

The stochastic, but not necessarily irreducible, matrix $[\Delta(\mu) - Q]^{-1} \cdot \Delta(\mu)$ has an invariant probability vector given by $\pi^* = (\pi\mu)^{-1}\pi\Delta(\mu)$. If the initial probability vector γ is chosen to be π^*, the corresponding service time is called the *average effective service time.* The mean number $\pi^*p'(1)$ of arrivals during it is equal to the traffic intensity ρ. The mean average effective service time E_s^* is equal to $(\pi\mu)^{-1}$.

Proof. Most of these statements follow by simple calculations. We show the details only for one of them. Clearly by (6.2.24), $p'(1) = [\Delta(\mu) - Q]^{-1}\lambda$, so that

$$\pi^*p'(1) = (\pi\mu)^{-1}\pi\Delta(\mu)[\Delta(\mu) - Q]^{-1}\lambda = (\pi\mu)^{-1}(\pi\lambda) = \rho,$$

since $\pi\Delta(\mu)[\Delta(\mu) - Q]^{-1} = \pi$.

The corresponding results for the interarrival times are given in the following.

Theorem 6.2.4. Given that an arrival at time 0 finds the Q-process in the state j, the elementary conditional probability that the next arrival occurs during $(t, t + dt]$ with the Q-process in the state h is given by the (j, h)-element of the matrix $\exp\{[Q - \Delta(\lambda)]t\}\Delta(\lambda)dt$.

For any initial probability vector γ on $\{1, \ldots, m\}$, the effective interarrival time has a PH-distribution with the representation $[\gamma, Q - \Delta(\lambda)]$, and its mean is given by $E_a = \gamma[\Delta(\lambda) - Q]^{-1}e$.

Given that there were an unlimited supply of customers at time 0, the probability generating functions of the number of departures during an effective interarrival time are given by the components of the vector

$$\tilde{p}(z) = [\Delta(\lambda) + (1 - z)\Delta(\mu) - Q]^{-1}\lambda.$$

For $\gamma = (\pi\lambda)^{-1}\pi\Delta(\lambda)$, the *average effective interarrival time* is obtained. The mean number of departures during it is given by ρ^{-1}, and its mean duration is $(\pi\lambda)^{-1}$.

Remark. Of particular interest are the interpretations of the traffic intensity ρ, found in theorems 6.2.3 and 6.2.4. With appropriate changes, ρ may for many queues be understood to be the mean number of arrivals during a suitably defined average service time. The quantity $1/\rho$ is often the mean number of departures during an appropriately defined average interarrival time, provided we have an unlimited supply of customers available so that no idle time will accrue during that interarrival interval. This is in contradistinction to the interpretation of ρ as the fraction of busy time in the stationary queue. The latter interpretation is valid only for some of the simpler classical queues.

The Multi-Server Case. It is entirely straightforward to extend the results of the preceding sections to the multi-server case. Let there be N_j servers of rate μ_j available whenever the Q-process is in the state j, $1 \leq j \leq m$. We set $N = \max_j N_j$, and assume in order to avoid triviality that $N \geq 1$, and that $\max_j(N_j\mu_j) > 0$. The vector μ^* is given by $\mu_j^* = N_j\mu_j$, $1 \leq j \leq m$. The arrival rates are given by the components of the vector $\boldsymbol{\lambda}$.

The resulting queueing model is again described by a QBD process, which differs from the single server case primarily in a more involved boundary behavior. The infinitesimal generator $\tilde{Q}$ is now given by

$$\tilde{Q} = \begin{vmatrix} A_{01} & A_{00} & & & & & & & \\ A_{12} & A_{11} & A_{10} & & & & & & \\ & A_{22} & A_{21} & A_{20} & & & & & \\ & & & \cdots & & & & & \\ & & & A_{N-2,2} & A_{N-2,1} & A_{N-2,0} & & & \\ & & & & A_{N-1,2} & A_{N-1,1} & A_0 & & \\ & & & & & A_2 & A_1 & A_0 & \cdots \\ & & & & & \vdots & \vdots & \vdots & \end{vmatrix}, \tag{6.2.25}$$

where

$$A_0 = \Delta(\boldsymbol{\lambda}), \qquad A_1 = Q - \Delta(\boldsymbol{\lambda} + \mu^*), \qquad A_2 = \Delta(\mu^*),$$

and for $1 \leq i \leq N - 1$,

$$A_{i0} = \Delta(\boldsymbol{\lambda}), \qquad A_{i1} = Q - A_{i0} - A_{i2}, \qquad A_{i2} = \Delta[\mu^*(i)],$$

with $\mu_j^*(i) = \mu_j \min(i, N_j)$, for $1 \leq j \leq m$. The matrices in the first row are given by $A_{01} = Q - \Delta(\lambda)$, $A_{00} = \Delta(\lambda)$.

The queue is *stable* if and only if $\pi\lambda < \pi\mu^*$, and in that case the stationary probability vector x of $\tilde{Q}$ is of a modified matrix-geometric form, with $x_i = x_{N-1}R^{i-N+1}$, for $i \geq N - 1$. The matrix R and the initial terms $x_0, \ldots, x_{N-1}$, are found by applying the general theory presented in this book. See also M. F. Neuts [230].

It is worth stressing that varying the number of servers requires a complete reimplementation of the algorithm. When N is large, each run may require much computer time. A practical time-saving approach is as follows. If the purpose of the interactive computation is to determine appropriate values of $N_1, N_2, \ldots, N_m$, to obtain desirable behavior of the queue, we first implement the algorithm for the single-server case for values of the service rates of the type $N_1\mu_1, \ldots, N_m\mu_m$. This is not an appropriate model for the multi-server queue, as it ignores the boundary behavior of the more complicated model. Except for queues that are highly underutilized we will so obtain a simpler model with substantially similar behavior. When values of $N_1, \ldots, N_m$ of approximately the right size have been identified, a number of additional searches with the more complicated algorithm can then be performed. For many practical purposes, the search for the m values of $N_1, \ldots, N_m$ is much simpler than it would appear. Only highly particular m-tuples $N_1, \ldots, N_m$ are usually practically meaningful, so that the search for desirable assignments of the numbers of servers to phases of the Q-process is limited to a fairly small set of alternatives.

In concluding the present discussion, we note that the infinite-server queue $M/M/\infty$ in a Markovian environment is surprisingly resistant to analytic solution. By methods similar to those in V. Ramaswami and M. F. Neuts [262] for the $PH/G/\infty$ queue, useful numerical information on the moments of the queue length in the various phases can be obtained. This, combined with a brute force numerical solution of a truncated version of the birth-and-death equations, enables one to solve this model for a wide range of parameter values in spite of the lack of a mathematically elegant solution. A similar approach to the $M/M/\infty$ queue with time-dependent arrival and service rates is discussed in T. Collings and C. Stoneman [63].

6.3. A QUEUE WITH SERVER BREAKDOWNS AND REPAIRS

The following model, which is a natural application of a queue in a Markovian environment, is discussed in greater detail in M. F. Neuts and

D. M. Lucantoni [234]. Consider a service system consisting of N parallel, exponential servers who process customers each at a rate μ. The servers are independently subject to breakdowns. For each server, the time until breakdown is exponentially distributed with parameter θ. Upon failure, a server enters a repair facility attended by $c \leq N$ repairmen. The actual repair times are exponentially distributed with parameter σ, and each repair is performed by one of the repair crews. A server who is not waiting for or in repair is said to be operative. The arrivals to the queue are according to a (modulated) Poisson process, whose rate may depend on the number of operative servers. The arrival rate, when j servers are operative, is denoted by λ_j, for $0 \leq j \leq N$.

This model has a number of variants which may be treated by exactly the same method. We shall briefly describe these without going into the details of their solutions.

a. In our discussion, servers may fail whether or not they are serving a customer. It is also possible to consider different failure rates θ' and θ'' for active and inactive servers. These failure rates may also be made to depend on the number of operative servers.

b. We discuss the case where services interrupted by the failure of the performing server are either resumed or repeated by an available operative server. The case where a customer whose service was interrupted must wait for the return of the performing server is much more involved. This variant requires a complicated description of the state space and is presently tractable only for very small values of N.

c. Free operative servers may be assigned to repair duty, with either repair or service receiving priority. We do not consider this option in our discussion. Servers and repairmen are treated as different entities who can only provide either service or repair.

In our model the subsystem formed by the N servers (or machines) and the c repairmen is the classical exponential machine interference model, treated by F. Benson and D. R. Cox [15], F. Benson [16], and P. Naor [213]. Textbook presentations are given, for example, in D. Gross and C. M. Harris [115], T. L. Saaty [272], and L. Takács [293]. The case where $c = N$ may provide a model for a queue of jobs handled by a pool of human operators. Server breakdowns are then interpreted as absences. This case was treated by transform methods by I. L. Mitrany and B. Avi-Itzhak [207].

The state space consists of the pairs (i, j), with $i \geq 0$ and $0 \leq j \leq N$. The index i denotes the number of customers in the system and j the number of operative servers. The model then readily leads to the QBD process with the generator $\tilde{Q}$, given by

$$\tilde{Q} = \begin{vmatrix} A_{00} & A_{01} & 0 & \cdots \\ A_{10} & A_{11} & A_{12} & \cdots \\ & & & \cdots \\ & & & & A_{N-2,0} & A_{N-2,1} & A_{N-2,2} \\ & & & & & A_{N-1,0} & A_{N-1,1} & A_0 \\ & & & & & & A_2 & A_1 & A_0 \\ & & & & & & & A_2 & A_1 & \cdots \\ & & & & & & & & \cdots \end{vmatrix}. \tag{6.3.1}$$

The square blocks of dimension $(N + 1) \times (N + 1)$ are given by

$$\begin{aligned} A_{00} &= Q - \Delta(\lambda), \\ A_{01} &= A_{12} = \cdots = A_{N-1,2} = A_0 = \Delta(\lambda), \\ A_{i0} &= \operatorname{diag}\{\mu \min(i, j), 0 \le j \le N\}, \quad \text{for} \quad 1 \le i \le N, \\ A_{i1} &= Q - \Delta(\lambda) - A_{i0}, \quad \text{for} \quad 1 \le i \le N - 1, \\ A_1 &= Q - \Delta(\lambda) - A_2, \\ A_2 &= \operatorname{diag}(0, \mu, 2\mu, \ldots, N\mu) = \Delta(\mu), \end{aligned} \tag{6.3.2}$$

where the matrix Q is the generator of the classical machine repair model and is given by

$$\begin{aligned} Q_{j,j-1} &= j\theta, && \text{for} \quad 1 \le j \le N, \\ Q_{jj} &= -j\theta - \sigma \min(c, N - j), && \text{for} \quad 0 \le j \le N, \\ Q_{j,j+1} &= \sigma \min(c, N - j), && \text{for} \quad 0 \le j \le N - 1, \\ Q_{jh} &= 0, && \text{for} \quad |j - h| > 1. \end{aligned} \tag{6.3.3}$$

The vector $\boldsymbol{\pi}$ is readily computed by solving the equations $\boldsymbol{\pi} Q = \mathbf{0}$, $\boldsymbol{\pi e} = 1$, although its components are explicitly given by

$$\begin{aligned} \pi_0 &= \left[\sum_{\nu=0}^{N-c+1} \frac{1}{\nu!} \left(\frac{c\sigma}{\theta}\right)^{\nu} + \sum_{\nu=N-c+2}^{N} \frac{1}{\nu!} \left(\frac{c\sigma}{\theta}\right)^{\nu} \prod_{r=1}^{\nu-N+c-1} \left(1 - \frac{r}{c}\right) \right]^{-1}, \\ \pi_j &= \frac{1}{j!} \left(\frac{c\sigma}{\theta}\right)^{j} \pi_0, \quad \text{for} \quad 0 \le j \le N - c + 1, \\ \pi_j &= \frac{1}{j!} \left(\frac{c\sigma}{\theta}\right)^{j} \prod_{r=1}^{j-N+c-1} \left(1 - \frac{r}{c}\right) \pi_0, \quad \text{for} \quad N - c + 2 \le j \le N. \end{aligned} \tag{6.3.4}$$

For $N = 1$, we obtain a fully explicit solution for the stationary probability vector x of $\tilde{Q}$. This solution is given in theorem 6.3.1. In the sequel, we then discuss the general case $N \geq 2$. We note that the case $N = 1$ is simply the $M/M/1$ queue in which the server is alternatingly up or down for exponential lengths of time with means θ^{-1} and σ^{-1} respectively. The arrival rates are λ_0 and λ_1, depending on whether the server is down or up.

Theorem 6.3.1. In the case $N = 1$, the queue is *stable* if and only if $\lambda_0\theta\sigma^{-1} + \lambda_1 < \mu$. The pair (π_0, π_1) is given by

$$\pi_0 = 1 - \pi_1 = \theta(\theta + \sigma)^{-1}.$$

The vector x, partitioned into the 2-vectors x_i, $i \geq 0$, is given by $x_i = \pi(I - R)R^i$, for $i \geq 0$, where the matrix R is explicitly given by

$$R = \begin{vmatrix} \lambda_0(\lambda_0 + \sigma)^{-1}(\theta + \mu)\mu^{-1} & \lambda_0\mu^{-1} \\ \lambda_1(\lambda_0 + \sigma)^{-1}\theta\mu^{-1} & \lambda_1\mu^{-1} \end{vmatrix}.$$

Proof. Since in this case, the matrix Q is given by

$$Q = \begin{vmatrix} -\sigma & \sigma \\ \theta & -\theta \end{vmatrix},$$

the expressions for π_0 and π_1 are obvious. The general equilibrium condition $\pi\lambda < \pi\mu$ clearly yields the stated inequality.

The equation for the matrix R, that is,

$$R^2\Delta(\mu) + R[Q - \Delta(\lambda + \mu)] + \Delta(\lambda) = 0,$$

leads to the system of equations

$$\begin{aligned} R_{00}(\sigma + \lambda_0) - R_{01}\theta - \lambda_0 = 0, \\ R_{01}(R_{00} + R_{11})\mu + R_{00}\sigma - R_{01}(\theta + \lambda_1 + \mu) = 0, \\ R_{10}(\sigma + \lambda_0) - R_{11}\theta = 0, \\ (R_{01}R_{10} + R_{11}{}^2)\mu + R_{10}\sigma - R_{11}(\theta + \lambda_1 + \mu) + \lambda_1 = 0. \end{aligned}$$

Recalling, however, that $R\mu = \lambda$, we obtain $R_{01}\mu = \lambda_0$ and $R_{11}\mu = \lambda_1$. Using the last two equations, the other two elements R_{00} and R_{10} are readily obtained. It is now a routine matter to verify that the vector $x_0 = \pi(I - R)$ is positive and satisfies the equations

$$x_0[Q - \Delta(\lambda) + R\Delta(\mu)] = \mathbf{0} \quad \text{and} \quad x_0(I - R)^{-1}e = 1.$$

In the general case $N \geq 2$, the queue is *stable* if and only if

$$\rho = \left(\sum_{j=0}^{N} \pi_j \lambda_j\right)\left(\mu \sum_{j=1}^{N} j\pi_j\right)^{-1} < 1. \tag{6.3.5}$$

The $(N + 1)$-vectors $\boldsymbol{x}_i$, $i \geq 0$, are then given by the general approach discussed in this book. For $i \geq N - 1$, we have $\boldsymbol{x}_i = \boldsymbol{x}_{N-1}R^{i-N+1}$, where R is the minimal nonnegative solution of the equation

$$R^2\Delta(\mu) + R[Q - \Delta(\lambda + \mu)] + \Delta(\lambda) = 0.$$

The j-th row of R is positive when $\lambda_j > 0$, and zero otherwise.

The vectors $\boldsymbol{x}_0, \ldots, \boldsymbol{x}_{N-1}$ are obtained by solving the linear equations

$$\begin{aligned} &\boldsymbol{x}_0 A_{00} + \boldsymbol{x}_1 A_{10} = \boldsymbol{0}, \\ &\boldsymbol{x}_{i-1}\Delta(\lambda) + \boldsymbol{x}_i A_{i1} + \boldsymbol{x}_{i+1} A_{i+1,0} = \boldsymbol{0}, \quad \text{for} \quad 1 \leq i \leq N-2, \\ &\boldsymbol{x}_{N-2}\Delta(\lambda) + \boldsymbol{x}_{N-1}[A_{N-1,1} + R\Delta(\mu)] = \boldsymbol{0}, \\ &\sum_{i=0}^{N-2} \boldsymbol{x}_i \boldsymbol{e} + \boldsymbol{x}_{N-1}(I - R)^{-1}\boldsymbol{e} = 1. \end{aligned} \tag{6.3.6}$$

In some applications the order of the system (6.3.6) may be very high, but its particular structure permits an efficient solution. Provided the vector $\boldsymbol{\lambda}$ is positive, we may solve the equations (6.3.6) recursively, starting with the penultimate one. This leads to

$$\boldsymbol{x}_i = \boldsymbol{x}_{N-1}C_i, \quad \text{for} \quad 0 \leq i \leq N-2, \tag{6.3.7}$$

where the matrices C_i are defined by

$$\begin{aligned} C_{N-1} &= I, \\ C_{N-2} &= -(A_{N-1,1} + RA_2)\Delta^{-1}(\lambda), \\ C_{N-\nu} &= -(C_{N-\nu+1}A_{N-\nu+1,1} + C_{N-\nu+2}A_{N-\nu+2,0})\Delta^{-1}(\lambda), \end{aligned} \tag{6.3.8}$$

for $3 \leq \nu \leq N$.

The vector $\boldsymbol{x}_{N-1}$ is then obtained by solving the system

$$\begin{aligned} &\boldsymbol{x}_{N-1}(C_0 A_{00} + C_1 A_{10}) = \boldsymbol{0}, \\ &\boldsymbol{x}_{N-1}\left[\sum_{\nu=0}^{N-2} C_\nu \boldsymbol{e} + (I - R)^{-1}\boldsymbol{e}\right] = 1. \end{aligned} \tag{6.3.9}$$

The recursion (6.3.8) may be implemented by storing only two of the matrices C_i. The vector $\sum_{\nu=0}^{N-2} C_\nu \boldsymbol{e}$ is accumulated as one goes along. At the final stage of the recursive computation, one has all the ingredients needed to compute the coefficient matrix and hence the solution of the system (6.3.9). The vector $\zeta = (I - R)^{-1}e$ is computed beforehand by solving the system $\zeta = e + R\zeta$ by Gauss-Seidel iteration.

With economy of storage, the matrices C_i, $0 \leq i \leq N-2$, need to be computed a second time for use in (6.3.7). This, however, requires only a small amount of additional processing time.

When some of the parameters λ_i are very small, the recursive computation of the matrices C_i may fail due to overflow. This procedure is a fortiori to be avoided when some λ_i are zero.

When either of these alternatives are likely to arise, the system (6.3.6), and others like it occurring in this book, may be solved by a more time-consuming, but safe, iterative method. To do so, let, for example, Δ_0 and Δ_i, $1 \leq i \leq N-1$, denote the matrices $-\mathrm{diag}(A_{00})$ and $-\mathrm{diag}(A_{i,1})$, $1 \leq i \leq N-1$, respectively.

The system (6.3.6) may then be written into the form

$$\begin{aligned} x_0 &= [x_0(A_{00} + \Delta_0) + x_1 A_{10}]\Delta_0^{-1}, \\ x_i &= [x_{i-1}\Delta(\lambda) + x_i(A_{i1} + \Delta_i) + x_{i+1}A_{i+1,0}]\Delta_i^{-1}, \\ &\qquad \text{for} \quad 1 \leq i \leq N-2, \\ x_{N-1} &= \{x_{N-2}\Delta(\lambda) + x_{N-1}[A_{N-1,1} + \Delta_{N-1} + R\Delta(\mu)]\}\Delta_{N-1}^{-1}, \end{aligned} \tag{6.3.10}$$

which is well suited, for example, for a Gauss-Seidel type iteration. The normalizing equation in (6.3.6) may be used at each step to keep the successive iterates within a compact set that contains the unique solution vector.

It should be noted that such numerical difficulties as may arise with the earlier recursive procedure are inherent in nearly all recursive methods for the solution of large systems of linear equations. The present system is, fortunately, always well suited for solution by iterative methods.

The Conditional Waiting Time Distributions. The details of the computation of the conditional waiting time distributions, given that an arriving customer finds j, $0 \leq j \leq N$, servers operative, are somewhat involved, because of the care needed in handling the boundary behavior of the queue. For the purpose of presenting a representative example, these details will now be discussed.

We denote by $W_j(x)$, $0 \leq j \leq N$, the conditional probability that a customer C arriving to the stationary queue will begin service after a wait no longer than x, given that upon his arrival j servers are operative. It is clear that

$$\begin{aligned} W_j(0+) &= 0, \qquad \text{for} \quad j = 0, \\ &= \pi_j^{-1} \sum_{\nu=0}^{j-1} x_{\nu j}, \quad \text{for} \quad 1 \leq j \leq N. \end{aligned} \tag{6.3.11}$$

An arriving customer does not have to wait if and only if there is at least one operative server and at most $j - 1$ (operative) servers are busy.

If the arriving customer must wait, he will do so until for the first time the number of customers ahead of him drops to one less than the number of operational servers. We shall construct a Markov process $\tilde{Q}_0$ with one absorbing state in which this waiting time becomes the time until absorption. For definiteness, we shall assume that customers who arrived ahead of the customer C, but suffered service interruptions due to server breakdowns, *retain* priority over the customer C. By minor modifications in the discussion that follows, one may also study the case where such customers lose their priority vis-à-vis the customer C. For the purpose of computing the conditional waiting time distributions, it suffices in the latter case to treat a breakdown of an *active* server also as a departure from the queue. The customer who was being processed by the failing server is now no longer ahead of the customer C and is, from his viewpoint, no longer in the queue.

The state space of the process $\tilde{Q}_0$ is given by

$$E_1 = \{*\} \cup \{(i, \nu): i \geq 0, 0 \leq \nu \leq \min(i, N)\}.$$

The state * is obtained by lumping together the states (i, j) with $0 \leq i \leq N - 1$, $i + 1 \leq j \leq N$, of the original process $\tilde{Q}$. The matrix $\tilde{Q}_0$ has a highly structured form, given by

$$\tilde{Q}_0 = \begin{array}{c|cccccccccc|}
* & 0 & \mathbf{0} & \mathbf{0} & \mathbf{0} & \cdots & & & & & \\
0 & c_0 & B_0 & 0 & 0 & \cdots & & & & & \\
1 & c_1 & D_1 & B_1 & 0 & \cdots & & & & & \\
2 & c_2 & 0 & D_2 & B_2 & \cdots & & & & & \\
\vdots & \vdots & \vdots & \vdots & \vdots & & \vdots & \vdots & \vdots & \vdots & \\
N-1 & c_{N-1} & 0 & 0 & 0 & \cdots & D_{N-1} & B_{N-1} & 0 & 0 & \cdots \\
N & c_N & 0 & 0 & 0 & \cdots & 0 & D_N & B_N & 0 & \cdots \\
N+1 & \mathbf{0} & 0 & 0 & 0 & \cdots & 0 & 0 & \Delta(\mu) & Q - \Delta(\mu) & \cdots \\
N+2 & \mathbf{0} & 0 & 0 & 0 & \cdots & 0 & 0 & 0 & \Delta(\mu) & \cdots \\
\vdots & \vdots & \vdots & \vdots & \vdots & & \vdots & \vdots & \vdots & \vdots &
\end{array}. \tag{6.3.12}$$

The vectors c_ν, $0 \leq \nu \leq N$, are of dimension $\nu + 1$. They are zero, except for their last component, which is given by $\nu\mu + \sigma \min(c, N - \nu)$. The matrices B_ν, $0 \leq \nu \leq N$, are Jacobi matrices of order $\nu + 1$. $B_0 = -c\sigma$, and the ν elements *below* the diagonal of B_ν, $1 \leq \nu \leq N$, are given by $r\theta$, $1 \leq r \leq \nu$. The ν elements above the diagonal of B_ν are all equal and given by $\sigma \min(c, N - r)$, $1 \leq r \leq \nu$. The matrices D_ν, $1 \leq \nu \leq N$, are of dimensions $(\nu + 1) \times \nu$. All their elements are zero, except for the elements $(D_\nu)_{rr} = r\mu$, for $0 \leq r \leq \nu - 1$. Finally, the diagonal elements of B_ν are chosen so that $B_\nu e + D_\nu e + c_\nu = \mathbf{0}$, for $1 \leq \nu \leq N$.

The waiting time distributions $W_j(\cdot)$, $0 \leq j \leq N$, are easily seen to be those of the time until absorption in the process $\tilde{Q}_0$ for different choices of the initial probability vectors. For $W_j(\cdot)$, we use the initial probability vector $y(0)$, given in partitioned form by

$$[W_j(0+), y_0(0), y_1(0), \ldots, y_N(0), y_{N+1}(0), \ldots].$$

The vectors $y_\nu(0)$ of dimension $\min(N, \nu) + 1$, are given by

$$[y_\nu(0)]_k = \delta_{jk}\pi_j^{-1}x_{\nu j}, \quad \text{for} \quad \nu \geq 0, \qquad 0 \leq k \leq \min(N, \nu). \qquad (6.3.13)$$

We see that the only nonabsorbing states *with positive initial probability* are those of the form (ν, j), $\nu \geq j \geq 0$.

In order to evaluate $W_j(x)$, $x \geq 0$, and its density portion $W_j'(x)$, we now solve the system of differential equations

$$y_\nu'(x) = y_\nu(x)B_\nu + y_{\nu+1}(x)D_{\nu+1}, \qquad \text{for} \quad 0 \leq \nu \leq N,$$

$$y_\nu'(x) = y_\nu(x)[Q - \Delta(\mu)] + y_{\nu+1}(x)\Delta(\mu), \quad \text{for} \quad \nu \geq N + 1, \quad (6.3.14)$$

where $D_{N+1} = \Delta(\mu)$, and evaluate

$$W_j(x) = 1 - \sum_{\nu=0}^{\infty} y_\nu(x)e, \quad \text{for} \quad x \geq 0,$$

$$W_j'(x) = \sum_{\nu=0}^{N} y_\nu(x)c_\nu, \qquad \text{for} \quad x > 0. \qquad (6.3.15)$$

The required numerical computations are performed by applying the first method discussed in section 3.9. It is advisable to compute the distributions $W_j(\cdot)$ for *decreasing* values of j, starting with $j = N$. For the lower indices j, these probability distributions may have slowly decaying exponential tails, which require long computation times, yet are not particularly informative.

The evaluation of the means of $W_j(\cdot)$ for $0 \leq j \leq N$ is discussed in detail in [234]. The second moments may be found by similar arguments, although rather involved matrix calculations are required.

A Numerical Example. The following numerical example deals with the computation of the stationary probability vector $\boldsymbol{x}$ for the model under discussion. This example and the interpretation of the numerical results supplement those given in [234].

We consider a queue with $N = 6$ servers, capable of serving customers at a rate $\mu = 1$. The breakdown and repair rates are chosen to be $\theta = \sigma = 0.05$. These four parameters will not be changed, but we shall discuss by examples the effect of varying the number c of repair crews or the rate of arrival λ_j, during periods of time when the number j of operational servers is low. The successive choices of c and λ are as follows:

Queue A: $c = 6$; $\lambda_j = 2.5$, for $0 \leq j \leq 6$.

Queue B: $c = 5$; $\lambda_j = 2.5$, for $0 \leq j \leq 6$.

Queue C: $c = 6$, $\lambda_0 = \lambda_1 = \lambda_2 = 1.25$, $\lambda_3 = \lambda_4 = \lambda_5 = \lambda_6 = 2.5$.

Queue D: $c = 5$, $\lambda_0 = \lambda_1 = \lambda_2 = 1.25$, $\lambda_3 = \lambda_4 = \lambda_5 = \lambda_6 = 2.5$.

Queue E: $c = 3$, $\lambda_0 = \lambda_1 = \lambda_2 = 1.25$, $\lambda_3 = \lambda_4 = \lambda_5 = \lambda_6 = 2.5$.

Queue F: $c = 2$, $\lambda_0 = \lambda_1 = \lambda_2 = 1.25$, $\lambda_3 = \lambda_4 = \lambda_5 = \lambda_6 = 2.5$.

The traffic intensities ρ for various values of c with the arrival rates as given in the queues A, B and C, D, E, and F respectively are given in table 6.5. If the arrival rate λ_j is maintained at 2.5 regardless of the number of

TABLE 6.5

c	A, B	C, D, E, F
1	2.50	1.35
2	1.28	0.84
3	0.95	0.73
4	0.86	0.70
5	0.84	0.69
6	0.83	0.69

operational servers, the value of c has to be at least *three* in order to achieve a stable queue. For the alternate choice of arrival rates, we see that c needs to be at least *two* and that the traffic intensities corresponding to the same values of c are considerably lower in this case. We see that in both cases the value of ρ becomes rapidly insensitive to increases in c. As we shall see, this is not the case for other important features of the queue.

In table 6.6 we list for each of the six queues the values of π_j, $0 \leq j \leq 6$. These are the stationary probabilities of the exponential machine interference model, obtained by ignoring the queue. These probabilities clearly depend only on the value of c and not on the λ_j. The corresponding values in the columns A and C, as well as in the columns B and D, are therefore equal to each other.

In table 6.7 the conditional means and standard deviations of the stationary queue length, given the number of operative servers, are presented.

TABLE 6.6

Number of Operational Servers	A	B	C	D	E	F
0	0.0156	0.0187	0.0156	0.0187	0.0557	0.1368
1	0.0937	0.0935	0.0937	0.0935	0.1671	0.2736
2	0.2344	0.2336	0.2344	0.2336	0.2507	0.2736
3	0.3125	0.3115	0.3125	0.3115	0.2507	0.1824
4	0.2344	0.2336	0.2344	0.2336	0.1880	0.0912
5	0.0937	0.0935	0.0937	0.0935	0.0752	0.0365
6	0.0156	0.0156	0.0156	0.0156	0.0125	0.0061

The stationary probabilities of the numbers of operational servers for the queues A, B, C, D, E, and F.

Our program evaluated the stationary probability vector $\mathbf{x} = [\mathbf{x}_0, \mathbf{x}_1, \ldots]$ by successively computing the vectors $\mathbf{x}_i$ until a probability less than 10^{-6} remained in the tail of the density $\mathbf{x}$. It then reported on the accuracy tests given by $R\boldsymbol{\lambda} = \boldsymbol{\mu}$, and $\Sigma_{i=0}^{\infty} \mathbf{x}_i = \boldsymbol{\pi}$, which were found to hold to at least five places of decimal accuracy.

Table 6.8 reports selected percentiles of the *conditional* queue length densities, given the number of operational servers.

As in the case discussed in section 6.2, a number of qualitative interpretations may be given for the numerical results presented here. It should be noted that the failure rate θ was chosen to be fairly high and the repair rate σ rather low as compared to the service rate $\mu = 1$. This was done to illustrate the oscillatory behavior of the queues by means of a small number of conceptually simple examples. Additional insight may be obtained by reducing θ and by increasing the arrival rates λ_j. The stationary probabilities π_j for the lower indices j would then be reduced to more realistic magnitudes, but with the increased rates of arrivals, the rare but long-lived build-ups of the queue would be as significant as or worse than those in the present examples.

Comparing the queues A and B, we see that the removal of one repairman results in an insignificant increase in ρ and in minor changes in the probabilities π_j. The increase in the means and standard deviations of the conditional queue length distributions is, on the other hand, substantial. We also note the overdispersion of the conditional queue length densities for the higher indices j. This is due to the occurrence of major queue build-ups that arise during periods when the queue is oversaturated. The effect of removing a repairman increases the *variability* of the periods of oversaturation, even though the means of such periods are not significantly affected. This increases the likelihood of deleterious build-ups, which take a long time to be cleared.

TABLE 6.7

Number of Servers	Mean						Standard Deviation					
	A	B	C	D	E	F	A	B	C	D	E	F
0	45.76	49.28	10.65	11.81	21.17	52.29	35.18	37.91	7.97	8.94	17.44	47.76
1	37.43	39.28	6.48	6.81	12.84	39.79	34.05	36.43	6.48	7.06	15.05	45.95
2	29.74	31.27	4.28	4.44	7.97	30.73	32.53	34.76	5.05	5.44	12.25	43.24
3	23.12	24.48	4.52	4.62	6.97	26.66	30.47	32.64	4.42	4.70	10.25	40.64
4	17.85	19.09	3.63	3.69	5.30	21.72	27.93	30.07	3.49	3.71	8.46	37.68
5	13.90	15.00	3.01	3.04	4.12	17.82	25.15	27.26	2.64	2.81	6.82	34.71
6	10.99	11.97	2.71	2.73	3.43	14.79	22.37	24.44	2.08	2.19	5.43	31.78

The conditional means and standard deviations of the queue length, given the number of operational servers.

TABLE 6.8

Number of Servers	90%						95%						99%					
	A	B	C	D	E	F	A	B	C	D	E	F	A	B	C	D	E	F
0	91	99	21	23	44	114	115	124	26	29	56	147	170	183	37	42	83	224
1	82	87	15	16	32	100	105	112	19	21	44	133	160	171	30	33	71	209
2	72	76	10	11	21	87	96	102	14	15	33	120	150	161	24	26	60	196
3	62	66	10	10	16	78	86	92	13	13	26	111	141	151	21	23	53	188
4	53	56	7	7	11	68	76	82	10	10	19	101	131	141	17	18	45	177
5	43	47	6	6	7	57	66	72	7	7	12	90	121	131	13	14	37	167
6	33	37	5	5	5	47	57	62	6	6	7	80	111	121	9	10	29	157

Selected percentiles of the conditional queue length distributions, given the number of operational servers.

The effect of reducing the input when many servers are down is dramatically illustrated by comparing the queues A and C. The effect of the build-up during the much rarer periods of oversaturation is greatly reduced. The prompt return of a failed server to operative status is here no longer so crucial. The queue lengths can now be kept within tolerable bounds when one (Queue D) or even three repairmen are removed, as was done in Queue E. The picture changes again radically when $c = 2$, as in the queue F. The large increase in ρ from 0.73 to 0.84 when c is reduced from three (E) to two (F) already suggests that the behavior of the queues E and F can be expected to be very different. Although the queues B and F have nearly the same traffic intensities, the queue F only rarely has more than three operative servers, and the slowness of repairs is causing a serious deterioration of service in the queue, in spite of the reduced arrival rates.

As a concluding comment, we note that the exponential assumptions, particularly for the times-to-failure and repair times, are a severe restriction of the model. There is computational evidence showing that the fluctuations of the queue are quite sensitive to the variability of these random variables. Except for very small values of N, even the generalization to PH-distributions results in an unmanageable explosion in the dimensionality, which makes an exact computational solution prohibitive. For practical purposes, a three-pronged attack by an exact algorithm, by diffusion approximation, and by simulation would be desirable. By itself, each of these methods clearly has limitations.

6.4. OVERFLOW MODELS

Some models for service systems in which a second service unit receives the overflow customers from a first unit of finite capacity may be viewed as queues in a (natural) Markovian environment.

We first consider a system in which the first unit is a *finite* $M/PH/1$ queue. Explicit formulas for the stationary probability vector of the finite $M/PH/1$ queue were obtained in section 3.2. As long as there are $K + 1$ customers in the first unit, all additional arrivals are diverted to the second unit, in which they are served by one of c exponential servers, each of rate μ. The queue in the second unit is unbounded. The *entire* system may be solved by considering a single $M/M/c$ queue in a Markovian environment. In order to avoid involved notation, we shall only discuss the case $c = 1$. The case with a general c is discussed in M. F. Neuts and S. Kumar [238]. Numerical examples and their interpretations are also given there.

The arrival rate to the first unit is λ. The PH-distribution of the service times in that unit has the irreducible representation $(\boldsymbol{\beta}, S)$ of order n and mean μ_1'. The matrix Q is of order $m = (K+1)n + 1$, and is given by

$$Q = \begin{vmatrix} -\lambda & \lambda\boldsymbol{\beta} & 0 & 0 & \cdots & 0 & 0 & 0 \\ \mathbf{S}^o & S-\lambda I & \lambda I & 0 & \cdots & 0 & 0 & 0 \\ \mathbf{0} & \mathbf{S}^o\mathbf{B}^o & S-\lambda I & \lambda I & \cdots & 0 & 0 & 0 \\ \mathbf{0} & 0 & \mathbf{S}^o\mathbf{B}^o & S-\lambda I & \cdots & 0 & 0 & 0 \\ \vdots & \vdots & \vdots & \vdots & & \vdots & \vdots & \vdots \\ \mathbf{0} & 0 & 0 & 0 & \cdots & \mathbf{S}^o\mathbf{B}^o & S-\lambda I & \lambda I \\ \mathbf{0} & 0 & 0 & 0 & \cdots & 0 & \mathbf{S}^o\mathbf{B}^o & S \end{vmatrix}. \tag{6.4.1}$$

Arrivals to the second unit can occur only when the Markov process Q is in one of the states $(K+1, j)$, $1 \le j \le n$. The *vector* $\boldsymbol{\lambda}$ of the Markovian environment model is therefore given by

$$\boldsymbol{\lambda} = (0, \mathbf{0}, \mathbf{0}, \ldots, \mathbf{0}, \lambda\mathbf{e}),$$

where $\boldsymbol{\lambda}$ is partitioned into a scalar and $K+1$ vectors of dimension n. If the service rate μ in the second unit does not depend on the state of the first unit, then the vector $\boldsymbol{\mu}$ is similarly given by

$$\boldsymbol{\mu} = \mu(1, \mathbf{e}, \mathbf{e}, \ldots, \mathbf{e}, \mathbf{e}).$$

The vector $\boldsymbol{\pi} = (\pi_0, \boldsymbol{\pi}_1, \ldots, \boldsymbol{\pi}_{K+1})$, similarly partitioned, is precisely the stationary probability vector of Q which was computed in section 3.2. The entire system will be *stable* if and only if

$$\mu > \lambda\boldsymbol{\pi}_{K+1}\mathbf{e}. \tag{6.4.2}$$

The inner product $\boldsymbol{\pi}_{K+1}\mathbf{e}$ is explicitly given by

$$\boldsymbol{\pi}_{K+1}\mathbf{e} = \boldsymbol{\beta}\hat{R}^K(-\lambda S^{-1}\mathbf{e})\left\{\boldsymbol{\beta}\left[\sum_{\nu=0}^{K}\hat{R}^\nu - \lambda\hat{R}^K S^{-1}\right]\mathbf{e}\right\}^{-1}, \tag{6.4.3}$$

in which $\hat{R} = \lambda(\lambda I - \lambda B^{oo} - S)^{-1}$.

The special form of the vector $\boldsymbol{\lambda}$ has far-reaching consequences for the computation of the matrix R. Although R is of order $(K+1)n + 1$, *only its last n rows are nonzero.* It is most convenient to partition the nonzero portion of R into a column vector $\mathbf{R}_0$ and $K+1$ square matrices R_ν, $1 \le \nu \le K+1$, of order n, that is,

$$R_0, R_1, R_2, \ldots, R_{K+1}.$$

The corresponding rows of R^2 are then given by

$$R_{K+1}R_0, R_{K+1}R_1, \ldots, R_{K+1}R_K, R_{K+1}^2.$$

The classical equation for the matrix R, that is,

$$R^2\Delta(\mu) + R[Q - \Delta(\lambda + \mu)] + \Delta(\lambda) = 0, \tag{6.4.4}$$

now leads to the system of matrix equations

$$\begin{aligned}
&[\mu R_{K+1} - (\lambda + \mu)I]\boldsymbol{R}_0 + \boldsymbol{R}_1 S^o = \mathbf{0},\\
&\mu R_{K+1}R_1 + R_1[S - (\lambda + \mu)I] + R_2 S^o B^o + \lambda \boldsymbol{R}_0 \boldsymbol{\beta} = 0,\\
&\mu R_{K+1}R_\nu + R_\nu[S - (\lambda + \mu)I] + R_{\nu+1}S^o B^o + \lambda R_{\nu-1} = 0, \quad \text{for} \quad 2 \le \nu \le K,\\
&\mu R_{K+1}^2 + R_{K+1}[S - (\lambda + \mu)I] + \lambda R_K + \lambda I = 0.
\end{aligned} \tag{6.4.5}$$

These equations may be conveniently solved by rewriting them as

$$\begin{aligned}
&\boldsymbol{R}_0 = [(\lambda + \mu)I - \mu R_{K+1}]^{-1}\boldsymbol{R}_1 S^o,\\
&R_1 = [\mu R_{K+1}R_1 + (R_2 S^o + \lambda \boldsymbol{R}^o)\boldsymbol{\beta}][(\lambda + \mu)I - S]^{-1},\\
&R_\nu = [\mu R_{K+1}R_\nu + \lambda R_{\nu-1} + (R_{\nu+1}S^o)\boldsymbol{\beta}][(\lambda + \mu)I - S]^{-1}, \quad \text{for} \quad 2 \le \nu \le K,\\
&R_{K+1} = [\mu R_{K+1}^2 + \lambda R_K + \lambda I][(\lambda + \mu)I - S]^{-1},
\end{aligned} \tag{6.4.6}$$

and by performing successive substitutions, starting with the zero solution. The matrix $[(\lambda + \mu)I - S]^{-1}$ needs to be computed only once, and the evaluation of the next value of $\boldsymbol{R}_0$ requires the solution of a system of n linear equations in n unknowns. The relation $R\boldsymbol{\mu} = \lambda$, or, equivalently,

$$\boldsymbol{R}_0 + \sum_{\nu=1}^{K+1} R_\nu \boldsymbol{e} = \frac{\lambda}{\mu}\boldsymbol{e}, \tag{6.4.7}$$

provides us with a useful accuracy check.

Once the matrix R has been computed, the vectors $\boldsymbol{x}(k)$, $k \ge 0$, can be easily evaluated from the formula $\boldsymbol{x}(k) = \boldsymbol{\pi}(I - R)R^k$, $k \ge 0$. Partitioning the vectors $\boldsymbol{x}(k)$ as $[x_0(k), \boldsymbol{x}_1(k), \ldots, \boldsymbol{x}_{K+1}(k)]$, we see that the ν-th component of $\boldsymbol{x}_i(k)$ is the stationary probability that there are i customers in the first queue and k in the second, and that the service phase of the first server is ν. Similarly, $x_0(k)$ is the steady-state probability that there are k customers in the second unit and that the first is empty.

A wealth of additional results can now be obtained by considering marginal and conditional distributions. From these, many inferences about the qualitative behavior can be drawn. If $\lambda\mu_1'$ is significantly less than one, only rare, clumped arrivals to Unit II will occur. If $\lambda\mu_1'$ is close to one, there will typically be long intervals during which arrivals to Unit II occur steadily, separated by long intervals of time with only rare arrivals. As $\lambda\mu_1'$ starts to exceed one, the arrivals to Unit II begin to be sustained over longer and longer intervals as fewer and fewer customers gain access to Unit I. The fluctuations of the second queue not only can be described informally but are, via the algorithmic solution, expressed in a precise quantitative form.

This overflow model may further be used to illustrate the danger of naive approximations and to point to the extreme care that is needed in studying even the simplest networks of queues with capacity constraints.

When $\lambda\mu_1'$ is much larger than one, it is tempting to replace the overflow process by a Poisson process with rate given by $\lambda\boldsymbol{\pi}_{K+1}\boldsymbol{e}$, and to treat the second service unit as an $M/M/1$ queue. For the reasons given in the discussion of the numerical examples in section 6.2, this approximation may lead to substantial underestimation of the queue lengths in the second queue. Similar comments apply to other overflow models. In A. Nilsson [241], the overflow from a finite loss system is processed by a secondary infinite-server queue. The substantial variability of the number of occupied channels in the second system is a direct consequence of the random fluctuations in the overflow stream.

This model also has several interesting generalizations and variants. The second unit may have its own independent Poisson arrival process of rate λ'. When this is the case, the vector λ in equation (6.4.4) is replaced by $\lambda + \lambda'\boldsymbol{e}$. The corresponding matrix $R(\lambda')$ is strictly positive when $\lambda' > 0$. The computational simplification inherent in the case $\lambda' = 0$ is no longer present, but the matrix R may still be partitioned in an analogous manner to make use of the special structure of the matrix Q. A detailed discussion is given in [238].

An interesting design problem deals with the optimal choice of K when there are costs associated with the number of unused places in the first queue and also with each customer who must go to the second unit. The value of K which minimizes, for example, the total expected cost per unit of time may be algorithmically determined by the approach described for a different model in section 4.3.

By an entirely similar analysis, we may also study a finite $M/M/r$ queue, whose overflow is directed to c identical exponential servers. A version of this with $r = c = 1$ and with both queues finite was studied by R. L. Disney [81]. As shown there, particular structural features of the generator may also be exploited for efficient computation in that case.

It is further feasible to study an $M/PH/1$ queue with finite waiting space, whose overflow is directed to a *single* server with a service time distribution *of phase type.* In order to study this model as a continuous-parameter Markov process, however, large matrices need to be manipulated, and careful computer coding is essential. For the sake of readers who wish to pursue the development of algorithms for this model, we shall display the generator $\tilde{Q}$ of the Markov process for that model. The following notation is used. The Poisson arrival rate to the system is λ; the maximum number of customers allowed at any one time in Unit I is $K + 1$; the service time distributions in Units I and II are of phase type, respectively with representations (β, S) of order n and (γ, L) of order r. The matrices $S^o B^o$ and $L^o \Gamma^o$ are defined as before.

The matrix $\tilde{Q}$ has the block-Jacobi structure

$$\tilde{Q} = \begin{vmatrix} B_0 & C_1 & 0 & 0 & 0 & \cdots \\ B_1 & A_1 & A_0 & 0 & 0 & \cdots \\ 0 & A_2 & A_1 & A_0 & 0 & \cdots \\ 0 & 0 & A_2 & A_1 & A_0 & \cdots \\ 0 & 0 & 0 & A_2 & A_1 & \cdots \\ \vdots & \vdots & \vdots & \vdots & \vdots & \end{vmatrix}, \tag{6.4.8}$$

where the dimensions of the blocks are given by

$$\begin{aligned} B_0&: \quad [(K+1)n+1] \times [(K+1)n+1], \\ C_1&: \quad [(K+1)n+1] \times [(K+1)nr+n], \\ B_1&: \quad [(K+1)nr+n] \times [(K+1)n+1], \\ A_0, A_1, A_2&: \quad [(K+1)nr+n] \times [(K+1)nr+n]. \end{aligned}$$

The states of $\tilde{Q}$ are written in lexicographic order and are given by

$(0, 0)$

$(0, i, h), \qquad 1 \le i \le K+1, \qquad 1 \le h \le n,$

$(k, 0, j), \qquad k \ge 1, \qquad 1 \le j \le r,$

$(k, i, j, h), \qquad k \ge 1, \qquad 1 \le i \le K+1, \qquad 1 \le j \le r, \qquad 1 \le h \le n,$

where i denotes the queue length in Unit I, k the queue length in Unit II, h the service phase in Unit I, and j the service phase in Unit II. A display of the highly particular and useful special structure of the matrices B_0,

C_1, B_1, A_0, A_1, and A_2 follows. For convenience, we have shown these matrices for $K = 3$; the general structure is then readily apparent.

$$B_0 = \begin{vmatrix} -\lambda & \lambda\beta & 0 & 0 & 0 \\ S^o & S - \lambda I & \lambda I & 0 & 0 \\ \mathbf{0} & S^o B^o & S - \lambda I & \lambda I & 0 \\ \mathbf{0} & 0 & S^o B^o & S - \lambda I & \lambda I \\ \mathbf{0} & 0 & 0 & S^o B^o & S - \lambda I \end{vmatrix}. \tag{6.4.9}$$

The matrix C_1 is zero except for the $n \times nr$ block given by $\gamma \otimes \lambda I$ in the lower right-hand corner.

$$B_1 = \begin{vmatrix} L^o & 0 & 0 & 0 & 0 \\ \mathbf{0} & L^o \otimes I & 0 & 0 & 0 \\ \mathbf{0} & 0 & L^o \otimes I & 0 & 0 \\ \mathbf{0} & 0 & 0 & L^o \otimes I & 0 \\ \mathbf{0} & 0 & 0 & 0 & L^o \otimes I \end{vmatrix}. \tag{6.4.10}$$

The symbol I stands for the identity matrix of order n when it occurs to the right of $\otimes$, and for the identity matrix of order r when it is to the left of $\otimes$. The matrix A_0 is zero except for the $nr \times nr$ block $\lambda I \otimes I$ in the lower right-hand corner.

$$A_1 = \begin{vmatrix} L-\lambda I & \lambda I \otimes \beta & 0 & 0 & 0 \\ I \otimes S^o & (L-\lambda I)\otimes I + I \otimes S & \lambda I \otimes I & 0 & 0 \\ 0 & I \otimes S^o B^o & (L-\lambda I)\otimes I + I \otimes S & \lambda I \otimes I & 0 \\ 0 & 0 & I \otimes S^o B^o & (L-\lambda I)\otimes I + I \otimes S & \lambda I \otimes I \\ 0 & 0 & 0 & I \otimes S^o B^o & (L-\lambda I)\otimes I + I \otimes S \end{vmatrix},$$

$$A_2 = \begin{vmatrix} L^o \Gamma^o & 0 & 0 & 0 & 0 \\ 0 & L^o \Gamma^o \otimes I & 0 & 0 & 0 \\ 0 & 0 & L^o \Gamma^o \otimes I & 0 & 0 \\ 0 & 0 & 0 & L^o \Gamma^o \otimes I & 0 \\ 0 & 0 & 0 & 0 & L^o \Gamma^o \otimes I \end{vmatrix}.$$

It is clear that, even for small values of K, n, and r, the particular structure of these matrices and of the matrix R need to be exploited in order to make the algorithmic solution of this model both feasible and efficient. For guidelines, we refer to the simpler case, discussed earlier. The overflow process from the $M/PH/1$ queue is also a particular case of the versatile Markovian point process, introduced in [232]. It is possible, in principle, to treat this overflow stream as the input to a single server unit with a *general* service time distribution by the techniques of V. Ramaswami [261]. However, for larger values of Kn, this represents a substantial computational effort.

6.5. TWO MODELS IN WHICH THE MATRIX A IS TRIANGULAR

The two models discussed in this section are related to interesting applications and may be viewed as queues in a random environment. In addition, they offer examples of block-partitioned matrices for which the matrix A is upper- or lower-triangular. The results, obtained in section 1.4, are germane to their solution.

The *first model* is for a queue in which customers who do not have to wait are served at a different rate than those who do not gain immediate access to a server. For single server queues, such as the $M/G/1$ and $GI/M/1$, this is a trivial modification. For multi-server queues, some interesting issues arise, even under the assumption of exponential services. Discussions of the $M/M/c$ queue, in which nonwaiting customers are served at a rate μ_0 and waiting customers at a rate μ_1, may be found in P. H. Brill [36, 37] and P. H. Brill and M. J. M. Posner [38]. These discussions are based on the *System Point Method,* developed in [36], or on transform techniques. In D. M. Lucantoni [195], it is shown that the $GI/M/c$ queue with the same distinction between customers is amenable to analysis by the methods presented in this book. For general interarrival time distributions, a fair amount of effort is needed to define and compute the matrices A_k, $k \geq 0$, and the boundary matrices that arise in the matrix $\tilde{P}$ of the Markov chain embedded at arrival epochs. We refer to [195] for the details and also for numerical examples.

This model is an instance of a queue in random environment. The presence of a number of nonwaiting customers causes the overall service rate to vary randomly. When μ_0 is much smaller than μ_1, the queue will exhibit large random fluctuations similar to those discussed for the preceding models. This may arise when a customer arriving at an idle server needs to spend some time activating the service unit.

The $M/M/c$ version of the model is described by a QBD process with generator $\tilde{Q}$, given by

$$\tilde{Q} = \begin{vmatrix} A_{01} & A_{02} & & & & & & \\ A_{10} & A_{11} & A_{12} & & & & & \\ & A_{20} & A_{21} & A_{22} & & & & \\ & & & \cdots & & & & \\ & & & A_{c,0} & A_{c,1} & A_0 & & \\ & & & & A_2 & A_1 & A_0 & \\ & & & & & A_2 & A_1 & A_0 \\ & & & & & & & \cdots \end{vmatrix}. \qquad (6.5.1)$$

The state space is given by $E = \{(i, j), i \geq 0, 0 \leq j \leq \min(i, c)\}$. The index i denotes the total number of customers in the system, and the index j the number of customers *in service, who have had to wait.*

The matrices A_0, A_1, and A_2 are of order $c + 1$ and are given by

$$A_0 = \lambda I,$$

$$A_1 = -\mathrm{diag}[\lambda + c\mu_0, \lambda + \mu_1 + (c-1)\mu_0, \lambda + 2\mu_1 + (c-2)\mu_0, \ldots, \lambda + c\mu_1]$$

$$A_2 = \begin{vmatrix} 0 & c\mu_0 & 0 & 0 & \cdots & 0 & 0 \\ 0 & \mu_1 & (c-1)\mu_0 & 0 & \cdots & 0 & 0 \\ 0 & 0 & 2\mu_1 & (c-2)\mu_0 & \cdots & 0 & 0 \\ \vdots & \vdots & \vdots & \vdots & & \vdots & \vdots \\ 0 & 0 & 0 & 0 & \cdots & (c-1)\mu_1 & \mu_0 \\ 0 & 0 & 0 & 0 & \cdots & 0 & c\mu_1 \end{vmatrix}. \qquad (6.5.2)$$

The boundary matrices are defined by

$$A_{01} = -\lambda, \qquad A_{02} = \lambda,$$

$$A_{k,1} = -\mathrm{diag}[\lambda + k\mu_0, \lambda + \mu_1 + (k-1)\mu_0, \ldots, \lambda + k\mu_1],$$

$$A_{k,0} = \begin{vmatrix} k\mu_0 & 0 & 0 & \cdots & 0 & 0 \\ \mu_1 & (k-1)\mu_0 & 0 & \cdots & 0 & 0 \\ 0 & 2\mu_1 & (k-2)\mu_0 & 0 & 0 & \\ \vdots & \vdots & \vdots & & \vdots & \vdots \\ 0 & 0 & 0 & \cdots & (k-1)\mu_1 & \mu_0 \\ 0 & 0 & 0 & \cdots & 0 & k\mu_1 \end{vmatrix}, \qquad (6.5.3)$$

for $1 \le k \le c$,

$$A_{k,2} = \begin{vmatrix} \lambda & 0 & 0 & \cdots & 0 & 0 \\ 0 & \lambda & 0 & \cdots & 0 & 0 \\ 0 & 0 & \lambda & \cdots & 0 & 0 \\ \cdot & \cdot & \cdot & & \cdot & \cdot \\ \cdot & \cdot & \cdot & & \cdot & \cdot \\ \cdot & \cdot & \cdot & & \cdot & \cdot \\ 0 & 0 & 0 & \cdots & 0 & 0 \\ 0 & 0 & 0 & \cdots & \lambda & 0 \end{vmatrix},$$

for $1 \le k \le c - 1$. We note that $A_{k,0}$ is of dimensions $(k + 1) \times k$ and $A_{k,2}$ of dimensions $(k + 1) \times (k + 2)$.

By adapting the result of theorem 1.4.1 to the present case, we see that the irreducible Markov process $\tilde{Q}$ is positive recurrent if and only if

$$\frac{(A_0)_{cc}}{(A_2)_{cc}} = \frac{\lambda}{c\mu_1} < 1. \tag{6.5.4}$$

The stationary probability vector $x = [y_0, y_1, \ldots, y_{c-1}, x_0, x_1, \ldots]$ of $\tilde{Q}$ satisfies

$$x_k = x_0 R^k, \quad \text{for} \quad k \ge 0. \tag{6.5.5}$$

The matrix R is upper-triangular and is the minimal nonnegative solution to the equation

$$R^2 A_2 + RA_1 + A_0 = 0. \tag{6.5.6}$$

All its elements on or above the diagonal are positive. We shall show in what follows how the matrix R may also be computed by an exact, non-iterative algorithm.

The numerical solution of the system

$$\begin{gathered} y_0 A_{01} + y_1 A_{10} = 0 \\ y_{r-1} A_{r-1,2} + y_r A_{r,1} + y_{r+1} A_{r+1,0} = \mathbf{0}, \quad \text{for} \quad 1 \le r \le c - 1, \\ y_{c-1} A_{c-1,2} + x_0 (A_{c,1} + RA_2) = \mathbf{0}, \\ y_0 + y_1 e + \cdots + y_{c-1} e + x_0 (I - R)^{-1} e = 1, \end{gathered} \tag{6.5.7}$$

is entirely routine. By judicious use of structure, its solution remains feasible for large values of c.

As discussed in detail in [195], the virtual waiting time distribution $W(\cdot)$ for this model and for its $GI/M/c$ version are easily computed by implementing the efficient algorithm discussed in section 3.9. The details of the algorithm are as given in theorem 3.9.2.

The computation of the mean waiting time presents a minor difficulty whose discussion is of interest to this and also to the next model.

The probability that a customer does not have to wait is given by

$$W(0) = y_0 + y_1 e + \cdots + y_{c-1} e, \tag{6.5.8}$$

and the Laplace-Stieltjes transform $w(s)$ of $W(\cdot)$ is given by

$$w(s) = W(0) + \sum_{k=0}^{\infty} x_0 R^k [(sI - D)^{-1} A_2]^{k+1} e, \tag{6.5.9}$$

where $D = -\text{diag}[c\mu_0, (c-1)\mu_0 + \mu_1, (c-2)\mu_0 + 2\mu_1, \ldots, c\mu_1]$.

The mean waiting time E_W is now given by

$$E_W = -w'(0) = x_0 \sum_{k=0}^{\infty} R^k \sum_{r=0}^{k} [-D^{-1}A_2]^r(-D^{-1}e). \tag{6.5.10}$$

The matrix $K = -D^{-1}A_2$ is upper-triangular, and the element in its right-hand lower corner is equal to one. A generalized inverse of the matrix $I - K$ may be simply obtained as follows. Let U be a matrix of order $c + 1$, given by $U = [\mathbf{0}, \mathbf{0}, \ldots, \mathbf{0}, e]$. The matrix $I - K + U$ is now clearly nonsingular and

$$\sum_{r=0}^{k} K^r(I - K + U) = I - K^{k+1} + (k+1)U, \quad \text{for} \quad k \geq 0. \tag{6.5.11}$$

Upon substitution in (6.5.10), we obtain

$$E_W = x_0[(I - R)^{-1} - \sum_{k=0}^{\infty} R^k K^{k+1} + (I - R)^{-2} U] \cdot (I - K + U)^{-1}(-D^{-1}e). \tag{6.5.12}$$

As shown in [195], the matrix $Y = \sum_{k=0}^{\infty} R^k K^{k+1}$ may also be efficiently computed by expressing the equation

$$Y = K + RYK \tag{6.5.13}$$

as a highly structured system of $(c + 1)^2$ linear equations by the use of the Kronecker product.

The *second model* is for a very interesting queue that arises in telecommunication studies. It deals with a c-channel system carrying both voice and data traffic. It was discussed by different methods in U. N. Bhat and M. J. Fischer [27]. Its numerical solution, even for large values of c (for example, up to one hundred), is entirely feasible and does not require extraordinary effort.

There are c parallel servers (transmission channels) and two Poisson arrival streams of rates λ_1 and λ_2 respectively. With respect to the second stream, the system operates as a *loss system*. A customer who cannot start service immediately upon arrival is lost. Customers from the first stream are allowed to queue up for service. They correspond to data packages that may be stored for transmission as soon as a channel becomes available. Customers in the second stream correspond to voice transmissions.

The service times are assumed to be mutually independent and exponentially distributed with parameters μ_1 and μ_2, depending on whether the customers come from the first or the second arrival stream. We note that, for consistency of notation, we have reversed the use of the indices 1 and 2 from that in [27].

This model again leads to a QBD process with the state space $E = \{(i, j): i \geq 0, 0 \leq j \leq c\}$. The index i is the number of customers of type 1 in the system. The index j denotes the number of channels serving customers of type 2.

The matrix $\tilde{Q}$ is of the form

$$\tilde{Q} = \begin{vmatrix} A_{01} & A_{02} & & & & & & \\ A_{10} & A_{11} & A_{12} & & & & & \\ & A_{20} & A_{21} & A_{22} & & & & \\ & & & \cdots & & & & \\ & & & A_{c-1,0} & A_{c-1,1} & A_0 & & \\ & & & & A_2 & A_1 & A_0 & \\ & & & & & A_2 & A_1 & \cdots \\ & & & & & & \cdots & \end{vmatrix}, \tag{6.5.14}$$

where all blocks are square matrices of order $c + 1$. The matrices $A_{\nu 2}$, $0 \leq \nu \leq c - 2$, and A_0 are all equal to $\lambda_1 I$. The matrix A_2 is a diagonal matrix, with diagonal entries $c\mu_1, (c-1)\mu_1, \ldots, \mu_1, 0$. The matrix A_1 is zero, except for the elements on the diagonal and the line immediately below the diagonal. These elements are given by

$$-\lambda_1 - c\mu_1, -\lambda_1 - (c-1)\mu_1 - \mu_2, \ldots,$$
$$-\lambda_1 - \mu_1 - (c-1)\mu_2, -\lambda_1 - c\mu_2$$

and

$$\mu_2, 2\mu_2, \ldots, (c-1)\mu_2, c\mu_2$$

respectively.

For $1 \leq \nu \leq c - 1$, the matrix $A_{\nu 0}$ is diagonal of order $c + 1$, with its diagonal elements given by $\nu\mu_1, \ldots, \nu\mu_1, (\nu-1)\mu_1, (\nu-2)\mu_1, \ldots, \mu_1, 0$.

To see this, we note that all ν customers of type 1 are in service if and only if there are at most $c - \nu$ channels occupied by customers of type 2. If j, with $c - \nu < j \leq c$, channels are busy with customers of type 2, then $c - j$ of the ν customers of type 1 are in service, and the remaining ones are waiting.

The matrices $A_{\nu 1}$, $0 \leq \nu \leq c - 1$, are Jacobi matrices of order $c + 1$. The elements on their inferior diagonal are given by $\mu_2, 2\mu_2, \ldots, c\mu_2$. The elements on the superior diagonal are equal to λ_2 in the rows labeled $0, 1, \ldots, c - \nu - 1$, and are zero in the remaining rows. The diagonal elements d_j of $A_{\nu 1}$ are chosen so that the row sums of $\tilde{Q}$ vanish and are therefore explicitly given by

$$\begin{aligned} d_j &= -(\lambda_1 + \lambda_2 + \nu\mu_1 + j\mu_2), \quad \text{for} \quad 0 \leq j \leq c - \nu - 1, \\ &= -[\lambda_1 + (c - j)\mu_1 + j\mu_2], \quad \text{for} \quad c - \nu \leq j \leq c. \end{aligned}$$

It follows from theorem 1.4.1 that the queue is *stable* if and only if

$$\lambda_1 < c\mu_1. \tag{6.5.15}$$

The numerical solution of this model proceeds exactly as in the preceding model, except that the solution of the equations

$$\begin{aligned} \boldsymbol{x}_0 A_{01} + \boldsymbol{x}_1 A_{10} &= \mathbf{0}, \\ \lambda_1 \boldsymbol{x}_{i-1} + \boldsymbol{x}_i A_{i1} + \boldsymbol{x}_{i+1} A_{i+1,0} &= \mathbf{0}, \quad \text{for} \quad 1 \leq i \leq c - 2, \\ \lambda_1 \boldsymbol{x}_{c-2} + \boldsymbol{x}_{c-1}(A_{c-1,1} + RA_2) &= \mathbf{0}, \\ \sum_{i=0}^{c-2} \boldsymbol{x}_i \boldsymbol{e} + \boldsymbol{x}_{c-1}(I - R)^{-1}\boldsymbol{e} &= 1, \end{aligned} \tag{6.5.16}$$

may be organized in a particularly efficacious manner due to the sparsity of the coefficient matrices.

The lower-triangular matrix R may be exactly computed in a finite number of steps as follows. The equation $R^2 A_2 + RA_1 + A_0 = 0$ leads to

$$(c - j)\mu_1 R_{jj}{}^2 - [\lambda_1 + (c - j)\mu_1 + j\mu_2]R_{jj} + \lambda_1 = 0, \tag{6.5.17}$$

for $0 \leq j \leq c$. Provided $\lambda_1 < c\mu_1$, the unique solutions to these quadratic equations in $(0, 1)$ are given by

$$\begin{aligned} R_{00} &= \lambda_1 (c\mu_1)^{-1}, \\ R_{jj} &= [2(c - j)\mu_1]^{-1}\{\lambda_1 + (c - j)\mu_1 + j\mu_2 \\ &\qquad - [(\lambda_1 + (c - j)\mu_1 + j\mu_2)^2 - 4(c - j)\lambda_1\mu_1]^{1/2}\}, \\ &\qquad \text{for} \quad 1 \leq j \leq c - 1, \\ R_{cc} &= \lambda_1(\lambda_1 + c\mu_2)^{-1}. \end{aligned} \tag{6.5.18}$$

Once the diagonal elements of R are known, the elements below the diagonal may be computed by recursively solving linear equations, since for $j > \nu$, we have

$$[R_{j\nu}(R_{\nu\nu} + R_{jj}) + \sum_{\nu<r<j} R_{jr}R_{r\nu}](A_2)_{\nu\nu}$$
$$+ R_{j\nu}(A_1)_{\nu\nu} + R_{j,\nu+1}(A_1)_{\nu+1,\nu} = 0. \quad (6.5.19)$$

For $j = \nu + 1$, $0 \leq \nu \leq c - 1$, we obtain

$$R_{\nu+1,\nu}[\lambda_1 + (c - \nu)\mu_1 + \nu\mu_2 - (c - \nu)\mu_2(R_{\nu\nu} + R_{\nu+1,\nu+1})]$$
$$= (\nu + 1)\mu_2 R_{\nu+1,\nu+1}. \quad (6.5.20)$$

We see that the quantities $R_{\nu+k,\nu}$, $0 \leq \nu \leq c - k$, may now be recursively computed for $1 \leq k \leq c$. At each stage only simple linear equations need to be solved.

The mean and the distribution of the virtual waiting time in the stationary queue may be computed as in [195]. It would be of interest to compare the heuristic approximation to the mean waiting time, proposed in [27], to its exact value for a wide range of parameter values. This model may again be viewed as a queue in a Markovian environment. The number of channels occupied by voice traffic *modulates* the total service rate for data traffic. As in earlier cases, the overall mean queue length or waiting time are not desirable descriptors of the behavior of this queue. When λ_1 and μ_1 are much larger than λ_2 and μ_2, it is to be expected that the data queue will be highly variable. For realistic parameter choices, excursions of the queue length into the thousands may arise. Numerical studies of this model are promising; they may suggest additional approximations and possibly new useful descriptors of highly oscillatory queues.

6.6. A FINITE-SOURCE PRIORITY QUEUE

A useful application of the $M/M/1$ queue in a Markovian environment is discussed in J. P. Colard [62]. A related model is discussed by S. Halfin [124] and S. Halfin and M. Segal [125]. One considers a finite internal source of N customers and an independent Poisson arrival stream of rate λ_1 of external customers. Each internal customer who is not in service issues a request for service after an exponentially distributed length of time of mean $1/\lambda_2$.

The service times of both types of customers are exponentially distributed, respectively with parameters μ_1 for external and μ_2 for internal customers.

The internal customers have *preemptive-resume priority* over the external customers. The classical independence assumptions on interarrival and service times are imposed. This model is proposed as a (simplified) description for the following two situations. In the first, the internal customers are thought of as N active computer terminals transmitting typically short jobs to a computer. The external customers are then batch jobs, requiring mostly longer processing times. In the second imaginative application, the server is viewed as a specialized technician who attends to N machines owned by a corporation. He may also service similar machines for customers outside, but only whenever none of the first N machines is down.

We shall present the definition of the matrices and the main results on this model only for the case of a single server. The extension to c exponential servers is routine.

The model leads to a QBD process on the state space $E = \{(i, j): i \geq 0, 0 \leq j \leq N\}$. The index i denotes the number of external customers in the system. The index j represents the number of internal customers requesting service. When the second index assumes the value j, the arrival rate of internal customers is given by $(N - j)\lambda_2$.

The matrix $\tilde{Q}$ is given by

$$\tilde{Q} = \left|\begin{array}{ccccc} A_2 + A_1 & A_0 & 0 & 0 & \cdots \\ A_2 & A_1 & A_0 & 0 & \cdots \\ 0 & A_2 & A_1 & A_0 & \cdots \\ 0 & 0 & A_2 & A_1 & \cdots \\ \vdots & \vdots & \vdots & \vdots & \end{array}\right|, \tag{6.6.1}$$

where A_0, A_1, A_2 are matrices of order $N + 1$. The matrix A_2 is zero, except for the entry μ_1 in the upper left-hand corner. The matrix A_1 is a Jacobi matrix with diagonal elements

$$-\lambda_1 - N\lambda_2 - \mu_1, -\lambda_1 - (N - 1)\lambda_2 - \mu_2,$$
$$-\lambda_1 - (N - 2)\lambda_2 - \mu_2, \ldots, -\lambda_1 - \lambda_2 - \mu_2, -\lambda_1 - \mu_2.$$

The elements on the superior diagonal are given by $N\lambda_2$, $(N - 1)\lambda_2$, $\ldots$, λ_2, and all elements on the inferior diagonal are equal to μ_2. The matrix A_0 is equal to $\lambda_1 I$.

The stationary probability vector $\boldsymbol{\pi}$ of $A = A_0 + A_1 + A_2$ is readily computed. One obtains

$$\pi_0 = \left[\sum_{k=0}^{N} \frac{N!}{(N-k)!} \left(\frac{\lambda_2}{\mu_2} \right)^k \right]^{-1},$$

$$\pi_j = \frac{N!}{(N-j)!} \left(\frac{\lambda_2}{\mu_2} \right)^j \pi_0, \quad \text{for} \quad 1 \leq j \leq N. \tag{6.6.2}$$

The equilibrium condition $\pi A_0 e < \pi A_2 e$ reduces to

$$\lambda_1 < \mu_1 \pi_0. \tag{6.6.3}$$

It also readily follows from the particular form of the boundary matrices in (6.6.1) that the stationary probability vector $x = [x_0, x_1, x_2, \ldots]$ of $\tilde{Q}$ is given by

$$x_i = \pi(I - R)R^i, \quad \text{for} \quad i \geq 0. \tag{6.6.4}$$

The present model is clearly a particular case of the $M/M/1$ queue in a Markovian environment, treated in section 6.2. It corresponds to the choice of parameters

$$\begin{aligned} \lambda &= \lambda_1 e, \\ \mu &= [\mu_1, 0, 0, \ldots, 0] \\ Q &= A_1 + \lambda_1 I + A_2. \end{aligned} \tag{6.6.5}$$

Its fundamental period may be studied with particular ease. Since all the columns of A_2 but the first vanish, equation (3.3.19) implies that the matrix $\hat{G}(z, s)$ is also zero, except for its first column, which we shall denote by $\gamma(z, s)$. The first column of $\hat{G}^2(z, s)$ is then given by $\gamma_0(z, s)\gamma(z, s)$, and it follows from (3.3.19) that

$$z[\mu_1, 0, \ldots, 0]' - (sI - A_1)\gamma(z, s) + \lambda_1 \gamma_0(z, s)\gamma(z, s) = \mathbf{0}. \tag{6.6.6}$$

From this equation, a number of expressions for the moments of the fundamental and busy periods may be derived by routine differentiations.

6.7. QUEUES WITH PAIRED CUSTOMERS

In this section we consider queues in which each service involves two customers, one from each of two arrival streams. Such waiting lines are not queues in a random environment. They represent, however, a novel and challenging class of queueing problems. Although their investigation has only recently been undertaken, such results as are available are germane to the general subject of this book.

In the simplest case, we consider two independent Poisson arrival streams of constant rates λ_1 and λ_2 respectively. Arrivals form a queue in front of a single server, but each service consists of the processing of a pair of customers, one from each stream. When no pairs are present, no service may be provided.

A convenient representation of the state of the queue is as follows. At any time, the queue length is represented by a pair (X, Y) of random variables. X denotes the *number of pairs* of customers present; Y is the *excess* number of customers of one type or the other who are currently in the queue. The range of X is the set of nonnegative integers, and Y may assume both nonnegative and negative values. The event $\{Y = j\}$, with $j \geq 0$, signifies that an excess of j customers from the stream 1 are present. With $j < 0$, this event refers to the presence of an excess of $-j$ customers from stream 2.

With only service by pairs permitted, we readily see that the excess Y will tend to either $+\infty$ or $-\infty$, depending on whether $\lambda_1 > \lambda_2$ or $\lambda_2 < \lambda_1$, or it will exhibit the behavior typical of the symmetric random walk, discussed, for example, in W. Feller [90, chap. 3]. With exponentially distributed service times, the process $[X(t), Y(t), t \geq 0]$ is a Markov process on the state space $E = \{(i, j): i \geq 0, -\infty < j < +\infty\}$. This process can only be transient or null-recurrent, yet stationary probability distributions for the marginal, non-Markovian process $\{X(t), t \geq 0\}$ may be well defined. This interesting feature of queues with paired customers is worthy of further study. It does not depend significantly on the assumptions of Poisson arrivals or exponential service times, which were made here for ease of illustration. It is also interesting to note that the marginal process $\{Y(t), t \geq 0\}$ is none other than the familiar *double-ended queue*, which has been studied by a number of authors [83, 151, 157–60].

For a satisfactory algorithmic solution, it is clear that upper and lower bounds J_1 and $-J_2$, $J_1 > 0$, $J_2 > 0$, must be imposed on the excess. As is often the case, this not only avoids certain mathematical difficulties inherent in the model, but introduces an additional degree of realism into it. As we shall discuss, this truncation also suggests how more complicated models with additional design variables may be investigated.

We now consider in detail the simplest model in which the services are exponentially distributed with parameter μ and in which the excess is constrained by $-J_2 \leq Y(t) \leq J_1$. When $Y(t)$ reaches its upper (respectively lower) bound, stream 1 (respectively stream 2) is interrupted until the excess no longer assumes the value specified by the bound.

The arrival process may also be viewed as a particular case of the point process, discussed in M. F. Neuts [232]. We consider the Markov process with generator Q, given by

$$Q = \begin{vmatrix} -\lambda_1 & \lambda_1 & & & & & \\ \lambda_2 & -\lambda_1 - \lambda_2 & \lambda_1 & & & & \\ & \lambda_2 & -\lambda_1 - \lambda_2 & \lambda_1 & & & \\ & & & \cdots & & & \\ & & & & \lambda_2 & -\lambda_1 - \lambda_2 & \lambda_1 \\ & & & & & \lambda_2 & -\lambda_2 \end{vmatrix}, \tag{6.7.1}$$

on the state space $\{-J_2, 1 - J_2, \ldots, -1, 0, 1, \ldots, J_1 - 1, J_1\}$. Transitions of the form $j \to j + 1$, for $-J_2 \leq j \leq -1$, or $j \to j - 1$, for $1 \leq j \leq J_1$, and only these transitions correspond to the formation of a pair. All other transitions correspond to a change in the excess only. From this viewpoint, the arrival process of pairs is a particular case of a Markovian input process as discussed in section 4.2b. Only selected transitions correspond to arrivals of pairs. It is clear that the same analysis as that which follows also applies to the case where the arrival rates are allowed to depend on the state of the process Q. The details of this model with excess-dependent arrival rates are given in G. Latouche [181].

The single server queue with exponential service is now readily studied as a QBD process on the state space $E = \{(i, j), i \geq 0, -J_2 \leq j \leq J_1\}$. The index i denotes the number of *pairs* in the system. The generator $\tilde{Q}$ is given by

$$\tilde{Q} = \begin{vmatrix} A_2 + A_1 & A_0 & & \\ A_2 & A_1 & A_0 & \\ & A_2 & A_1 & A_0 \\ & & \cdots & \end{vmatrix}, \tag{6.7.2}$$

where all blocks are matrices of order $J_1 + J_2 + 1$. The matrix $A_2 = \mu I$ and the matrices A_1 and A_0 are Jacobi matrices constructed as follows. The diagonal elements of A_1 are given by $-\mu - \lambda_1, -\mu - \lambda_1 - \lambda_2, \ldots, -\mu - \lambda_1 - \lambda_2, -\mu - \lambda_2$. On the inferior diagonal, the element λ_2 occurs in the rows with indices $1 - J_2, \ldots, -1, 0$. On the superior diagonal, the element λ_1 occurs in the rows labeled $0, 1, \ldots, J_1 - 1$. All other elements are zero. The diagonal elements of A_0 are zero. The element λ_1 occurs on the superior diagonal in the rows labeled $-J_2, \ldots, -1$. The element λ_2 occurs on the inferior diagonal in the rows labeled

$1, \ldots, J_1$. All other elements are zero. We note in particular that in the matrix A_0, the row with index 0 vanishes.

Clearly $A_0 + A_1 + A_2 = Q$, and the stationary probability vector $\boldsymbol{\pi}$ of Q is given by

$$\pi(-J_2) = \left[\sum_{\nu=0}^{J_1+J_2} \left(\frac{\lambda_1}{\lambda_2} \right)^{\nu} \right]^{-1}$$

$$\pi(j) = \left(\frac{\lambda_1}{\lambda_2} \right)^{j+J_2} \pi(-J_2), \quad \text{for} \quad -J_2 \le j \le J_1. \tag{6.7.3}$$

The queue is stable if and only if

$$\mu > \lambda_1 \sum_{\nu=-J_2}^{-1} \pi(\nu) + \lambda_2 \sum_{\nu=1}^{J_1} \pi(\nu). \tag{6.7.4}$$

It is easily seen that the quantity on the right-hand side of (6.7.4) is the stationary *arrival rate of pairs.*

When (6.7.4) holds, the stationary probability vector $\boldsymbol{x} = [\boldsymbol{x}_0, \boldsymbol{x}_1, \boldsymbol{x}_2, \ldots]$ of $\tilde{Q}$ is given by

$$\boldsymbol{x}_i = \boldsymbol{\pi}(I - R)R^i, \quad \text{for} \quad i \ge 0. \tag{6.7.5}$$

The matrix R, the minimal solution of $R^2 A_2 + RA_1 + A_0 = 0$, is positive except for the row with index 0. As in the matrix A_0, this row also vanishes in R.

Remarks

a. For $\lambda_1 = \lambda_2 = \lambda$, the inequality (6.7.4) simplifies to

$$\mu > \lambda \left(1 - \frac{1}{J_1 + J_2 + 1} \right). \tag{6.7.6}$$

This suggests that as $J_1 + J_2 \to \infty$, the marginal density $\{\boldsymbol{x}_i \boldsymbol{e}, i \ge 0\}$ of the number of pairs in the system will approach a limit density, provided $\mu > \lambda$. The bivariate Markov process with unbounded excess cannot have a stationary probability density. If it did, the marginal process $\{Y(t), t \ge 0\}$, which is Markovian, would be ergodic, but this is clearly not the case.

This example raises interesting open problems regarding the existence of stationary measures for the queue with unbounded excess and regarding their relationship to the limit of the density $\{\boldsymbol{x}_i \boldsymbol{e}, i \ge 0\}$ as

$J_1 + J_2$ tends to infinity. The latter question is related to the behavior of R as the *order* of the matrices increases and is inherently difficult.

b. For $\lambda_1 > \lambda_2$, the inequality (6.7.4) may be rewritten as

$$\mu > \lambda_2 \frac{\left(\frac{\lambda_1}{\lambda_2}\right)^{J_1+J_2+1} - \left(\frac{\lambda_1}{\lambda_2}\right)}{\left(\frac{\lambda_1}{\lambda_2}\right)^{J_1+J_2+1} - 1}. \tag{6.7.7}$$

As $J_1 + J_2$ tends to infinity, the arrival rate of pairs approaches the smaller of the two rates λ_1 and λ_2. This is, of course, highly intuitive, and suggests that, provided $\mu > \min(\lambda_1, \lambda_2)$, the number of pairs in the system will have a stationary probability density. With an overabundance of customers from stream 1, the queue *of pairs* should eventually behave as an $M/M/1$ queue with arrival rate λ_2 and service rate μ.

Generalizations. The simple queue with paired customers and exponential service has a large number of possible generalizations of interest. For some of these a detailed algorithmic solution is possible, but others lead to a state description of such a high dimensionality that their analysis remains prohibitive.

a. In addition to the bounds J_1 and $-J_2$, we may introduce parameters $K_1 \geq 0$ and $K_2 \geq 0$, such that $-J_2 < -K_2 \leq 0 \leq K_1 < J_1$, with the following significance. When the excess reaches the bound J_1, the input stream 1 is shut off. It is not started up again until the excess drops to the value K_1. The parameter K_2 has a similar significance with respect to the second input stream. Our original model then corresponds to the choices $K_1 = J_1 - 1$, $K_2 = J_2 - 1$. It may be advantageous to choose smaller values of K_1 and K_2, particularly when there are costs associated with each interruption of an arrival stream.

With exponentially distributed service times, this more versatile model may again be studied as a QBD process of the form (6.7.2), but with a slightly more complicated state description. To each index j satisfying $-J_2 < j < -K_2$, or $K_1 < j < J_1$, there now correspond two states in the Markov process Q. We shall denote these two states by j and $\bar{j}$ respectively. The index j signifies that both arrival streams are uninterrupted, while the index $\bar{j}$ indicates that one of the arrival streams is interrupted. (Which one is clear from the value of j.) With this "splitting" of appropriate states, the arrival process may again be considered as a Markovian input in which certain transitions lead to pair formation. This is similar to the state description needed in model C of section 5.2.

For the sake of illustration, we display the matrix Q corresponding to $J_1 = 5, K_1 = 1, K_2 = 3, J_2 = 6$.

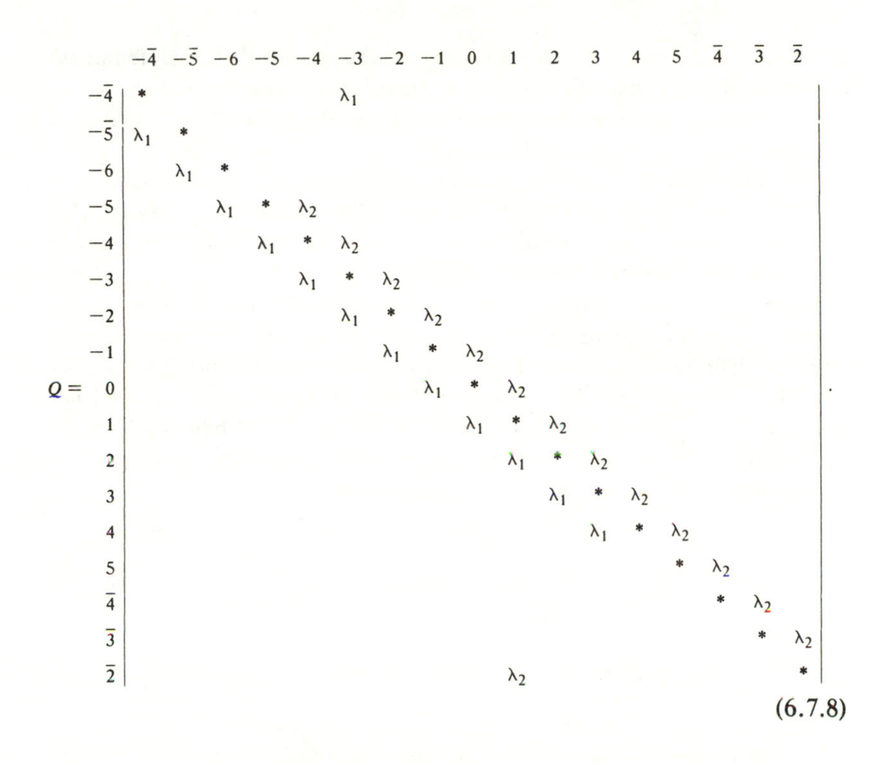

The diagonal elements, represented for ease of display by asterisks, are such that the row sums of Q are equal to zero. The matrices A_0, A_1, and A_2 are constructed in a similar manner, and the analysis proceeds exactly as in the simpler case discussed earlier.

b. The generalization of the model in (*a*) to c exponential servers is routine and results in a QBD process with more complicated boundary behavior. A discussion may be found in G. Latouche [181].

c. The case of paired customers from two Poisson streams queueing at a single server with *general* service time distribution may be solved, even for the more complex model described in (*a*) above, by implementing the general approach in V. Ramaswami [261]. A Markov chain of the $M/G/1$ type is embedded at departure epochs. The overhead computations needed to evaluate the blocks arising in the partitioned stochastic matrix $\tilde{P}$ of the embedded chain are quite substantial and require careful programming. Once this is accomplished, the main algorithm itself is straightforward. Its discussion falls outside the scope of this book.

d. The case where $r_1 \geq 1$ and $r_2 \geq 1$ customers from streams 1 and 2 are required to initiate a service may *in principle* be solved by the same methods of analysis as just discussed. The state description, however, grows so rapidly in complexity that an actual algorithmic implementation is of prohibitively high dimension. Although the formal structure of the solution is clear, practical numerical methods are presently not available except for very small values of r_1 and r_2.

e. The sensitivity of the solution of the preceding models to the assumption of Poisson arrival processes is well worth investigating. Useful algorithms may be developed for the case where the input streams are independent PH-renewal processes.

We shall present some detail on the generalization of the simple model to such input streams. The streams 1 and 2 are PH-renewal processes with interarrival time distributions having representations (α, T) and (β, S) respectively. The matrices T and S are of order m_1 and m_2 respectively, $\alpha_{m_1+1} = \beta_{m_2+1} = 0$, and T^oA^o and S^oB^o have their usual significance. The parameters J_1 and J_2 play the same roles as before.

If we define the matrices C_0, C_1, and C_2 by

$$\begin{aligned} C_0 &= I \otimes S^oB^o, \\ C_1 &= T \otimes I + I \otimes S, \\ C_2 &= T^oA^o \otimes I, \end{aligned} \tag{6.7.9}$$

then the analogue of the matrix Q of formula (6.7.1) is given by

$$Q = \left|\begin{array}{cccccc} C_0 + C_1 & C_2 & & & & \\ C_0 & C_1 & C_2 & & & \\ & C_0 & C_1 & C_2 & & \\ & & \cdots & & & \\ & & & C_0 & C_1 & C_2 \\ & & & & C_0 & C_1 + C_2 \end{array}\right|. \tag{6.7.10}$$

The states of the Markov chain Q are represented by the triplets (j, ν_1, ν_2), $-J_2 \le j \le J_1$, $1 \le \nu_1 \le m_1$, $1 \le \nu_2 \le m_2$, written in lexicographic order. The construction of the matrices A_0, A_1, and A_2 from Q is completely analogous to the simple case. These matrices are, however, now of order $(J_1 + J_2 + 1)m_1m_2$. The formal solution of the queue with exponential service is similar to that of the simple case. The m_1m_2 rows of A_0 and R, corresponding to the indices $(0, \nu_1, \nu_2)$, $1 \le \nu_1 \le m_1$, $1 \le \nu_2 \le m_2$, are now equal to zero.

The stationary density of the matrix $\tilde{Q}$ has an interesting invariance property, which follows from the particular form of the matrix Q. The joint stationary marginal density of the number of pairs in the system and of the phases of both arrival processes depends on J_1 and J_2 *only* through their sum $J_1 + J_2$, that is, in two queueing models with $-J_2 < 0 < J_1$ and $-J_2' < 0 < J_1'$ respectively, for which $J_1 + J_2 = J_1' + J_2'$, the joint marginal densities of the number of pairs and the two arrival phases are identical.

The Stationary Probability Vector of Q. The computation of the stationary probability vector $\boldsymbol{\pi}$ of Q is of some independent interest. It provides us with the joint steady-state probabilities of the queue length and the two arrival phases in a *bounded double-ended queue* with PH-renewal input processes. Once the vector $\boldsymbol{\pi}$ is known, many other steady-state distributions of the bounded double-ended queue may easily be computed. Except for the particular case discussed in the following, the vector $\boldsymbol{\pi}$ cannot be explicitly computed.

It is convenient to index the rows of blocks in the matrix Q, given in (6.7.10), by $0, 1, \ldots, J = J_1 + J_2$, and to partition the vector $\boldsymbol{\pi}$ accordingly into $m_1 m_2$-vectors $\boldsymbol{\pi}(0), \boldsymbol{\pi}(1), \ldots, \boldsymbol{\pi}(J)$. The equation $\boldsymbol{\pi} Q = \mathbf{0}$ then leads to

$$\boldsymbol{\pi}(0)[T \otimes I + I \otimes S + I \otimes S^o B^o] + \boldsymbol{\pi}(1)(I \otimes S^o B^o) = \mathbf{0},$$

$$\boldsymbol{\pi}(j-1)(T^o A^o \otimes I) + \boldsymbol{\pi}(j)[T \otimes I + I \otimes S] + \boldsymbol{\pi}(j+1)(I \otimes S^o B^o) = \mathbf{0},$$
$$\text{for} \quad 1 \leq j \leq J-1,$$

$$\boldsymbol{\pi}(J-1)(T^o A^o \otimes I) + \boldsymbol{\pi}(J)[T \otimes I + T^o A^o \otimes I + I \otimes S] = \mathbf{0}. \tag{6.7.11}$$

Upon addition of the $J + 1$ equations in (6.7.11), we obtain

$$\sum_{\nu=0}^{J} \boldsymbol{\pi}(\nu)[(T + T^o A^o) \otimes I + I \otimes (S + S^o B^o)] = \mathbf{0}. \tag{6.7.12}$$

This, together with the normalizing equation $\boldsymbol{\pi}\mathbf{e} = 1$, implies that

$$\sum_{\nu=0}^{J} \boldsymbol{\pi}(\nu) = \boldsymbol{\theta} \otimes \boldsymbol{\phi}, \tag{6.7.13}$$

where the probability vectors $\boldsymbol{\theta}$ and $\boldsymbol{\phi}$ are uniquely determined by the equations

$$\boldsymbol{\theta}(T + T^o A^o) = \mathbf{0}, \qquad \boldsymbol{\theta}\mathbf{e} = 1,$$

$$\boldsymbol{\phi}(S + S^o B^o) = \mathbf{0}, \qquad \boldsymbol{\phi}\mathbf{e} = 1. \tag{6.7.14}$$

It appears to be most convenient for the application at hand to solve the system (6.7.11) for successively increasing values of $J \geq 2$ by successive substitution. After a small amount of overhead computation and after rewriting the equations (6.7.11) in a more convenient form, the successive substitutions may be performed in a particularly efficacious manner.

To this effect, we partition the row vectors $\pi(j)$ further into m_2-vectors $\pi_1(j)$, $\pi_2(j)$, $\ldots$, $\pi_{m_1}(j)$. The equations (6.7.11) may then be written as

$$\pi_r(0)(T_{rr}I + S) + \sum_{\substack{\nu=1\\ \nu\neq r}}^{m_1} \pi_\nu(0)T_{\nu r} + [\pi_r(0)S^o + \pi_r(1)S^o]\beta$$

$$= \mathbf{0}, \quad \text{for} \quad 1 \le r \le m_1,$$

$$\pi_r(j)(T_{rr}I + S) + \alpha_r \sum_{\nu=1}^{m_1} \pi_\nu(j-1)T_\nu{}^o + \sum_{\substack{\nu=1\\ \nu\neq r}}^{m_1} \pi_\nu(j)T_{\nu r} + [\pi_r(j+1)S^o]\beta$$

$$= \mathbf{0}, \quad \text{for} \quad 1 \le r \le m_1, \qquad 1 \le j \le J-1,$$

$$\pi_r(J)(T_{rr}I + S) + \sum_{\substack{\nu=1\\ \nu\neq r}}^{m_1} \pi_\nu(J)T_{\nu r} + \alpha_r \sum_{\nu=1}^{m_1} [\pi_\nu(J-1) + \pi_\nu(J)]T_\nu{}^o$$

$$= \mathbf{0}, \quad \text{for} \quad 1 \le r \le m_1. \tag{6.7.15}$$

The m_1 inverses $K_r = -(T_{rr}I + S)^{-1}$, $1 \le r \le m_1$, exist and are nonnegative matrices. In the form

$$\pi_r(0) = \sum_{\substack{\nu=1\\ \nu\neq r}}^{m_1} \pi_\nu(0)T_{\nu r} \cdot K_r + [\pi_r(0)S^o + \pi_r(1)S^o] \cdot \beta K_r,$$

$$\text{for} \quad 1 \le r \le m_1,$$

$$\pi_r(j) = \left[\sum_{\substack{\nu=1\\ \nu\neq r}}^{m_1} \pi_\nu(j)T_{\nu r} + \alpha_r \sum_{\nu=1}^{m_1} \pi_\nu(j-1)T_\nu{}^o\right]K_r \tag{6.7.16}$$

$$+ [\pi_r(j+1)S^o] \cdot \beta K_r,$$

$$\text{for} \quad 1 \le r \le m_1, \qquad 1 \le j \le J-1,$$

$$\pi_r(J) = \left[\sum_{\substack{\nu=1\\ \nu\neq r}}^{m_1} \pi_\nu(J)T_{\nu r} + \alpha_r \sum_{\nu=1}^{m_1} \pi_\nu(J-1)T_\nu{}^o + \alpha_r \sum_{\nu=1}^{m_1} \pi_\nu(J)T_\nu{}^o\right] K_r,$$

$$\text{for} \quad 1 \le r \le m_1,$$

the equations are particularly well suited for solution by iterative methods.

The m_1 matrices K_r and the corresponding vectors βK_r are computed only once.

It is indicated to let the distribution with the smaller of the two mean interarrival times correspond to the representation (α, T). As we increase J, this leads to rapid numerical stabilization of the vectors $\pi(j)$ with low indices j. These components of the vector π then typically contain most of the probability mass. The vector π corresponding to a given value of J may also be used as the starting solution for the next value of J. This results in major savings in computation time.

When one of the two interarrival time distributions is *exponential*, the vector π may be explicitly expressed in terms of an easily computed matrix. We shall give the details for the case where $T = \lambda_1$, $T^o = \lambda_1$, $\alpha = 1$. The solution for the alternate case is obtained by symmetry.

The equations (6.7.11) now become

$$\pi(0)(S + S^oB^o - \lambda_1 I) + \pi(1)S^oB^o = \mathbf{0},$$

$$\lambda_1\pi(j-1) + \pi(j)(S - \lambda_1 I) + \pi(j+1)S^oB^o = \mathbf{0}, \quad \text{for} \quad 1 \le j \le J-1,$$

$$\lambda_1\pi(J-1) + \pi(J)S = \mathbf{0}. \tag{6.7.17}$$

As in the proof of theorem 3.2.1, we obtain

$$\lambda_1\pi(j)B^{oo} = \pi(j+1)S^oB^o, \quad \text{for} \quad 0 \le j \le J-1, \tag{6.7.18}$$

and in turn

$$\pi(0) = [\pi(0)S^o]\beta(\lambda_1 I - \lambda_1 B^{oo} - S)^{-1},$$

$$\pi(j) = \pi(j-1)\lambda_1(\lambda_1 I - \lambda_1 B^{oo} - S)^{-1}, \quad \text{for} \quad 1 \le j \le J-1,$$

$$\pi(J) = \pi(J-1)(-\lambda_1 S^{-1}). \tag{6.7.19}$$

The equations (6.7.19) determine the vector π up to a multiplicative constant that is readily obtained from the normalizing equation.

It is also a matter of easy verification that in the general case, when $T = S$, and $\alpha = \beta$, the vectors $\pi(j)$ are given by

$$\pi(j) = \frac{1}{J+1}\,\theta \otimes \theta, \quad \text{for} \quad 0 \le j \le J, \tag{6.7.20}$$

where θ is defined in formula (6.7.14).

It is clear that in the computation of the matrix R, which is typically of high order, it is imperative to take all available special structure of the matrices A_0, A_1, and A_2 into account. How this may be done should at this stage be evident.

References†

[1] Adiri, I.; Hofri, M.; and Yadin, M. A multiprogramming queue. *J. Assoc. Comp. Mach.*, 20, 589-603, 1973.

[2] Ali Khan, M. S., and Gani, J. Infinite dams with inputs forming a Markov chain. *J. Appl. Prob.*, 5, 72-84, 1968.

[3] Al-Khayyal, F. A., and Gross, D. On approximating and bounding $GI/M/c$ queues. In *Algorithmic Methods in Probability,* TIMS Studies in the Management Sciences, no. 7. London: North-Holland Publishing Co., 233-45, 1977.

[4] Arora, K. L. Time dependent solution of the two server queue fed by general arrival and exponential service time distributions. *Opns. Res.*, 10, 327-34, 1962.

[5] Arora, K. L. Two-server bulk-service queueing process. *Opns. Res.*, 12, 286-294, 1964.

[6] Avi-Itzhak, B., and Yadin, M. A sequence of two queues with no intermediate queue. *Mgmt. Sci.*, 11, 553-64, 1965.

[7] Avi-Itzhak, B., and Heyman, D. P. Approximate queueing models for multiprogramming computer systems. *Opns. Res.*, 21, 1212-30, 1973.

[8] Avis, D. M. Computing waiting times in $GI/E_k/c$ queueing systems. In *Algorithmic Methods in Probability,* TIMS Studies in the Management Sciences, no. 7. London: North-Holland Publishing Co., 215-32, 1977.

† In addition to the references cited in the text, this bibliography also lists papers dealing with other aspects of the models under discussion in this book, or, in a few instances, with additional stochastic processes to which the proposed methodology could be fruitfully brought to bear. Every effort has been made to make the present list of references as complete and informative as possible. With the literature, notably on the theory of queues, being as extensive and widespread as it is, some omissions are both certain and wholly unintentional.

[9] Bailey, N. T. J. On queueing processes with bulk service. *J. Roy. Statist. Soc. Ser. B,* 16, 80–87, 1954.

[10] Baily, D. E., and Neuts, M. F. Algorithmic methods for multi-server queues with group arrivals and exponential servers. Tech. rept. no. 78/14. Applied Mathematics Institute, University of Delaware, Newark, 1978. (To appear in *Euro. J. of Oper. Res.*)

[11] Barnett, A. Optimal control of the $B/B/C$ queue. *Interfaces,* 8, 49–52, 1978.

[12] Barnett, V. A simple random walk on parallel axes moving at different rates. *J. Appl. Prob.,* 12, 466–76, 1975.

[13] Bellman, R. *Dynamic programming.* Princeton, N.J.: Princeton University Press, 1957.

[14] Bellman, R. *Introduction to matrix analysis.* New York: McGraw-Hill Book Co., 1960.

[15] Benson, F., and Cox, D. R. The productivity of machines requiring attention at random intervals. *J. Roy. Statist. Soc. Ser. B,* 13, 65–82, 1951.

[16] Benson, F. Further notes on the productivity of machines requiring attention at random intervals. *J. Roy. Statist. Soc. Ser. B,* 14, 200–10, 1952.

[17] Bhat, U. N. On single-server bulk-queueing processes with binomial input. *Opns. Res.,* 12, 527–33, 1964.

[18] Bhat, U. N. Imbedded Markov chain analysis of single server bulk queues. *J. Austral. Math. Soc.,* 4, 244–63, 1964.

[19] Bhat, U. N. Customer overflow in queues with finite waiting space. *Austral. J. Statist.,* 7, 15–19, 1965.

[20] Bhat, U. N. On a stochastic process occurring in queueing systems. *J. Appl. Prob.,* 2, 467–69, 1965.

[21] Bhat, U. N. The queue $GI/M/2$ with service rate depending on the number of busy servers. *Ann. Inst. Statist. Math.,* 18, 211–21, 1966.

[22] Bhat, U. N. Some explicit results for the queue $GI/M/1$ with group service. *Sankhyā,* ser. A, 29, 199–206, 1967.

[23] Bhat, U. N. *A study of the queueing systems* $M/G/1$ *and* $GI/M/1$. Lecture Notes in Opns. Res. and Math. Econ., no. 2. New York: Springer-Verlag, 1968.

[24] Bhat, U. N. Transient behavior of multi-server queues with recurrent input and exponential service times. *J. Appl. Prob.,* 5, 158–68, 1968.

[25] Bhat, U. N. Two measures for describing queue behavior. *Opns. Res.,* 20, 357–72, 1972.

[26] Bhat, U. N. Imbedded Markov chains in queueing systems $M/G/1$ and $GI/M/1$ with limited waiting room. *Metrika,* 22, 153–60, 1975.

[27] Bhat, U. N., and Fischer, M. J. Multichannel queueing systems with heterogeneous classes of arrivals. *Nav. Res. Logist. Quart.*, 23, 271-82, 1976.

[28] Bhat, U. N. The value of queueing theory: A rejoinder. *Interfaces*, 8, 27-28, 1978.

[29] Bogdanoff, J. L. A new cumulative damage model: Part 1. *J. Appl. Mech.*, 45, 246-50, 1978.

[30] Bogdanoff, J. L., and Krieger, W. A new cumulative damage model: Part 2. *J. Appl. Mech.*, 45, 251-57, 1978.

[31] Bogdanoff, J. L. A new cumulative damage model: Part 3. *J. Appl. Mech.*, 45, 733-39, 1978.

[32] Boudreau, P. E.; Griffin, J. S.; and Kac, M. An elementary queueing problem. *Amer. Math. Monthly*, 69, 713-24, 1962.

[33] Boxma, O. J.; Cohen, J. W.; and Huffels, N. Approximations of the mean waiting time in an $M/G/s$ queueing system. *Opns. Res.*, 27, 1115-27, 1979.

[34] Boxma, O. J., and Konheim, A. G. Approximate analysis of exponential queueing systems with blocking. *Acta Informatica*, 15, 19-66, 1981.

[35] Brandwajn, A. An iterative solution of two-dimensional birth and death processes. *Opns. Res.*, 27, 595-605, 1979.

[36] Brill, P. H. System point theory in exponential queues. Ph.D. diss., Department of Industrial Engineering, University of Toronto, 1975.

[37] Brill, P. H. An embedded level crossing technique for dams and queues. *J. Appl. Prob.*, 16, 174-86, 1979.

[38] Brill, P. H., and Posner, M. J. M. A two server queue with non-waiting customers receiving specialized service. Tech. rept., Department of Industrial Engineering, University of Toronto, Toronto, Canada.

[39] Brockwell, P. J. The transient behaviour of a single server queue with batch arrivals. *J. Austral. Math. Soc.*, 3, 241-48, 1963.

[40] Brockwell, P. J. The transient behaviour of the queueing system $GI/M/1$. *J. Austral. Math. Soc.*, 3, 249-56, 1963.

[41] Burrows, C. Some numerical results for waiting times in the queue $E_k/M/1$. *Biometrika*, 47, 202-3, 1960.

[42] Bux, W., and Herzog, U. Approximation von Verteilungsfunktionen-Ein wichtiger Schritt bei der Modellbildung für Rechensysteme. Workshop über Modelle für Rechensysteme, Institut für Informatik, University of Bonn, Bonn, Germany, March, 1977.

[43] Bux, W., and Herzog, U. The phase concept: Approximation of measured data and performance analysis. Paper presented at the International Symposium

on Computer Performance Modeling, Measurement and Evaluation, Yorktown Heights, N.Y., August, 1977.

[44] Buzen, J. P. Queueing network models of multiprogramming. Ph.D. diss., Department of Computer Sciences, Harvard University, 1971.

[45] Buzen, J. P. Computational algorithms for closed queueing networks with exponential servers. *Comm. A.C.M.*, 16, 527-31, 1973.

[46] Byrd Jr, J. The value of queueing theory. *Interfaces*, 8, 22-26, 1978.

[47] Carson, C. C. Computational methods for single server queues with interarrival and service time distributions of phase type. Ph.D. diss., Statistics Department, Purdue University, 1975.

[48] Caseau, P., and Pujolle, G. Throughput capacity of a sequence of queues with blocking due to finite waiting room. *I.E.E.E. Trans. Software Eng.* (to appear).

[49] Chaudhry, M. L. Some queueing problems with phase-type service. *Opns. Res.*, 14, 466-77, 1966.

[50] Chaudhry, M. L., and Templeton, J. G. C. Bounds for the least positive root of a characteristic equation in the theory of queues. *INFOR*, 11, 177-79, 1973.

[51] Chung, K. L. *Markov chains with stationary transition probabilities*. 2d ed. New York: Springer-Verlag, 1967.

[52] Çinlar, E. Time dependence of queues with semi-Markovian services. *J. Appl. Prob.*, 4, 356-64, 1967.

[53] Çinlar, E. Queues with semi-Markovian arrivals. *J. Appl. Prob.*, 4, 365-79, 1967.

[54] Çinlar, E., and Disney, R. L. Streams of overflows from a finite queue. *Opns. Res.*, 15, 131-34, 1967.

[55] Çinlar, E. Decomposition of a semi-Markov process under a state-dependent rule. *SIAM J. Appl. Math.*, 15, 252-63, 1967.

[56] Çinlar, E. Markov renewal theory. *Adv. Appl. Prob.*, 1, 123-87, 1969.

[57] Clarke, A. B. Markovian queues with servers in tandem. Math. rept. no. 49. Western Michigan University, Kalamazoo, 1977.

[58] Clarke, A. B. A two-server tandem queueing system with storage between servers. Math. rept. no. 50. Western Michigan University, Kalamazoo, 1977.

[59] Clarke, A. B. A multiserver general service time queue with servers in series. Math. rept. no. 51. Western Michigan University, Kalamazoo, 1978.

[60] Clarke, A. B. Waiting times for Markovian queue with servers in series. Math. rept. no. 52. Western Michigan University, Kalamazoo, 1978.

[61] Cohen, J. W. *The single server queue.* London: North-Holland Publishing Co., 1969.

[62] Colard, J. P. Analyse Algorithmique d'un Modèle de Systeme d'Attente à Source Prioritaire Finie. Mémoire de Licence, Faculté des Sciences, Université Libre de Bruxelles, Brussels, Belgium, 1979.

[63] Collings, T., and Stoneman, C. The $M/M/\infty$ queue with varying arrival and departure rates. *Opns. Res.*, 24, 760–73, 1976.

[64] Conolly, B. W. A difference equation technique applied to the simple queue. *J. Roy. Statist. Soc. Ser. B*, 20, 165–67, 1958.

[65] Conolly, B. W. A difference equation technique applied to the simple queue with arbitrary arrival interval distribution. *J. Roy. Statist. Soc. Ser. B*, 20, 167–75, 1958.

[66] Conolly, B. W. The busy period in relation to the queueing process $GI/M/1$. *Biometrika*, 46, 246–51, 1959.

[67] Conolly, B. W. The busy period in relation to the single-server queueing system with general independent arrivals and Erlangian service. *J. Roy. Statist. Soc. Ser. B*, 22, 89–96, 1960.

[68] Conolly, B. W. Queueing at a single serving point with group arrival. *J. Roy. Statist. Soc. Ser. B*, 22, 285–98, 1960.

[69] Cosmetatos, G. P. Approximate explicit formulae for the average queueing time in the processes $(M/D/r)$ and $(D/M/r)$. *INFOR*, 13, 328–32, 1975.

[70] Cosmetatos, G. P. Some approximate equilibrium results for the multiserver queue $(M/G/r)$. *Oper. Res. Quart.*, 27, 615–20, 1976.

[71] Cosmetatos, G. P. Some approximate equilibrium results for the $E_k/M/r$ no-queue system. *R.A.I.R.O. Informatique/Computer Science*, 11, 355–61, 1977.

[72] Cox, D. R. A use of complex probabilities in the theory of stochastic processes. *Proc. Camb. Phil. Soc.*, 51, 313–19, 1955.

[73] Cox, D. R. The analysis of non-Markovian stochastic processes by the inclusion of supplementary variables. *Proc. Camb. Phil. Soc.*, 51, 433–41, 1955.

[74] Cox, D. R., and Smith, W. L. *Queues.* London: Methuen's Monographs on Applied Probability and Statistics, 1961.

[75] Cox, D. R. *Renewal theory.* London: Methuen's Monographs on Applied Probability and Statistics, 1962.

[76] Daley, D. J. Monte Carlo estimation of the mean queue size in a stationary $GI/M/1$ queue. *Opns. Res.*, 16, 1002–5, 1968.

[77] Descloux, A. On overflow processes of trunk groups with Poisson inputs and exponential service times. *Bell Syst. Tech. J.*, 42, 383–97, 1963.

[78] De Smit, J. H. A. On the many server queue with exponential service times. *Adv. Appl. Prob.*, 5, 170-82, 1973.

[79] Disney, R. L. Analytic studies of stochastic networks using methods of network decomposition. *J. of Indust. Eng.*, 18, 140-45, 1967.

[80] Disney, R. L., and Solberg, J. The effect of three switching rules on queueing networks. *J. of Indust. Eng.*, 19, 584-90, 1968.

[81] Disney, R. L. A matrix solution for the two server queue with overflow. *Mgmt. Sci.*, 19, 254-65, 1972.

[82] Disney, R. L. Random flow in queueing networks: A review and a critique. *A.I.I.E. Trans.*, 7, 268-88, 1975.

[83] Dobbie, J. M. A double-ended queueing problem of Kendall. *Opns. Res.*, 9, 755-57, 1961.

[84] Eisen, M. M. Effects of slow-downs and failure on stochastic service systems. *Technometrics*, 5, 385-92, 1963.

[85] Eisen, M., and Tainiter, M. Stochastic variations in queueing processes. *Opns. Res.*, 11, 922-27, 1963.

[86] Elsner, L. Verfahren zur Berechnung des Spektralradius nicht-negativer, irreduzibler Matrizen I. *Computing*, 8, 32-39, 1971.

[87] Elsner, L. Verfahren zur Berechnung des Spektralradius nicht-negativer, irreduzibler Matrizen II. *Computing*, 9, 69-73, 1972.

[88] Erlang, A. K. Solution of some problems in the theory of probabilities of significance in automatic telephone exchanges. *The Post Office Electrical Engineer's Journal*, 10, 189-97, 1917-18.

[89] Evans, R. V. Geometric distribution in some two-dimensional queueing systems. *Opns. Res.*, 15, 830-46, 1967.

[90] Feller, W. *An introduction to probability theory and its applications.* Vol. 1. 3d ed. New York: John Wiley and Sons, 1968.

[91] Finch, P. D. The effect of the size of the waiting room on a simple queue. *J. Roy. Statist. Soc. Ser. B*, 20, 182-86, 1958.

[92] Finch, P. D. Balking in the queueing system $GI/M/1$. *Acta Math. Acad. Sci. Hungar.*, 10, 241-47, 1959.

[93] Finch, P. D. The transient behaviour of a coincidence variate in telephone traffic. *Ann. Math. Statist.*, 32, 230-34, 1961.

[94] Finch, P. D. On the busy period in the queueing system $GI/G/1$. *J. Austral. Math. Soc.*, 2, 217-28, 1961.

[95] Finch, P. D. The single server queueing system with non-recurrent input process and Erlang service time. *J. Austral. Math. Soc.*, 3, 220-36, 1963.

[96] Finch, P. D. A coincidence problem in telephone traffic with non-recurrent arrival process. *J. Austral. Math. Soc.*, 3, 237-40, 1963.

[97] Finch, P. D. A note on the queueing system $GI/E_k/1$. *J. Appl. Prob.*, 6, 708-10, 1969.

[98] Fischer, M. J. The waiting time in the $E_k/M/1$ queueing system. *Opns. Res.*, 22, 898-902, 1974.

[99] Foster, F. G. On stochastic matrices associated with certain queueing processes. *Ann. Math. Statist.*, 24, 355-60, 1953.

[100] Foster, F. G. Queues with batch arrivals I. *Acta Math. Acad. Sci. Hungar.*, 12, 1-10, 1961.

[101] Foster, F. G., and Perera, A. G. A. D. Queues with batch arrivals II. *Acta Math. Acad. Sci. Hungar.*, 16, 275-87, 1965.

[102] Gani, J., and Prabhu, N. U. Stationary distributions of the negative exponential type for the infinite dam. *J. Roy. Statist. Soc. Ser. B*, 19, 342-51, 1957.

[103] Gani, J. A note on the first emptiness of dams with Markovian inputs. *J. Math. Anal. Appl.*, 26, 270-74, 1969.

[104] Gantmacher, F. R. *The theory of matrices.* New York: Chelsea, 1959.

[105] Gaver, D. P. The influence of servicing times in queueing processes. *Opns. Res.*, 2, 139-49, 1954.

[106] Gaver, D. P. Probability models for multiprogramming computer systems. *J. Assoc. Comp. Mach.*, 14, 423-38, 1967.

[107] Gaver, D. P., and Shedler, G. S. Processor utilization in multiprogramming systems via diffusion approximations. *Opns. Res.*, 21, 569-76, 1973.

[108] Gaver, D. P., and Humfeld, G. Multitype multiprogramming models. *Acta Informatica*, 7, 111-21, 1976.

[109] Gergely, T., and Török, T. L. On the busy period of discrete-time queues. *J. Appl. Prob.*, 11, 853-57, 1974.

[110] Gibson, A. E., and Conolly, B. W. On certain unrestricted, linear, unit step, continuous time random walks. *J. Appl. Prob.*, 8, 374-80, 1971.

[111] Grandell, J. *Doubly stochastic Poisson processes.* New York: Springer-Verlag, 1976.

[112] Greenberg, I. Some duality results in the theory of queues. *J. Appl. Prob.*, 6, 99-121, 1969.

[113] Grenander, U. Mathematical experiments on the computer. Manuscript, Division of Mathematical Sciences, Brown University, Providence, R.I., April, 1979.

[114] Grinstein, J., and Rubinovitch, M. Queues with random service output: The case of Poisson arrivals. *J. Appl. Prob.*, 11, 771–84, 1974.

[115] Gross, D., and Harris, C. M. *Fundamentals of queueing theory.* New York: John Wiley and Sons, 1974.

[116] Gue, R. L. Signal flow graphs and analog computation in the analysis of finite queues. *Opns. Res.*, 14, 342–50, 1966.

[117] Gupta, S. K., and Goyal, J. K. Queues with Poisson input and hyper-exponential output with finite waiting space. *Opns. Res.*, 12, 75–81, 1964.

[118] Gupta, S. K. Queues with batch Poisson arrivals and a general class of service time distributions. *J. Indust. Eng.*, 15, 319–20, 1964.

[120] Gupta, S. K. Queues with Poisson input and mixed-Erlangian service time distribution with finite waiting space. *J. Opns. Res. Soc. Japan*, 8, 24–31, 1965.

[121] Gupta, S. K. Queues with mixed-Erlangian input and negative exponential service time distributions with finite waiting space. *Canad. Opns. Res. Soc. J.*, 3, 22–28, 1965.

[122] Gupta, S. K. Queues fed by Poisson input and hyper-mixed Erlangian service time distribution with finite waiting space. *Unternehmungsforschung*, 9, 80–90, 1965.

[123] Gupta, S. K., and Goyal, J. K. Queues with batch Poisson arrivals and hyperexponential service. *Nav. Res. Log. Quart.*, 12, 323–29, 1965.

[124] Halfin, S. Steady-state distribution for the buffer content of an $M/G/1$ queue with varying service rate. *SIAM J. Appl. Math.*, 23, 356–63, 1972.

[125] Halfin, S., and Segal, M. A priority queueing model for a mixture of two types of customers. *SIAM J. Appl. Math.*, 23, 369–79, 1972.

[126] Hall, W. K., and Disney, R. L. Finite queues in parallel under a generalized channel selection rule. *J. Appl. Prob.*, 8, 413–16, 1971.

[127] Harris, T. J. Duality of finite Markovian queues. *Opns. Res.*, 15, 575–76, 1967.

[128] Harrison, J. M. The heavy traffic approximation for single server queues in series. *J. Appl. Prob.*, 10, 613–29, 1973.

[129] Harrison, J. M., and Lemoine, A. J. Limit theorems for periodic queues. *J. Appl. Prob.*, 14, 566–76, 1977.

[130] Harrison, J. M. The diffusion approximation for tandem queues in heavy traffic. *Adv. Appl. Prob.*, 10, 886–905, 1978.

[131] Healy, T. L. Queues with "Exponential-type" service-time distributions. *Opns. Res.*, 8, 719–21, 1960.

[132] Heathcote, C. R., and Winer, P. An approximation for the moments of waiting times. *Opns. Res.*, 17, 175–86, 1969.

[133] Heffer, J. C. Steady-state solution of the $M/E_k/c$ (∞, FIFO) queueing system. *INFOR*, 7, 16–30, 1969.

[134] Heffes, H. Analysis of first-come, first-served queueing systems with peaked inputs. *Bell Syst. Tech. J.*, 52, 1215–28, 1973.

[135] Heffes, H., and Holtzman, J. M. Peakedness of traffic carried by a finite trunk group with renewal input. *Bell Syst. Tech. J.*, 52, 1617–42, 1973.

[136] Henderson, J., and Finch, P. D. A note on the queueing system $E_k/G/1$. *J. Appl. Prob.*, 7, 473–75, 1970.

[137] Heyde, C. C. On the stationary waiting time distribution in the $GI/G/1$ queue. *J. Appl. Prob.*, 1, 175–78, 1964.

[138] Heyman, D. P. Optimal operating policies for $M/G/1$ queueing systems. *Opns. Res.*, 16, 362–82, 1968.

[139] Hildebrand, D. K. Stability of finite queue, tandem service systems. *J. Appl. Prob.*, 4, 571–83, 1967.

[140] Hildebrand, D. K. On the capacity of tandem server, finite queue, service systems. *Opns. Res.*, 16, 72–82, 1968.

[141] Hillier, F. S., and Boling, R. W. Finite queues in series with exponential or Erlang service times: A numerical approach. *Opns. Res.*, 15, 286–303, 1971.

[142] Hillier, F. S., and Lo, F. D. Tables for multi-server queueing systems involving Erlang distributions. Tech. rept. no. 31. Department of Operations Research, Stanford University, Stanford, Calif., 1971.

[143] Hokstad, P. The $GI/M/m$ queue with finite waiting room. *J. Appl. Prob.*, 12, 779–92, 1975.

[144] Hokstad, P. Asymptotic behaviour of the $E_k/G/1$ queue with finite waiting room. *J. Appl. Prob.*, 14, 358–66, 1977.

[145] Hokstad, P. On the steady-state solution of the $M/G/2$ queue. *Adv. Appl. Prob.*, 11, 240–55, 1979.

[146] Hunt, G. C. Sequential arrays of waiting lines. *Opns. Res.*, 4, 674–83, 1956.

[147] Hunter, J. J. On the moments of Markov renewal processes. *Adv. Appl. Prob.*, 1, 188–210, 1969.

[148] Jackson, R. R. P. Queueing system with phase type service. *Opns. Res. Quart.*, 5, 109–20, 1954.

[149] Jackson, R. R. P. Queueing processes with phase-type service. *J. Roy. Statist. Soc. Ser. B*, 18, 129–32, 1956.

[150] Jackson, R. R. P., and Nickols, D. J. Some equilibrium results for the queueing process $E_k/M/1$. *J. Roy. Statist. Soc. Ser. B*, 18, 275–79, 1956.

[151] Jain, H. C. A double ended queueing system. *Defense Sci. J.*, 12, 327–32, 1962.

[152] Jansson, B. Choosing a good appointment system: A study of queues of the type (*D, M,* 1). *Opns. Res.,* 14, 292-312, 1966.

[153] Jensen, A. *A distribution model applicable to economics.* Copenhagen: Munksgaard, 1954.

[154] Kabak, I. W. Blocking and delays in $M^{(n)}/M/c$ queueing systems. *Opns. Res.,* 16, 830-40, 1968.

[155] Kapadia, A. S. A k-server queue with phase input and service distribution. *Opns. Res.,* 21, 623-28, 1973.

[156] Karlin, S., and Taylor, H. M. *A first course in stochastic processes.* 2d ed. New York: Academic Press, 1975.

[157] Kashyap, B. R. K. A double-ended queueing system with limited waiting space. *Proc. Natl. Inst. Sci. India,* 31, 559-70, 1965.

[158] Kashyap, B. R. K. The double-ended queue with bulk service and limited waiting space. *Opns. Res.,* 14, 822-34, 1966.

[159] Kashyap, B. R. K. Further results for the double ended queue. *Metrika,* 11, 168-86, 1967.

[160] Kashyap, B. R. K. The double-ended queue with batch departures. In *Advancing frontiers in operational research,* Delhi, India: Hindustan Publishing Co., 139-43, 1969.

[161] Kawamura, T. Single server queue with Erlangian input and holding time. *Yokohama Math. J.,* 12, 39-61, 1964.

[162] Keilson, J. *Green's function methods in probability theory.* Griffin's Statistical Monographs and Courses, no. 17. London: Charles Griffin & Co., 1965.

[163] Keilson, J., and Nunn, W. R. Laguerre transformation as a tool for the numerical solution of integral equations of convolution type. *Appl. Math. and Comp.,* 5, 313-59, 1979.

[164] Kemeny, J., and Snell, J. L. *Finite Markov chains.* Princeton, N.J.: Van Nostrand Publishing Co., 1960.

[165] Kendall, D. G. On the role of variable generation time in the development of a stochastic birth process. *Biometrika,* 35, 316-30, 1948.

[166] Kendall, D. G. Some problems in the theory of queues. *J. Roy. Statist. Soc. Ser. B,* 13, 151-85, 1951.

[167] Kendall, D. G. Stochastic processes occurring in the theory of queues and their analysis by the method of imbedded Markov chains. *Ann. Math. Statist.,* 24, 338-54, 1953.

[168] Kendall, D. G. Some recent work and further problems in the theory of queues. *Theory of Prob. and Its Appl.,* 9, 1-15, 1964.

[169] Khinchin, A. Ya. Mathematical theory of a stationary queue (in Russian). *Math. Sbornik,* 39, 73-84, 1932.

[170] Kingman, J. F. C. A convexity property of positive matrices. *Quart. J. Math.*, 12, 283–84, 1961.

[171] Kingman, J. F. C. The use of Spitzer's identity in the investigation of the busy period and other quantities in the queue *GI/G/*1. *J. Austral. Math. Soc.*, 2, 345–56, 1962.

[172] Kleinrock, L. *Queueing systems.* Vol. 1, *Theory.* Vol. 2, *Computer applications.* New York: John Wiley and Sons, 1975.

[173] Kolesar, P. A quick and dirty response to the quick and dirty crowd; particularly to Jack Byrd's "The value of queueing theory." *Interfaces*, 9, 77–82, 1979.

[174] Konheim, A. G., and Reiser, M. A queueing model with finite waiting room and blocking. *J. Assoc. Comp. Mach.*, 23, 328–41, 1976.

[175] König, D., and Stoyan, D. *Methoden der Bedienungstheorie.* Braunschweig: Vieweg, 1976.

[176] Kotiah, T. C. T.; Thompson, J. W.; and Waugh, W. A. O'N. Use of Erlangian distributions for single-server queueing systems. *J. Appl. Prob.*, 6, 584–93, 1969.

[177] Kotiah, T. C. T., and Slater, N. B. On two-server Poisson queues with two types of customers. *Opns. Res.*, 21, 597–603, 1973.

[178] Labetoulle, J., and Pujolle, G. A study of queueing networks with deterministic service and applications to computer networks. *Acta Informatica*, 7, 183–96, 1976.

[179] Latouche, G. Algorithmic analysis of a multiprogramming-multiprocessor computer system. Tech. rept. no. 78/13. Applied Mathematics Institute, University of Delaware, Newark, 1978.

[180] Latouche, G. Queueing systems with heterogeneous customers and group services. Tech. rept. no. 76. Lab. Informatique Théorique, Université Libre de Bruxelles, Brussels, Belgium, 1978.

[181] Latouche, G. A queue with paired customers. Tech. rept. no. 79. Université Libre de Bruxelles, Brussels, Belgium, January, 1979.

[182] Latouche, G. On a Markovian queue with weakly correlated interarrival times. Tech. rept. no. 86. Université Libre de Bruxelles, Brussels, Belgium, May, 1979.

[183] Latouche, G., and Neuts, M. F. Efficient algorithmic solutions to exponential tandem queues with blocking. *SIAM J. Algebraic and Discrete Meth.*, 1, 93–106, 1980.

[184] Lavenberg, S. S. Stability and maximum departure rate of certain open queueing networks having finite capacity constraints. *R.A.I.R.O. Informatique/ Computer Science*, 12, 353–70, 1978.

[185] Lehoczky, J. P. A note on the first emptiness time of an infinite reservoir with inputs forming a Markov chain. *J. Appl. Prob.*, 8, 276–84, 1971.

[186] Lehoczky, J. P., and Gaver, D. P. Approximate models for central server systems with two job types. Tech. rept. no. 130. Department of Statistics, Carnegie-Mellon University, Pittsburgh, Pa., 1977.

[187] Lemoine, A. J. Networks of queues: A survey of equilibrium analysis. *Mgmt. Sci.*, 24, 464–81, 1977.

[188] Lemoine, A. J. Networks of queues: A survey of weak convergence results. *Mgmt. Sci.*, 24, 1175–93, 1978.

[189] Lewis, P. A. W., and Shedler, G. S. A cyclic-queue model of system overhead in multiprogrammed computer systems. *J. Assoc. Comp. Mach.*, 18, 199–220, 1971.

[190] Lindley, D. V. The theory of queues with a single server. *Proc. Camb. Phil. Soc.*, 48, 277–89, 1952.

[191] Loris-Teghem, J. Un Traitement Algébrique du Modèle d'Attente $GI/M/2$. *Cahiers du Centre de Recherche Opérationnelle*, 13, 57–62, 1971.

[192] Loris-Teghem, J. An algebraic approach to the waiting time process in $GI/M/S$. *J. Appl. Prob.*, 10, 181–91, 1973.

[193] Love, R. F. Steady-state solution of the queueing system $E_w/M/s$ with batch service. *Opns. Res.*, 18, 160–71, 1970.

[194] Lucantoni, D. M., and Neuts, M. F. Numerical methods for a class of Markov chains arising in queueing theory. Tech. rept. no. 78/10. Applied Mathematics Institute, University of Delaware, Newark, 1978.

[195] Lucantoni, D. M. A $GI/M/c$ queue with a different service rate for customers who need not wait: An algorithmic solution. Tech. rept. no. 48B. Applied Mathematics Institute, University of Delaware, Newark, 1979.

[196] Luchak, G. The solution of the single channel queueing equations characterized by a time-dependent Poisson distributed arrival rate and a general class of holding times. *Opns. Res.*, 4, 711–32, 1956.

[197] Luchak, G. The distribution of the time required to reduce to some preassigned level a single-channel queue characterized by a time-dependent Poisson distributed arrival rate and a general class of holding times. *Opns. Res.*, 5, 205–9, 1957.

[198] Luchak, G. The continuous-time solution of the equations of the single-channel queue with a general class of service-time distribution by the method of generating functions. *J. Roy. Statist. Soc. Ser. B*, 20, 176–81, 1958.

[199] Maaløe, E. Approximation formulae for estimation of waiting time in multiple-channel queueing system. *Mgmt. Sci.*, 19, 703–10, 1973.

[200] Marcus, M., and Minc, H. *A survey of matrix theory and matrix inequalities.* Boston: Allyn and Bacon, 1964.

[201] Marks, B. I. State probabilities of $M/M/1$ priority queues. *Opns. Res.*, 21, 974–87, 1973.

[202] Mayhugh, J. O., and McCormick, R. E. Steady-state solution of the queue $M/E_k/r$. *Mgmt. Sci.*, 14, 692–712, 1968.

[203] Meisling, T. Discrete-time queueing theory. *Opns. Res.*, 6, 96–105, 1958.

[204] Mercer, A. A queueing problem in which the arrival times of the customers are scheduled. *J. Roy. Statist. Soc. Ser. B*, 22, 108–13, 1960.

[205] Miller, L. W. A note on the busy period of an $M/G/1$ finite queue. *Opns. Res.*, 23, 1179–82, 1975.

[206] Mitrani, I. Nonpriority multiprogramming systems under heavy demand conditions: Customer's viewpoint. *J. Assoc. Comp. Mach.*, 19, 445–52, 1972.

[207] Mitrany, I. L., and Avi-Itzhak, B. A many-server queue with service interruptions. *Opns. Res.*, 16, 628–38, 1968.

[208] Mittwoch, D. Z. The complete steady-state solution of the $E_k/M/r$ queueing system and its uniqueness proof. In *Proceedings of the Twentieth International Meeting, Institute Management Sciences.* Vol. 2. Jerusalem: Jerusalem Academic Press, 509–16, 1973.

[209] Moore, S. C., and Bhat, U. N. On a computational approach to some value functions in an $M/E_z/1$ queue. *A.I.I.E. Trans.*, 7, 73–76, 1975.

[210] Morse, P. M. Stochastic processes of waiting lines. *Opns. Res.*, 3, 255–61, 1955.

[211] Morse, P. M. *Queues, inventories and maintenance.* New York: John Wiley and Sons, 1958.

[212] Nance, R. E., and Bhat, U. N. A processor utilization model for a multiprocessor computer system. *Opns. Res.*, 26, 881–95, 1978.

[213] Naor, P. On machine interference. *J. Roy. Statist. Soc. Ser. B*, 18, 280–87, 1956.

[214] Naor, P., and Yechiali, U. Queueing problems with heterogeneous arrivals and service. *Opns. Res.*, 19, 722–34, 1971.

[215] Neuts, M. F. The single server queue with Poisson input and semi-Markov service times. *J. Appl. Prob.*, 3, 202–30, 1966.

[216] Neuts, M. F. Two queues in series with a finite intermediate waiting room. *J. Appl. Prob.*, 5, 123–42, 1968.

[217] Neuts, M. F. Two servers in series, treated in terms of a Markov renewal branching process. *Adv. Appl. Prob.*, 2, 110–49, 1970.

[218] Neuts, M. F. A queue subject to extraneous phase changes. *Adv. Appl. Prob.*, 3, 78–119, 1971.

[219] Neuts, M. F. The Markov renewal branching process. In *Proceedings of the Conference on Mathematical Methodology in the Theory of Queues, Kalamazoo, Mich.* New York: Springer-Verlag, 1–21, 1974.

[220] Neuts, M. F. Probability distributions of phase type. In *Liber Amicorum Prof. Emeritus H. Florin,* Department of Mathematics. Belgium: University of Louvain, 173–206, 1975.

[221] Neuts, M. F. Computational uses of the method of phases in the theory of queues. *Computers and Math. with Appl.,* 1, 151–66, 1975.

[222] Neuts, M. F. Statement in "Applied probability: Its nature and scope." Edited by N. U. Prabhu. *Stoch. Proc. and Appl.,* 3, 223–57, 1975.

[223] Neuts, M. F. Moment formulas for the Markov renewal branching process. *Adv. Appl. Prob.,* 8, 690–711, 1976.

[224] Neuts, M. F. Some explicit formulas for the steady-state behavior of the queue with semi-Markovian service times. *Adv. Appl. Prob.,* 9, 141–57, 1977.

[225] Neuts, M. F. Algorithms for the waiting time distributions under various queue disciplines in the $M/G/1$ queue with service time distributions of phase type. In *Algorithmic Methods in Probability,* TIMS Studies in the Management Sciences, no. 7. London: North-Holland Publishing Co., 177–97, 1977.

[226] Neuts, M. F. Renewal processes of phase type. *Nav. Res. Logist. Quart.,* 25, 445–54, 1978.

[227] Neuts, M. F. Markov chains with applications in queueing theory, which have a matrix-geometric invariant vector. *Adv. Appl. Prob.,* 10, 185–212, 1978.

[228] Neuts, M. F. The second moments of the absorption times in the Markov renewal branching process. *J. Appl. Prob.,* 15, 707–14, 1978.

[229] Neuts, M. F. The $M/M/1$ queue with randomly varying arrival and service rates. *Opsearch,* 15, 139–57, 1978.

[230] Neuts, M. F. Further results on the $M/M/1$ queue with randomly varying rates. *Opsearch,* 15, 158–68, 1978.

[231] Neuts, M. F. An algorithmic solution to the $GI/M/c$ queue with group arrivals. *Cahiers du Centre de Recherche Opérationnelle,* 21, 109–19, 1979.

[232] Neuts, M. F. A versatile Markovian point process. *J. Appl. Prob.,* 16, 764–79, 1979.

[233] Neuts, M. F. Some algorithms for queues with group arrivals or group services. *Proceedings of the Tenth Annual Pittsburgh Conference. Modeling and Simulation,* Vol. 10, pt. 2. Pittsburgh, Pa.: Instrument Society of America, 311–14, 1979.

[234] Neuts, M. F., and Lucantoni, D. M. A Markovian queue with N servers subject to breakdowns and repairs. *Mgmt. Sci.,* 25, 849–61, 1979.

[235] Neuts, M. F. Computational problems related to the Galton-Watson process. In *Computational Probability,* Proceedings of the 1975 Brown Actuarial Research Conference on Computational Probability, ed. P. M. Kahn. New York: Academic Press, 11–37, 1980.

[236] Neuts, M. F. The probabilistic significance of the rate matrix in matrix-geometric invariant vectors. *J. Appl. Prob.*, 17, 291-96, 1980.

[237] Neuts, M. F., and Chakravarthy, S. The single server queue with platooned arrivals and service of phase type. Tech. rept. no. 50B. Applied Mathematics Institute, University of Delaware, Newark, 1980. (To appear in *Euro. J. of Oper. Res.*)

[238] Neuts, M. F., and Kumar, S. Algorithmic solution of some queues with overflows. Tech. rept. no. 51B. Applied Mathematics Institute, University of Delaware, Newark, 1980.

[239] Newell, G. F. Approximate behavior of tandem queues. Research rept. Inst. Transp. and Traffic Engin., University of California, Berkeley, chaps. 1-4, 1975.

[240] Newell, G. F. Approximate behavior of tandem queues. Research rept. Inst. of Transp. and Traffic Engin., University of California, Berkeley, chaps. 5-8, 1977.

[241] Nilsson, A. On overflow systems in telephone networks: General service times in the secondary group. *Ericsson Technics*, 34, 46-70, 1978.

[242] Nishida, T. On the multiple exponential channel queueing systems with hyper-Poisson arrivals. *J. Oper. Res. Soc. Japan*, 5, 57-65, 1962.

[243] Nozaki, S. A., and Ross, S. M. Approximations in finite-capacity multi-server queues with Poisson arrivals. *J. Appl. Prob.*, 15, 826-34, 1978.

[244] Odoom, S., and Lloyd, E. H. A note on the equilibrium distribution of levels in a semi-infinite reservoir subject to Markovian inputs and unit withdrawals. *J. Appl. Prob.*, 2, 215-22, 1965.

[245] Ortega, J. M., and Rheinboldt, W. C. *Iterative solution of nonlinear equations in several variables.* New York: Academic Press, 1970.

[246] Pack, C. D. The output of multiserver queueing systems. *Opns. Res.*, 26, 492-509, 1978.

[247] Paige, C. C.; Styan, G. P. H.; and Wachter, P. G. Computation of the stationary distribution of a Markov chain. *J. Statist. Comput. Simul.*, 4, 173-86, 1975.

[248] Pakes, A. G., and Phatarfod, R. M. The limiting distribution for the infinitely deep dam with a Markovian input. *J. Stoch. Proc. Appl.*, 8, 199-209, 1978.

[249] Pakes, A. G. Discrete dams with Markovian inputs. Preprint, Department of Mathematics, Monash University, Clayton, Victoria, Australia, 1979.

[250] Phatarfod, R. M. The bottomless dam. *J. Hydrology*, 40, 337-63, 1979.

[251] Phatarfod, R. M. A solution to the dam problem. Preprint, Department of Mathematics, Monash University, Clayton, Victoria, Australia.

[252] Prabhu, N. U. Transient behavior of a tandem queue. *Mgmt. Sci.*, 13, 631-39, 1966.

[253] Pujolle, G., and Potier, D. Réseaux de Files d'Attente à Capacité Limitée avec des Applications aux Systemes Informatiques. *R.A.I.R.O. Informatique/ Computer Science*, 13, 175-97, 1979.

[254] Purdue, P. The $M/M/1$ queue in a Markovian environment. *Opns. Res.*, 22, 562-69, 1974.

[255] Pyke, R. Markov renewal processes: Definitions and preliminary properties. *Ann. Math. Statist.*, 32, 1231-42, 1961.

[256] Pyke, R. Markov renewal processes with finitely many states. *Ann. Math. Statist.*, 32, 1243-59, 1961.

[257] Rajeswari, A. R. Nonpreemptive priority queue with binomial input. *Opns. Res.*, 16, 416-21, 1968.

[258] Ramaswami, V. The $N/G/\infty$ queue. Tech. rept., Department of Mathematics, Drexel University, Philadelphia, Pa., October, 1978.

[259] Ramaswami, V., and Lucantoni, D. M. On the merits of an approximation to the busy period of the $GI/G/1$ queue. *Mgmt. Sci.*, 25, 285-89, 1979.

[260] Ramaswami, V. A note on Neuts' versatile Markovian point process. Tech. rept. no. 7-79. Department of Mathematics, Drexel University, Philadelphia, Pa., 1979.

[261] Ramaswami, V. The $N/G/1$ queue and its detailed analysis. *Adv. Appl. Prob.*, 12, 222-61, 1980.

[262] Ramaswami, V., and Neuts, M. F. Some explicit formulas and computational methods for infinite server queues with phase type arrivals. *J. Appl. Prob.*, 17, 498-514, 1980.

[263] Ramaswami, V. The busy period of queues which have a matrix-geometric steady-state probability vector. Tech. rept. no. 3-80, Department of Mathematics, Drexel University, Philadelphia, Pa., 1980.

[264] Rana, R. K. Queueing problems with arrivals in general stream and phase type service. *Metrika*, 18, 69-80, 1971.

[265] Reiser, M., and Konheim, A. G. Finite capacity queueing systems with applications in computer modeling. *SIAM J. on Computing*, 7, 210-29, 1978.

[266] Restrepo, R. A. A queue with simultaneous arrivals and Erlang service distribution. *Opns. Res.*, 13, 375-81, 1965.

[267] Roes, P. B. M. A many server bulk queue. *Opns. Res.*, 14, 1037-44, 1966.

[268] Rolski, T., and Stoyan, D. On the comparison of waiting times in $GI/G/1$ queues. *Opns. Res.*, 24, 197-200, 1976.

[269] Rosenlund, S. I. On the length and number of served customers of the busy period of a generalised $M/G/1$ queue with finite waiting room. *Adv. Appl. Prob.*, 5, 379-89, 1973.

[270] Rossa, G. Die Analyse von empirisch verteilten Zufallsgrössen auf dem Analogrechner. *Zastosowania Matematyki*, 12, 135–51, 1971.

[271] Rothkopf, M. H., and Oren, S. S. A closure approximation for the nonstationary $M/M/s$ queue. *Mgmt. Sci.*, 25, 522–34, 1979.

[272] Saaty, T. L. *Elements of queueing theory with applications*. New York: McGraw-Hill Book Co., 1961.

[273] Sahin, I. On the least positive root of the equation $z = K(z)$ in queueing theory. *J. Canad. Oper. Res. Soc.*, 8, 109, 1970.

[274] Sakasegawa, H. Numerical tables of the queueing system $E_k/E_2/s$. Institute of Statistical Mathematical Sciences, Comp. Sc. monograph no. 10, Tokyo, Japan, 1978.

[275] Saksena, V. N. Discrete time analysis of a deteriorating structure. *Sankhyā*, 40, 61–75, 1978.

[276] Schassberger, R. *Warteschlangen*. New York: Springer-Verlag, 1973.

[277] Schassberger, R. Insensitivity of steady-state distributions of generalized semi-Markov processes, I. *Ann. Prob.*, 5, 87–99, 1977.

[278] Schassberger, R. Insensitivity of steady-state distributions of generalized semi-Markov processes, II. *Ann. Prob.*, 6, 85–93, 1978.

[279] Schassberger, R. Insensitivity of steady-state distributions of generalized semi-Markov processes with speeds. *Adv. Appl. Prob.*, 10, 836–51, 1978.

[280] Schassberger, R. The insensitivity of stationary probabilities in networks of queues. *Adv. Appl. Prob.*, 10, 906–12, 1978.

[281] Shanbhag, D. N. A note on queueing systems with Erlangian service time distributions. *Ann. Math. Statist.*, 36, 1574–78, 1965.

[282] Shanbhag, D. N. On a duality principle in the theory of queues. *Opns. Res.*, 14, 947–49, 1966.

[283] Shapiro, S. The M-server queue with Poisson input and gamma-distributed service of order two. *Opns. Res.*, 14, 685–94, 1966.

[284] Shyu, K. On the queueing processes in the system $GI/M/n$ with bulk service. *Chinese Math.*, 1, 196–204, 1962.

[285] Singh, V. P. Two-server Markovian queues with balking: Heterogeneous vs. homogeneous servers. *Opns. Res.*, 18, 145–59, 1970.

[286] Smith, W. L. Distribution of queueing times. *Proc. Camb. Phil. Soc.*, 49, 449–61, 1953.

[287] Soriano, A. Comparison of two scheduling systems. *Opns. Res.*, 14, 388–97, 1966.

[288] Soriano, A. On the problem of batch arrivals and its application to a scheduling system. *Opns. Res.*, 14, 398-408, 1966.

[289] Suzuki, T. On a tandem queue with blocking. *J. Opns. Res. Soc. Japan*, 6, 137-57, 1964.

[290] Syski, R. Phase-type distributions and perturbation model. *Zastosowania Matematyki* (to appear).

[291] Takács, L. Transient behavior of single-server queueing processes with recurrent input and exponentially distributed service times. *Opns. Res.*, 8, 231-45, 1960.

[292] Takács, L. The transient behavior of a single server queueing process with recurrent input and gamma service time. *Ann. Math. Statist.*, 32, 1286-98, 1961.

[293] Takács, L. *Introduction to the theory of queues.* New York: Oxford University Press, 1962.

[294] Takahashi, Y. A lumping method for numerical calculations of stationary distributions of Markov chains. Research rept., Information Sciences no. B18, Tokyo Institute of Technology, Tokyo, Japan, 1975.

[295] Takahashi, Y., and Takami, Y. A numerical method for the steady-state probabilities of a $GI/G/c$ queueing system in a general class. *J. Oper. Res. Soc. Japan*, 19, 147-57, 1976.

[296] Takahashi, Y. On computation of the waiting-time distribution of a queueing system with phase-type interarrival and service distributions. Manuscript, Tohoku University, Faculty of Economics, Sendai, Japan, 1979.

[297] Truslove, A. L. The busy period of the $E_k/G/1$ queue with finite waiting room. *Adv. Appl. Prob.*, 7, 416-30, 1975.

[298] Truslove, A. L. Queue length for the $E_k/G/1$ queue with finite waiting room. *Adv. Appl. Prob.*, 7, 215-26, 1975.

[299] Ushijima, T. A queueing system with Markov arrivals. *J. Oper. Res. Soc. Japan*, 15, 167-93, 1972.

[300] Vaszonyi, A. To queue or not to queue: A rejoinder. *Interfaces*, 9, 83-86, 1979.

[301] Wachter, P. G. Solving certain systems of homogeneous equations with special reference to Markov chains. M.Sc. Thesis, Department of Mathematics, McGill University, 1973.

[302] Wallace, V. The solution of quasi birth and death processes arising from multiple access computer systems. Ph.D. diss., Systems Engineering Laboratory, University of Michigan, Tech. rept. no. 07742-6-T, 1969.

[303] Washburn, A. A multiserver queue with no passing. *Opns. Res.*, 22, 428-34, 1974.

[304] Winsten, C. B. Geometric distributions in the theory of queues. *J. Roy. Stat. Soc.*, 21, 1-35, 1959.

[305] Wishart, D. M. G. A queueing system with χ^2-service time distribution. *Ann. Math. Statist.*, 27, 768-79, 1956.

[306] Wishart, D. M. G. A queueing system with service time distribution of mixed chi-squared type. *Opns. Res.*, 7, 174-79, 1959.

[307] Wong, B.; Giffin, W.; and Disney, R. L. Two finite $M/M/1$ queues in tandem: A matrix solution for the steady state. *Opsearch*, 14, 1-18, 1977.

[308] Wu, F. On the queueing process $GI/M/n$. *Chinese Math.*, 2, 333-43, 1961.

[309] Xu, G. The transient behavior of the queueing process $GI/M/n$. *Chinese Math.*, 6, 393-421, 1965.

[310] Yadin, M. On Markov processes with randomly varying intensities with application to a queueing model. Opns. Res., Statist. and Econ. mimeograph series no. 209, Technion, Haifa, Israel, February, 1978.

[311] Yadin, M., and Frostig, E. On numerical methods for the evaluation of finite horizon cost functions for Markov systems with varying intensities. Opns. Res., Statist. and Econ. mimeograph series no. 220, Technion, Haifa, Israel, August, 1978.

[312] Yadin, M., and Syski, R. Randomization of intensities in a Markov chain. *Adv. Appl. Prob.*, 11, 397-421, 1979.

[313] Yechiali, U. A queueing-type birth-and-death process defined on a continuous-time Markov chain. *Opns. Res.*, 21, 604-9, 1973.

[314] Yu, O. S. Stochastic bounds for heterogeneous-server queues with Erlang service times. *J. Appl. Prob.*, 11, 785-96, 1974.

[315] Yu, O. S. The steady-state solution of a heterogeneous-server queue with Erlang service times. In *Algorithmic Methods in Probability*, TIMS Studies in the Management Sciences, no. 7. London: North-Holland Publishing Co., 199-213, 1977.

Index

A CATALOG OF SELECTED

DOVER BOOKS

IN SCIENCE AND MATHEMATICS

RELATIVITY, THERMODYNAMICS AND COSMOLOGY, Richard C. Tolman. Landmark study extends thermodynamics to special, general relativity; also applications of relativistic mechanics, thermodynamics to cosmological models. 501pp. 5⅜ × 8½. 65383-8 Pa. $12.95

APPLIED ANALYSIS, Cornelius Lanczos. Classic work on analysis and design of finite processes for approximating solution of analytical problems. Algebraic equations, matrices, harmonic analysis, quadrature methods, much more. 559pp. 5⅜ × 8½. 65656-X Pa. $13.95

SPECIAL RELATIVITY FOR PHYSICISTS, G. Stephenson and C.W. Kilmister. Concise elegant account for nonspecialists. Lorentz transformation, optical and dynamical applications, more. Bibliography. 108pp. 5⅜ × 8½. 65519-9 Pa. $4.95

INTRODUCTION TO ANALYSIS, Maxwell Rosenlicht. Unusually clear, accessible coverage of set theory, real number system, metric spaces, continuous functions, Riemann integration, multiple integrals, more. Wide range of problems. Undergraduate level. Bibliography. 254pp. 5⅜ × 8½. 65038-3 Pa. $7.95

INTRODUCTION TO QUANTUM MECHANICS With Applications to Chemistry, Linus Pauling & E. Bright Wilson, Jr. Classic undergraduate text by Nobel Prize winner applies quantum mechanics to chemical and physical problems. Numerous tables and figures enhance the text. Chapter bibliographies. Appendices. Index. 468pp. 5⅜ × 8½. 64871-0 Pa. $11.95

ASYMPTOTIC EXPANSIONS OF INTEGRALS, Norman Bleistein & Richard A. Handelsman. Best introduction to important field with applications in a variety of scientific disciplines. New preface. Problems. Diagrams. Tables. Bibliography. Index. 448pp. 5⅜ × 8½. 65082-0 Pa. $12.95

MATHEMATICS APPLIED TO CONTINUUM MECHANICS, Lee A. Segel. Analyzes models of fluid flow and solid deformation. For upper-level math, science and engineering students. 608pp. 5⅜ × 8½. 65369-2 Pa. $13.95

ELEMENTS OF REAL ANALYSIS, David A. Sprecher. Classic text covers fundamental concepts, real number system, point sets, functions of a real variable, Fourier series, much more. Over 500 exercises. 352pp. 5⅜ × 8½. 65385-4 Pa. $10.95

PHYSICAL PRINCIPLES OF THE QUANTUM THEORY, Werner Heisenberg. Nobel Laureate discusses quantum theory, uncertainty, wave mechanics, work of Dirac, Schroedinger, Compton, Wilson, Einstein, etc. 184pp. 5⅜ × 8½. 60113-7 Pa. $5.95

INTRODUCTORY REAL ANALYSIS, A.N. Kolmogorov, S.V. Fomin. Translated by Richard A. Silverman. Self-contained, evenly paced introduction to real and functional analysis. Some 350 problems. 403pp. 5⅜ × 8½. 61226-0 Pa. $9.95

PROBLEMS AND SOLUTIONS IN QUANTUM CHEMISTRY AND PHYSICS, Charles S. Johnson, Jr. and Lee G. Pedersen. Unusually varied problems, detailed solutions in coverage of quantum mechanics, wave mechanics, angular momentum, molecular spectroscopy, scattering theory, more. 280 problems plus 139 supplementary exercises. 430pp. 6½ × 9¼. 65236-X Pa. $12.95